T0130991

ORGANIC SYNTHESES

ORGANIC SYNTHESES

AN ANNUAL PUBLICATION OF SATISFACTORY
METHODS FOR THE PREPARATION
OF ORGANIC CHEMICALS
VOLUME 81
2005

A JOHN WILEY & SONS, INC., PUBLICATION

The procedures in this text are intended for use only by persons with prior training in the field of organic chemistry. In the checking and editing of these procedures, every effort has been made to identify potentially hazardous steps and to eliminate as much as possible the handling of potentially dangerous materials; safety precautions have been inserted where appropriate. If performed with the materials and equipment specified, in careful accordance with the instructions and methods in this text, the Editors believe the procedures to be very useful tools. However, these procedures must be conducted at one's own risk. Organic Syntheses, Inc., its Editors, who act as checkers, and its Board of Directors do not warrant or guarantee the safety of individuals using these procedures and hereby disclaim any liability for any injuries or damages claimed to have resulted from or related in any way to the procedures herein.

For general information on our other products and services please contact our Customer Care Department within the U.S. at 877-762-2974, outside the U.S. at 317-572-3993 or fax 317-572-4002.

Wiley also publishes its books in a variety of electronic formats. Some content that appears in print, however, may not be available in electronic format.

"John Wiley & Sons, Inc. is pleased to publish this volume of Organic Syntheses on behalf of Organic Syntheses, Inc. Although Organic Syntheses, Inc. has assured us that each preparation contained in this volume has been checked in an independent laboratory and that any hazards that were uncovered are clearly set forth in the write-up of each preparation, John Wiley & Sons, Inc. does not warrant the preparations against any safety hazards and assumes no liability with respect to the use of the preparations."

For ordering and customer service, call 1-800-CALL-WILEY.

Library of Congress Catalog Card Number: 21-17747
ISBN 0-471-68257-8

Printed in the United States of America
10 9 8 7 6 5 4 3 2 1

ORGANIC SYNTHESES

VOLUME	EDITOR-IN-CHIEF	PAGES
I*	†ROGER ADAMS	84
II*	†JAMES BRYANT CONANT	100
III*	†HANS THACHER CLARKE	105
IV*	†OLIVER KAMM	89
V*	†CARL SHIPP MARVEL	110
VI*	†HENRY GILMAN	120
VII*	†FRANK C. WHITMORE	105
VIII*	†ROGER ADAMS	139
IX*	†JAMES BRYANT CONANT	108
Collective Vol. I	A revised edition of Annual Volumes I–IX	580
	†HENRY GILMAN, *Editor-in-Chief*	
	2nd Edition revised by †A. H. BLATT	
X*	†HANS THACHER CLARKE	119
XI*	†CARL SHIPP MARVEL	106
XII*	†FRANK C. WHITMORE	96
XIII*	†WALLACE H. CAROTHERS	119
XIV*	†WILLIAM W. HARTMAN	100
XV*	†CARL R. NOLLER	104
XVI*	†JOHN R. JOHNSON	104
XVII*	†L. F. FIESER	112
XVIII*	†REYNOLD C. FUSON	103
XIX*	†JOHN R. JOHNSON	105
Collective Vol. II	A revised edition of Annual Volumes X–XIX	654
	†A. H. BLATT, *Editor-in-Chief*	
20*	†CHARLES F. H. ALLEN	113
21*	†NATHAN L. DRAKE	120
22*	†LEE IRVIN SMITH	114
23*	†LEE IRVIN SMITH	124
24*	†NATHAN L. DRAKE	119
25*	†WERNER E. BACHMANN	120
26*	†HOMER ADKINS	124
27*	†R. L. SHRINER	121
28*	†H. R. SNYDER	121
29*	†CLIFF S. HAMILTON	119
Collective Vol. III	A revised edition of Annual Volumes 20–29	890
	†E. C. HORNING, *Editor-in-Chief*	
30*	†ARTHUR C. COPE	115
31*	†R. S. SCHREIBER	122

*Out of print.
†Deceased.

VOLUME	EDITOR-IN-CHIEF	PAGES
32*	†Richard T. Arnold	119
33*	†Charles C. Price	115
34*	†William S. Johnson	121
35*	†T. L. Cairns	122
36*	N. J. Leonard	120
37*	†James Cason	109
38*	†John C. Sheehan	120
39*	†Max Tishler	114
Collective Vol. IV	A revised edition of Annual Volumes 30–39 †Norman Rabjohn, *Editor-in-Chief*	1036
40*	†Melvin S. Newman	114
41*	John D. Roberts	118
42*	†Virgil Boekelheide	118
43*	B. C. McKusick	124
44*	†William E. Parham	131
45*	†William G. Dauben	118
46*	E. J. Corey	146
47*	†William D. Emmons	140
48*	†Peter Yates	164
49*	Kenneth B. Wiberg	124
Collective Vol. V	A revised edition of Annual Volumes 40–49 Henry E. Baumgarten, *Editor-in-Chief*	1234
Cumulative Indices to Collective Volumes, I, II, III, IV, V †Ralph L. and †Rachel H. Shriner, *Editors*		
50*	Ronald Breslow	136
51*	†Richard E. Benson	209
52*	Herbert O. House	192
53*	Arnold Brossi	193
54*	Robert E. Ireland	155
55*	†Satoru Masamune	150
56*	†George H. Büchi	144
57*	Carl R. Johnson	135
58*	†William A. Sheppard	216
59*	Robert M. Coates	267
Collective Vol. VI	A revised edition of Annual Volumes 50–59 Wayland E. Noland, *Editor-in-Chief*	1208
60*	†Orville L. Chapman	140
61*	†Robert V. Stevens	165

Out of print.
†*Deceased.*

VOLUME	EDITOR-IN-CHIEF	PAGES
62*	MARTIN F. SEMMELHACK	269
63*	GABRIEL SAUCY	291
64	ANDREW S. KENDE	308
Collective Vol. VII	A revised edition of Annual Volumes 60–64 JEREMIAH P. FREEMAN, *Editor-in-Chief*	602
65	EDWIN VEDEJS	278
66*	CLAYTON H. HEATHCOCK	265
67	BRUCE E. SMART	289
68	JAMES D. WHITE	318
69	LEO A. PAQUETTE	328
Collective Vol. VIII	A revised edition of Annual Volumes 65–69 JEREMIAH P. FREEMAN, *Editor-in-Chief*	696
70	ALBERT I. MEYERS	305
71	LARRY E. OVERMAN	285
72	†DAVID L. COFFEN	333
73	ROBERT K. BOECKMAN, JR.	352
74	ICHIRO SHINKAI	341
Collective Vol. IX	A revised edition of Annual Volumes 70–74 JEREMIAH P. FREEMAN, *Editor-in-Chief*	840
75	AMOS B. SMITH III	257
76	STEPHEN F. MARTIN	340
77	DAVID J. HART	312
78	WILLIAM R. ROUSH	326
79	LOUIS S. HEGEDUS	328
Collective Vol. X	A revised edition of Annual Volumes 75–79 JEREMIAH P. FREEMAN, *Editor-in-Chief*	
80	STEVEN WOLFF	259
81	RICK L. DANHEISER	296

Collective Volumes, Collective Indices to Collective Volumes I–VIII, Volumes 75–81 and Reaction Guide are available from John Wiley & Sons, Inc.

*Out of print.
†Deceased.

NOTICE

With Volume 62, the Editors of *Organic Syntheses* began a new presentation and distribution policy to shorten the time between submission and appearance of an accepted procedure. The soft cover edition of this volume is produced by a rapid and inexpensive process, and is sent at no charge to members of the Organic Division of the American Chemical Society, Gesellschaft Deutscher Chemiker, Polskie Towarzystwo Chemiczne, Royal Society of Chemistry, and The Society of Synthetic Organic Chemistry, Japan. The soft cover edition is intended as the personal copy of the owner and is not for library use. The hard cover edition is published by John Wiley and Sons, Inc., in the traditional format, and it differs in content primarily by the inclusion of an index. The hard cover edition is intended primarily for library collections and is available for purchase through the publisher. Incorporation of graphical abstracts into the Table of Contents began with Volume 77. Annual Volumes 70–74 and 75–79 have been incorporated into five-year versions of the collective volumes of *Organic Syntheses* that appeared as *Collective Volume IX and X* in the traditional hard cover format, available for purchase from the publishers. The Editors hope that the new *Collective Volume* series, appearing twice as frequently as the previous decennial volumes, will provide a permanent and timely edition of the procedures for personal and institutional libraries. The Editors welcome comments and suggestions from users concerning the new editions.

Organic Syntheses, Inc., joined the age of electronic publication in 2001 with the release of its free web site (www.orgsyn.org) followed in 2003 with the completion of a commercially available electronic database (www.mrw.interscience.wiley.com/osdb). Organic Syntheses, Inc., fully funded the creation of the free website at www.orgsyn.org in a partnership with CambridgeSoft Corporation, and Data-Trace Publishing Company. The success of this site in its first full year of operation was overwhelming, with an average of nearly 48,000 site hits/day and more than 27,000 page views/day. The site is accessible to most internet browsers using Macintosh and Windows operating systems and may be used with or without a ChemDraw plugin or Active X control. Because of continually evolving system requirements, users should review software compatibility at the website prior to use. John Wiley & Sons, Inc., and Accelrys, Inc., partnered with Organic Syntheses, Inc., to develop the new database (www.mrw.interscience.wiley.com/osdb) that is available for license with internet solutions from John Wiley & Sons, Inc., and intranet solutions from Accelrys, Inc.

Both the commercial database and the free website contain all annual and collective volumes and indices of *Organic Syntheses*. Chemists can draw structural queries and combine structural or reaction transformation queries with full-text and bibliographic search terms, such as chemical name, reagents, molecular formula, apparatus, or even hazard warnings or phrases. The preparations are categorized into reaction types, allowing search by category. The contents of individual or collective volumes can be browsed by lists of titles, submitters' names, and volume and page references, without or with reaction equations.

The commercial database at www.mrw.interscience.wiley.com/osdb also enables the user to choose his/her preferred chemical drawing package, or to utilize several freely available plug-ins for entering queries. The user is also able to cut and paste existing structures and reactions directly into the structure search query or their preferred chemistry editor, streamlining workflow. Additionally, this database contains links to the full text of primary literature references via CrossRef, ChemPort, Medline, and ISI Web of Science. Links to local holdings for institutions using open url technology can also be enabled. The database user can limit his/her search to, or order the search results by, such factors as reaction type, percentage yield, temperature, and publication date, and can create a customized table of reactions for comparison. Connections to other Wiley references are currently made via text search, with cross-product structure and reaction searching to be added in the coming year. Incorporation of new preparations will occur as new material becomes available.

INFORMATION FOR AUTHORS OF PROCEDURES

Organic Syntheses welcomes and encourages submission of experimental procedures that lead to compounds of wide interest or that illustrate important new developments in methodology. Proposals for *Organic Syntheses* procedures will be considered by the Editorial Board upon receipt of an outline proposal as described below. A full procedure will then be invited for those proposals determined to be of sufficient interest. These full procedures will be evaluated by the Editorial Board, and if approved, assigned to a member of the Board for checking. In order for a procedure to be accepted for publication, each reaction must be successfully repeated in the laboratory of a member of the Editorial Board at least twice, with similar yields (generally $\pm 5\%$) and selectivity to that reported by the submitters.

Organic Syntheses Proposals

A cover sheet should be included providing full contact information for the principal author and including a scheme outlining the proposed reactions (an Organic Syntheses Proposal Cover Sheet can be downloaded from the *Organic Syntheses* websites). Attach an outline proposal describing the utility of the methodology and/or the usefulness of the product. Identify and reference the best current alternatives. For each step, indicate the proposed scale, yield, method of isolation and purification, and how the purity of the product is determined. Describe any unusual apparatus or techniques required, and any special hazards associated with the procedure. Identify the source of starting materials. Enclose copies of relevant publications (attach pdf files if this an electronic submission).

Submit proposals by mail or as email attachments to:

Professor Charles K. Zercher
Associate Editor, *Organic Syntheses*
Department of Chemistry
University of New Hampshire
23 College Road, Parsons Hall
Durham, NH 03824

For electronic submissions: org.syn@unh.edu

Submission of Procedures

Authors invited by the Editorial Board to submit full procedures should prepare their manuscripts in accord with the Instructions to Authors which may be obtained from the Associate Editor or downloaded from the *Organic Syntheses* websites. Submitters are also encouraged to consult earlier volumes of *Organic Syntheses* for models with regard to style, format, and the level of experimental detail expected in *Organic Syntheses* procedures. Manuscripts should be submitted in triplicate to the Associate Editor. Electronic submissions are encouraged; procedures will be accepted as email attachments in the form of Microsoft Word files with all schemes and graphics also sent separately as ChemDraw files.

Procedures that do not conform to the Instructions to Authors with regard to experimental style and detail will be returned to authors for correction. Authors will be notified when their manuscript is approved for checking by the Editorial Board and it is the goal of the Board to complete the checking of procedures within a period of no more than six months.

Additions, corrections, and improvements to the preparations previously published are welcomed; these should be directed to the Associate Editor. However, checking of such improvements will only be undertaken when new methodology is involved. Substantially improved procedures have been included in the Collective Volumes in place of a previously published procedure.

NOMENCLATURE

Both common and systematic names of compounds are used throughout this volume, depending on which the Editor-in-Chief felt was more appropriate. The *Chemical Abstracts* indexing name for each title compound, if it differs from the title name, is given as a subtitle. Systematic *Chemical Abstracts* nomenclature, used in the Collective Indexes for the title compound and a selection of other compounds mentioned in the procedure, is provided in an appendix at the end of each preparation. Registry numbers, which are useful in computer searching and identification, are also provided in these appendixes. Whenever two names are concurrently in use and one name is the correct *Chemical Abstracts* name, that name is preferred.

ACKNOWLEDGMENT

Organic Syntheses wishes to acknowledge the contributions of Discovery Partners Intl., Hoffmann-La Roche, Inc., Merck & Co., and Pfizer, Inc. to the success of this enterprise through their support, in the form of time and expenses, of members of the Boards of Directors and Editors.

HANDLING HAZARDOUS CHEMICALS
A Brief Introduction

General Reference: *Prudent Practices in the Laboratory*; National Academy Press; Washington, DC, 1995.

Physical Hazards

Fire. Avoid open flames by use of electric heaters. Limit the quantity of flammable liquids stored in the laboratory. Motors should be of the nonsparking induction type.

Explosion. Use shielding when working with explosive classes such as acetylides, azides, ozonides, and peroxides. Peroxidizable substances such as ethers and alkenes, when stored for a long time, should be tested for peroxides before use. Only sparkless "flammable storage" refrigerators should be used in laboratories.

Electric Shock. Use 3-prong grounded electrical equipment if possible.

Chemical Hazards

Because all chemicals are toxic under some conditions, and relatively few have been thoroughly tested, it is good strategy to minimize exposure to all chemicals. In practice this means having a good, properly installed hood; checking its performance periodically; using it properly; carrying out all operations in the hood; protecting the eyes; and, since many chemicals can penetrate the skin, avoiding skin contact by use of gloves and other protective clothing at all times.

a. Acute Effects. These effects occur soon after exposure. The effects include burn, inflammation, allergic responses, damage to the eyes, lungs, or nervous system (e.g., dizziness), and unconsciousness or death (as from over-exposure to HCN). The effect and its cause are usually obvious and so are the methods to prevent it. They generally arise from inhalation or skin contact, so should not be a problem if one follows the admonition "work in a hood and keep chemicals off your hands." Ingestion is a rare route, being generally the result of eating in the laboratory or not washing hands before eating.

b. Chronic Effects. These effects occur after a long period of exposure or after a long latency period and may show up in any of numerous organs. Of the chronic effects of chemicals, cancer has received the most attention lately. Several dozen chemicals have been demonstrated to be carcinogenic in man and hundreds to be carcinogenic to animals. Although there is no simple correlation between carcinogenicity in animals and in man, there is little doubt that a significant proportion of the chemicals used in laboratories have some potential for carcinogenicity in man. For this and other reasons, chemists should employ good practices at all times.

The key to safe handling of chemicals is a good, properly installed hood, and the referenced book devotes many pages to hoods and ventilation. It recommends that in a laboratory where people spend much of their time working with chemicals there should be a hood for each two people, and each should have at least 2.5 linear feet (0.75 meter) of working space at it. Hoods are more than just devices to keep undesirable vapors from the laboratory atmosphere. When closed they provide a protective barrier between chemists and chemical operations, and they are a good containment device for spills. Portable shields can be a useful supplement to hoods, or can be an alternative for hazards of limited severity, e.g., for small-scale operations with oxidizing or explosive chemicals.

Specialized equipment can minimize exposure to the hazards of laboratory operations. Impact resistant safety glasses are basic equipment and should be worn at all times. They may be supplemented by face shields or goggles for particular operations, such as pouring corrosive liquids. Because skin contact with chemicals can lead to skin irritation or sensitization or, through absorption, to effects on internal organs, protective gloves should be worn at all times.

Laboratories should have fire extinguishers and safety showers. Respirators should be available for emergencies. Emergency equipment should be kept in a central location and must be inspected periodically.

MSDS (Materials Safety Data Sheets) are available from the suppliers of commercially available reagents, solvents, and other chemical materials; anyone performing an experiment should check these data sheets before initiating an experiment to learn of any specific hazards associated with the chemicals being used in that experiment.

DISPOSAL OF CHEMICAL WASTE

General Reference: *Prudent Practices in the Laboratory*, National Academy Press, Washington, D.C. 1995

Effluents from synthetic organic chemistry fall into the following categories:

1. **Gases**

    1a. Gaseous materials either used or generated in an organic reaction.
    1b. Solvent vapors generated in reactions swept with an inert gas and during solvent stripping operations.
    1c. Vapors from volatile reagents, intermediates and products.

2. **Liquids**

    2a. Waste solvents and solvent solutions of organic solids (see item 3b).
    2b. Aqueous layers from reaction work-up containing volatile organic solvents.
    2c. Aqueous waste containing non-volatile organic materials.
    2d. Aqueous waste containing inorganic materials.

3. **Solids**

    3a. Metal salts and other inorganic materials.
    3b. Organic residues (tars) and other unwanted organic materials.
    3c. Used silica gel, charcoal, filter aids, spent catalysts and the like.

The operation of industrial scale synthetic organic chemistry in an environmentally acceptable manner* requires that all these effluent categories be dealt with properly. In small scale operations in a research or

---

*An environmentally acceptable manner may be defined as being both in compliance with all relevant state and federal environmental regulations *and* in accord with the common sense and good judgement of an environmentally aware professional.

academic setting, provision should be made for dealing with the more environmentally offensive categories.

1a. Gaseous materials that are toxic or noxious, e.g., halogens, hydrogen halides, hydrogen sulfide, ammonia, hydrogen cyanide, phosphine, nitrogen oxides, metal carbonyls, and the like.
1c. Vapors from noxious volatile organic compounds, e.g., mercaptans, sulfides, volatile amines, acrolein, acrylates, and the like.
2a. All waste solvents and solvent solutions of organic waste.
2c. Aqueous waste containing dissolved organic material known to be toxic.
2d. Aqueous waste containing dissolved inorganic material known to be toxic, particularly compounds of metals such as arsenic, beryllium, chromium, lead, manganese, mercury, nickel, and selenium.
3. All types of solid chemical waste.

Statutory procedures for waste and effluent management take precedence over any other methods. However, for operations in which compliance with statutory regulations is exempt or inapplicable because of scale or other circumstances, the following suggestions may be helpful.

Gases

Noxious gases and vapors from volatile compounds are best dealt with at the point of generation by "scrubbing" the effluent gas. The gas being swept from a reaction set-up is led through tubing to a (large!) trap to prevent suck-back and on into a sintered glass gas dispersion tube immersed in the scrubbing fluid. A bleach container can be conveniently used as a vessel for the scrubbing fluid. The nature of the effluent determines which of four common fluids should be used: dilute sulfuric acid, dilute alkali or sodium carbonate solution, laundry bleach when an oxidizing scrubber is needed, and sodium thiosulfate solution or diluted alkaline sodium borohydride when a reducing scrubber is needed. Ice should be added if an exotherm is anticipated.

Larger scale operations may require the use of a pH meter or starch/iodide test paper to ensure that the scrubbing capacity is not being exceeded.

When the operation is complete, the contents of the scrubber should be handled as aqueous waste, as outlined in the "Liquids" section that follows. In many instances, this will require neutralization, followed by concentration to a minimum volume, or concentration to dryness before disposal as concentrated liquid or solid chemical waste.

Liquids

Every laboratory should be equipped with a waste solvent container in which *all* waste organic solvents and solutions are collected. The contents of these containers should be periodically transferred to properly labeled waste solvent drums and arrangements made for contracted disposal in a regulated and licensed incineration facility.**

Aqueous waste containing dissolved toxic organic material should be decomposed *in situ*, when feasible, by adding acid, base, oxidant, or reductant. Otherwise, the material should be concentrated to a minimum volume and added to the contents of a waste solvent drum.

Aqueous waste containing dissolved toxic inorganic material should be evaporated to dryness and the residue handled as a solid chemical waste.

Solids

Soluble organic solid waste can usually be transferred into a waste solvent drum, provided near-term incineration of the contents is assured.

Inorganic solid wastes, particularly those containing toxic metals and toxic metal compounds, used Raney nickel, manganese dioxide, etc. should be placed in glass bottles or lined fiber drums, sealed, properly labeled, and arrangements made for disposal in a secure landfill.** Used mercury is particularly pernicious and small amounts should first be amalgamated with zinc or combined with excess sulfur to solidify the material.

Other types of solid laboratory waste including used silica gel and charcoal should also be packed, labeled, and sent for disposal in a secure landfill.

Special Note

Since local ordinances may vary widely from one locale to another, one should always check with appropriate authorities. Also, professional disposal services differ in their requirements for segregating and packaging waste.

**If arrangements for incineration of waste solvent and disposal of solid chemical waste by licensed contract disposal services are not in place, a list of providers of such services should be available from a state or local office of environmental protection.

PREFACE

Beginning in 1921, and continuing for more than 80 years, *Organic Syntheses* has provided the chemistry community with annual collections of detailed, reliable, and carefully checked procedures for organic synthesis. There is good reason to believe that the value of *Organic Syntheses* to the synthetic community will only increase in the future. In recent years the amount of experimental detail in most journals has decreased significantly, and in many publications nearly all experimental information is now relegated to supporting information. It is common that such "supplementary material" does not receive the same scrutiny from referees and journal editors that is accorded the text of journal articles. With this trend increasing, the importance of *Organic Syntheses*, with its carefully checked and extremely detailed experimental procedures, will undoubtedly only increase.

This volume of *Organic Syntheses* provides 28 checked and edited experimental procedures which describe the preparation of useful chemicals and/or illustrate important synthetic methods. The procedures in this volume have been grouped in several broad categories based on the type of chemistry employed and the nature of the target molecules described.

The volume begins with four procedures that highlight the application of transition-metal organometallic chemistry in organic synthesis. The first procedure describes the preparation of **ACETIC ACID 2-METHYLENE-3-PHENETHYLBUT-3-ENYL ESTER** and illustrates the "cross enyne metathesis" reaction for the synthesis of conjugated dienes. The example given here features the use of the Grubbs "second-generation" ruthenium carbene as the metathesis catalyst. In this procedure, the alkyne substrate is prepared by an application of the Corey-Fuchs alkyne synthesis. The next procedure details the synthesis of **N,N-DIBENZYL-N-(2-ETHENYLCYCLOPROPYL)AMINE** and provides an example of an important variant of the Kulinkovich reaction, which in this case produces cyclopropylamines rather than cyclopropanols. The application of a vanadium catalyst for the reductive coupling of aldehydes is illustrated in the next procedure on the preparation of **d,l-1,2-DICYCLOHEXYLETHANEDIOL**. A transition-metal catalyzed cross-coupling reaction is the subject of the fourth procedure in this series. A procedure for the preparation of **4-NONYLBENZOIC ACID** provides an example of a "ligand-free" iron catalyst system for effecting the coupling of alkyl Grignard reagents with aryl chlorides.

Continuing in the area of transition-metal catalyzed reactions, the next five procedures describe important variants of palladium-catalyzed carbon-carbon bond-forming processes, a class of reactions that has revolutionized organic synthesis. The first two procedures describe methods for the effective hydroarylation of alkynes to form (E) or (Z) styrene derivatives. Both procedures feature the application of the palladium-catalyzed coupling of an alkenylsilanol derivative with 4-iodoanisole, and these procedures provide efficient routes to (E)- and (Z)-1-(HEPTENYL)-4-METHOXYBENZENE. In the first procedure, the alkenylsilanol intermediate is accessed via iodination of 1-heptyne followed by hydroboration, halogen-metal exchange to generate an alkenyllithium species, and silylation. Platinum-catalyzed hydrosilylation is employed in the second procedure to prepare the requisite intermediate for cross coupling.

Recently, several highly active ligand systems have been developed that allow the extension of palladium-catalyzed carbon-carbon bond-forming reactions to classes of substrates that previously did not participate smoothly in these reactions. The next procedure describes the application of one such system, a palladium-tri-*tert*-butylphosphine catalyst, in Heck reactions of aryl chlorides leading to **(E)-2-METHYL-3-PHENYLACRYLIC ACID BUTYL ESTER** and **(E)-4-(2-PHENYLETHENYL) BENZONITRILE**. The next procedure details the preparation of **N-(*tert*-BUTOXYCARBONYL)-β-IODOALANINE METHYL ESTER** and its application as a useful building block for the synthesis of non-natural amino acids via palladium-catalyzed Negishi-type coupling reactions. The final procedure in this series features the application of palladium-catalyzed carbon-carbon bond-forming reactions in the synthesis of heterocyclic compounds. Specifically, a procedure is described for the synthesis of **3-PYRIDYLBORONIC ACID** and its **PINACOL ESTER**, as well as the Suzuki coupling of the latter with 3-bromoquinoline to form **3-PYRIDIN-3-YLQUINOLINE**.

Three additional procedures concerned with the synthesis of heterocyclic compounds follow. An improved protocol for the electrophilic bromination of isoquinoline with NBS is demonstrated in the synthesis of **5-BROMOISOQUINOLINE** and **5-BROMO-8-NITROISOQUINOLINE**. Optimized conditions for the condensation of α-bromo ketones with amidines leading to 2,4-disubstituted imidazoles are illustrated in the preparation of **4-(4-METHOXYPHENYL)-2-PHENYL-1H-IMIDAZOLE**. The third procedure focuses on the preparation of N-aryl-5-hydroxymethyl-2-oxazolidinones via the alkylation of metalated carbamates with (R)-glycidyl butyrate, as illustrated with the synthesis of **N-PHENYL-(5R)-HYDROXYMETHYL-2-OXAZOLIDINONE**.

Two valuable methods based on organolithium chemistry are presented next. The generation and cyclization of 5-hexenyllithium to form **2-CYCLOPENTYLACETOPHENONE** illustrates an important method for ring formation. The next procedure describes the synthesis of ortho-substituted arylboronic esters such as **2-(5,5-DIMETHYL-1,3,2-DIOXABORINAN-2-YL)BENZOIC ACID ETHYL ESTER** via a novel *in situ* trapping of an unstable aryllithium intermediate generated by orthometalation of ethyl benzoate with LiTMP.

Four useful synthetic building blocks are the subject of the next procedures. The first of these procedures details the synthesis of **1,2:5,6-DIANHYDRO-3,4-O-ISOPROPYLIDENE-L-MANNITOL** via a two-step sequence beginning with L-mannonic acid δ-lactone and featuring the application of Mitsunobu chemistry for the conversion of 1,2-diols to terminal epoxides. The preparation of **(R)-4-PHENYL-3-(1,2-PROPADIENYL)OXAZOLIDIN-2-ONE** illustrates a general method for the synthesis of chiral allenamides. Next, the preparation of **NONRACEMIC 2-(*tert*-BUTYLDIMETHYLSILYLOXY)-2-PENTYN-1-OL** from (S)-ethyl lactate illustrates a valuable method for the synthesis of alkynyllithium compounds from aldehydes. Finally, the fourth procedure in this series describes the generation of 6-spiroepoxy-2,4-cyclohexadienone and its trapping in situ with cyclopentadiene to form **9-SPIROEPOXY-endo-TRICYCLO[5.2.2.0^{2,6}]UNDECA-4,10-DIEN-8-ONE**.

Four procedures featuring oxidation and reduction processes are next. A powerful method for the asymmetric reduction of ketones is the subject of the first procedure, which illustrates the synthesis and Ru(II)-BINAP catalyzed hydrogenation of a keto ester derived from hydroxyproline leading to **2(S)-(β-tert-BUTOXYCARBONYL-α-(R)- and α-(S)-HYDROXYETHYL-4(R)-HYDROXYPYRROLIDINE-1-CARBOXYLIC ACID, *tert*-BUTYL ESTER**. The indium-mediated reduction of aromatic nitro compounds is illustrated in the preparation of **ETHYL 4-AMINOBENZOATE**. Next, a procedure for the preparation of **4-METHOXYPHENYLACETIC ACID** demonstrates the utility of a method for the oxidation of primary alcohols to carboxylic acids using sodium chlorite and catalyzed by TEMPO and bleach. Finally, the synthesis of *N*-**BENZYLIDENE-BENZYLAMINE** *N*-**OXIDE** illustrates the application of the methyltrioxorhenium-catalyzed oxidation of amines to nitrones.

The volume concludes with five miscellaneous procedures. First, a method for the synthesis of orthogonally protected C-α, C-α-disubstituted amino acids is exemplified with the preparation of **1-*tert-*BUTYL-CARBONYL-4-((9-FLUORENYLMETHOXY CARBONYL) AMINO)PIPERIDINE-4-CARBOXYLIC ACID**. The next procedure provides conditions for the synthesis and use of glycosyl phosphates as glycosyl donors in the preparation of oligosaccharides. This chemistry is illustrated by its application in the synthesis of **3,4,6-TRI-O-BENZYL-2-O-PIVALOYL-β-D-GLUCOPYRANOSYL-(1 → 6)-1,2:3,4-DI-O-ISOPROPYLIDENE-α-D-GALACTO-PYRANOSIDE**. The next procedure illustrates the use of polystyrylsulfonyl chloride resin as a solid-supported condensation reagent for the preparation of esters from alcohols and carboxylic acids. The utility of this method is exemplified with its application to the synthesis of **N-[(9-FLUORENYLMETHOXY)CARBONYL]-L-ASPARTIC ACID, α-*tert*-BUTYL ESTER, β-(2-ETHYL[(1E)-(4-NITROPHENYL)-AZO]PHENYL]ETHYL ESTER**. The next procedure details the preparation of **2-PHENYLTHIO-5-HEPTANOL**, thus illustrating a general method for the activation of non-activated δ-carbon atoms via photolysis of alkyl benzensulfenate esters. Conditions for the preparation of **PIVALOYL HYDRAZIDE** by the reaction of pivaloyl chloride with hydrazine in aqueous media are described next, and finally the volume concludes with the synthesis of **N-BENZYL-4-PHENYLBUTYR-AMIDE** via the application of a procedure for the boric acid-catalyzed condensation of amines and carboxylic acids.

The efforts of many dedicated individuals have contributed to the making of this volume of *Organic Syntheses*. First, I would like to especially thank my colleagues on the Board of Editors for their hard work, thoughtful intellectual contributions, and overall commitment to excellence. The responsibilities of members of the Editorial Board include identifying important new chemistry for inclusion in *Organic Syntheses*, carefully reviewing submitted procedures to ensure that they are adequately detailed and clear, and last but not least, the important task of carefully checking the procedures being considered for publication.

In this connection, I would also like to acknowledge the very significant contributions of the actual "Checkers", that is, the members of the research groups of Editorial Board members who often devote considerable time and effort to the checking of procedures.

Finally, I would like to conclude this Preface by acknowledging the enormous contributions of Professor Jeremiah P. Freeman, Secretary to the Board of Editors of *Organic Syntheses* for the past 25 years. The success of *Organic Syntheses* and its value to the organic chemistry community is due in no small part to the skill and dedication of Professor Freeman. Regrettably, Dr. Freeman will be retiring in September, and so this volume marks the end of the "Freeman Era" of *Organic Syntheses*. It has been a privilege to work with Jerry, and on behalf of all of the members of the Editorial Board I would like to thank him for his outstanding service to *Organic Syntheses* and to the organic chemistry community.

RICK L. DANHEISER

Cambridge, Massachusetts
April, 2004

Virgil Boekelheide
1919–2003

Virgil Boekelheide died in his sleep on September 24, 2003, in Eugene, Oregon. He was 84, and although he had been retired from his position at the University of Oregon for many years, he retained an active interest in both chemistry and the musical arts to the end of his life. Virgil was born in Chelsea, South Dakota, to parents of German descent, and his early education was at a small school in Northville, South Dakota, from which he graduated at the age of fifteen. He completed an A.B. degree *Magna cum Laude* from Dakota Wesleyan University in 1939 and then attended the University of Minnesota where he received his Ph.D. under C.F. Koelsch in 1943. Virgil began his independent academic career that year as an Instructor at the University of Illinois, and from there he moved in 1946 to the University of Rochester where he remained until 1960. His final move was to the University of Oregon where he spent a highly productive 24 years until his retirement in 1984.

Virgil Boekelheide's early research focused on alkaloid chemistry, but this merely served as the entry point to the main theme which occupied him through the rest of his career – the synthesis of structurally intriguing molecules with interesting chemical properties. To a large extent, this interest centered on non-benzenoid aromatic systems, and his 1964 PNAS paper describing the synthesis of a 15,16-dihydropyrene is acknowledged as one of the classics of the organic literature. This work led to a fundamental advance in the annulene field by creating a [14]-annulene with substituents positioned inside the π cloud, and Boekelheide soon capitalized on this brilliant accomplishment by demonstrating remarkable electronic and chemical properties of his new system. The dihydropyrene synthesis illustrated a second structural feature associated with annulenes, namely that transannular steric interactions could be removed and annulenes stabilized by replacing internal hydrogens with carbon-carbon bonds. This important principle has become a pivotal component of annulene design and has been used to great advantage, for example by Emmanuel Vogel in his syntheses of methano-bridged annulenes.

Boekelheide's success in this area led seamlessly to a second major research endeavor, the construction of stacked aromatic systems in the form of cyclophanes. The culmination of this effort was the spectacular preparation of "superphane," a compound in which two benzene rings are clamped together by six ethano bridges. This unique molecule in which the pair of benzene decks are separated by less than the distance between graphite layers was long considered among the "holy grails" of organic synthesis, and the Boekelheide accomplishment must surely rank among the major achievements in the field.

Virgil Boekelheide's long and distinguished career in science was recognized with numerous awards including election to the National Academy of Sciences in 1962, the first Oregonian to be so honored. He also received a Guggenheim Fellowship (1953), an Alfred P. Sloan Fellowship (1958), a Fulbright Distinguished Fellowship (1953), and Senior Alexander von Humboldt Fellowships in 1974 and 1982. These last two awards enabled Virgil to build extremely fruitful and congenial collaborations with colleagues in Germany who shared his research interests. Foremost among these individuals was Professor Henning Hopf at the Technical University of Braunschweig who remained a lifelong friend of Virgil's. Virgil's welcoming disposition encouraged a large number of international scholars to join his research group, and many of these continue to practice the special brand of novel and exciting chemistry that Virgil has left as his legacy. Virgil Boekelheide joined the Board of Editors of *Organic Syntheses* in 1956 and edited Volume 62. He retired from the Board in 1964.

In his later years and especially after his retirement, Virgil became an active patron of the arts, and he contributed substantial donations to the opera, ballet and symphonic organizations in his hometown of Eugene. For his philanthropic participation in these activities, he was elected President of the Eugene Ballet Society where he served from

1988–1991. Virgil was deservedly proud of his role in stimulating artistic as well as scientific accomplishment, and in 1994, he and his wife Caroline established an endowment fund at the University of Oregon with the unusual mission of supporting "teaching and research in chemistry, music, and dance." Virgil Boekelheide is survived by his wife of 58 years, Caroline, his sons, Erich and Karl, his daughter, Anne, his brother Irving, his sister Dorothy Anderson, and his grandchildren, Amanda and David.

JAMES D. WHITE
Corvallis, Oregon

Orville L. Chapman
June 26, 1932–January 22, 2004

Orville L. Chapman was born on June 26, 1932, in New London, Connecticut. The son of a Naval officer, he grew up in several cities in the United States and Central America. He attended high school in San Diego and received his undergraduate degree at Virginia Polytechnic Institute in Blacksburg, Virginia with a double major in Chemistry and English. Orville received his Ph.D. with Jerrold Meinwald at Cornell University in 1957. He became an Instructor at Iowa State in 1957 and moved up the ranks to Professor in 1964. Orville was an early pioneer in the emerging field of organic photochemistry.

Chapman moved from Iowa State to UCLA in 1974, on the heels of his exciting successes in applying matrix isolation spectroscopy to the characterization of cyclobutadiene and benzyne. The years 1975–85 were an extremely productive period for the investigation of a wide variety of organic reactive intermediates: carbenes, nitrenes, propadienones, silenes, carbonyl oxides, strained alkynes, etc. At UCLA, Orville's ideas concerning the novel molecule, C_{60}, germinated in ca. 1979/80. In 1981, Orville initiated efforts directed at the chemical synthesis of C_{60}. This work was but one part of a new effort in the synthesis and characterization of various types of strained, non-planar aromatic compounds. In retrospect, these efforts are now recognized as pioneering contributions to materials chemistry.

In 1984, Orville and his colleague, Arlene Russell, formed a company that offered in-house short courses in technical writing. Arlene and Orville also collaborated in the production of a laser videodisc for teaching NMR spectroscopy to undergraduates (ca. 1986–88). This was perhaps the first effort in Orville's emerging interest in revamping the undergraduate curriculum. It led to the idea of using ^{13}C NMR spectroscopy as a method for introducing the topic of organic chemistry. In 1989, he became Associate Dean for Educational Innovation at UCLA, a position that he held until his death. Orville spearheaded the UCLA proposal for curriculum reform in chemistry that was funded by the NSF.

Orville joined the Board of Editors of Organic Syntheses in 1975. He actively encouraged and solicited procedures illustrating the synthetic applications of photochemistry. He edited Volume 60 (1981) in this series, which was in large part devoted to photochemical processes.

Chapman received many national and international awards, including the Pure Chemistry Award and the Arthur C. Cope Medal from the American Chemical Society, the Havinga Medal from the Stichtung Havinga, Leiden, the Netherlands, and the Texas Instruments Foundation Founders' Prize. He was elected to the National Academy of Sciences in 1974. Professor Chapman received the ComputerWorld Smithsonian Institute Award for the best use of computers in education and academia in 1995. He was a long-term consultant for Mobil chemical, and was involved in the invention and development of a significant number of their processes.

Professor Chapman was internationally recognized as a brilliant, creative scholar and an intellectual leader in various fields of endeavor. He was a trailblazer and innovator in photochemistry, matrix isolation spectroscopy, reaction intermediates, chemical communication, the mechanism of olfactory perception, polymers, and materials design. He also achieved a worldwide reputation for bringing the best of information technology to higher education.

Orville is survived by his mother, his wife Susan, his two sons, Kevin and Kenneth, and three grandsons. A fellowship in his name will be established at UCLA.

CHRISTOPHER S. FOOTE
University of California, Los Angeles

David Llewellyn Coffen
March 2, 1939–December 18, 2002

David Llewellyn Coffen was born in St. John's, Newfoundland, and as a "Newfie" was proud of his Canadian heritage, even though he spent his entire professional life in the United States. He took his undergraduate degree in 1961 at the University of Toronto where he obtained a B.Sc. with First Class Honors in Chemistry. As a result of his distinguished record at Toronto, he was awarded a Woodrow Wilson Fellowship to begin graduate study at MIT. He chose George Büchi as his mentor, and the influence of "GB" on David Coffen's approach to organic chemistry remained strong throughout his life. David's synthetic work on the Iboga alkaloids, which formed the major portion of his Ph.D. thesis, is still cited as the seminal contribution in this area.

Following completion of his Ph.D. in 1965, David Coffen took a Postdoctoral Fellowship at ETH where he worked in Albert Eschenmoser's laboratory. This was a time of great excitement at both ETH and in the Woodward group at Harvard where collaborative efforts towards the synthesis of vitamin B_{12} were underway, and David found this heady atmosphere exhilarating. His year in Switzerland was not entirely consumed with academic activity, however, and among his several recreational ventures he learned to cook a delicious fondue. David returned to the U.S. in September 1966 to take up a position as Assistant Professor in the Chemistry Department at the University of

Colorado. Boulder, with its strong tradition in physical organic chemistry, turned out to be a difficult setting to build a program in synthesis, but one of his accomplishments was the first synthesis of tetrathiafulvalene, a molecule of much interest at the time to those in search of organic semiconductors.

In mid-1971, David left the University of Colorado to begin a long and illustrious career in industry. His first position was as a Senior Scientist on the Chemical Research Staff of Hoffman-La Roche in Nutley, NJ, and over the course of twenty-four years with the company, he rose through the ranks to become Vice President for Chemistry Research. David's expertise in synthetic organic chemistry and his inclination towards application of his skills in the creation and development of pharmaceutical products made him a highly valued member of the Roche staff. During his career with Roche Nutley, he led a program that resulted in the first U.S. multi-kilo production of β-lactam antibiotics, and he devised a new process for large scale vitamin D metabolite production. He was also a pioneer in establishing an industrial basis for using enzyme catalysis as a method for the synthesis of enantiomerically pure drugs. Towards the end of his career with Roche, David's responsibilities encompassed a very large segment of the company's research and development portfolio, including the entire process research operation as well as programs dedicated to the synthesis of prostaglandins, leukotrienes, steroidal anti-inflammatories and CNS agents. Alongside these endeavors, he succeeded in developing a broad-based program in combinatorial synthesis which resulted in a large number of both peptide and non-peptidic libraries. Lead candidates for the treatment of HIV and other diseases have emerged from these programs.

David retired from Hoffmann-La Roche in 1995 and joined ArQule, at that time a fledgling combinatorial chemistry company in Medford, Massachusetts, as its Senior Vice President. With David's help, ArQule grew to be one of the foremost practitioners of automated parallel synthesis, and the company rapidly expanded its research and development operations through collaborations with the pharmaceutical, agrochemical, and bioseparation industries. The move from New Jersey to Massachusetts was not without some anguish, the large disparity in housing costs being a particular concern, but David and his wife, Charlotte, were fortunate to find a delightful older home in Cambridge. In 1998, an opportunity presented itself for David to move to Southern California as the Chief Scientific Officer for Discovery Partners International and Vice President of Chem Rx, a combinatorial chemistry company acquired by DPI from Axys Pharmaceuticals when the latter merged with Celera Genomics. Although not instinctively a Californian, David relished the ambiance and climate of La Jolla after six decades of life in the northeastern U.S. and Canada. A heavy regimen of travel and a demanding workload had begun to take its toll, however, and David's health began to decline in the late spring of 2002. Shortly after a visit to Oregon State University as a guest lecturer, he was diagnosed with colon cancer, and in spite of a valiant effort to survive the disease, he died a few days before Christmas of that year.

David Coffen's service to the organic community took many forms. In addition to adjunct faculty appointments at Seton Hall University and at Rutgers University in Newark, he was in frequent demand as a visiting speaker. His carefully prepared lectures, lucid style, and wry humor made him an entertaining speaker who could translate the missions and constraints of industrial chemistry into a form that was immediately comprehensible to an audience of both students and professionals. David was above all a practical scientist who saw chemistry as the means to an end rather than an edifice of scholarly accomplishment. He was, nevertheless, acutely aware of the history of the subject, and he had a remarkable ability to recall incidents and events from his past, which had shaped his thinking about science in general and chemistry in particular.

David Coffen served on a number of scientific advisory boards during his career, including that of the Dow Chemical Company, IRIX Pharmaceuticals, Linden Technologies, the ACS Petroleum Research Fund, and the National Research Council of Canada's Steacie Institute. He was a member of the board of Editors of Organic Syntheses from 1988 until 1996 and was Editor-in-Chief of Volume 72. He was subsequently elected to the Board of Directors of Organic Syntheses and remained a director until his death. He gave generously of his time to all these activities and often provided a perspective, which would not otherwise have been appreciated. He was good at thinking "outside the box," an attribute which undoubtedly contributed to his success in the world of industrial chemistry, and although his opinions on many subjects were unshakeable, his faith in his own judgement combined with a dry wit gave him an authority that left an indelible impression. To his close friends he was intensely loyal, and to his family he was a dedicated husband and father.

David Coffen is survived by his wife, Charlotte, his son, Charles, and his daughter, Kirsten.

JAMES D. WHITE
Corvallis, OR

Satoru Masamune
July 24, 1928–November 9, 2003

Satoru Masamune, Editor-in-Chief of Volume 55 of *Organic Syntheses*, passed away on November 9, 2003. He was 75 years old.

Sat Masamune was born in Fukuoka, Japan, and received his undergraduate education at Tohoku University. In Sendai, he carried out undergraduate research in the laboratory of his future father-in-law, Professor Tetsuo Nozoe. After receiving his bachelor's degree in 1952, he moved to the United States to pursue graduate studies at the University of California, Berkeley as one of the first Fulbright Fellows. He received his Ph.D. in 1957, working under the direction of Henry Rapoport.

Masamune spent the period 1956 to 1961 at the University of Wisconsin, Madison, first as a postdoctoral fellow in the laboratory of Eugene Van Tamelen, and then as a lecturer. In 1961, he received an appointment as a Fellow of the Mellon Institute in Pittsburgh, where he was to stay for the next three years.

Masamune's syntheses of the diterpene kaurene and the diterpene alkaloids atisine and garryine catapulted him to the first rank of synthetic organic chemists upon their publication in four back-to-back-to-back-to-back communications in the *Journal of the American Chemical Society* in 1964. These diterpene alkaloids possess exceedingly complex hexacyclic structures consisting of an intricate network of six interconnected rings, and their structures appear formidable and intimidating even to a synthetic organic chemist today. Perhaps most astonishing is the fact that Sat accomplished this tour de force singlehandedly, that is, literally with only his own hands, thus beating out more than 20 competing research laboratories around the world who were in hot pursuit of the same target molecules. This boded well for what Sat might accomplish were he to have the benefit of a group of coworkers!

In 1964, Masamune joined the faculty of the University of Alberta in Edmonton where he was promoted to full professor in 1967. He was recruited to join the faculty of MIT in 1978, and was named the Arthur C. Cope Professor of Chemistry in 1991. He became an emeritus professor at MIT in 2000, but remained actively interested in research, visiting his office nearly every day until his death.

Sat was a great friend, an extraordinary scientist, and an inspiration as a colleague. His creativity, high standard of excellence, and his incredible work ethic were an inspiration to his coworkers and colleagues alike. Among the many remarkable features of Sat's scientific career, two qualities particularly stand out as characterizing his great body of work: breadth and depth.

The breadth of Sat's interests and accomplishments was extraordinary. His selection of research projects was guided, always, by what he identified as *the* most important problems of the day, regardless of whether that might require him to foray out of his current comfortable and familiar sphere of research, thereby abandoning an area in which he might recently have established pre-eminence. As a result, his program spanned an astonishingly broad range of problems, bordering on one hand research in physics, as in his collaboration with the late Professor Tanaka of MIT, and bordering, at the other end of the chemistry spectrum, the interface where chemistry meets biology, as exemplified in his research on enzymatic reactions and on catalytic antibodies.

With regard to depth, I allude to the way in which, with every scientific problem that attracted his interest, Sat attacked the problem with a thoroughness and intensity unsurpassed by any chemist in my personal experience. These characteristics, together with his uncompromisingly high standards of scientific rigor, served him well as on more than one occasion he entered a research area rife with controversial results and competing claims, and time and again in short order carried out the definitive experiments that solved the key problems of the field and clarified the vexing ambiguities from earlier research.

In addition to his pioneering work on the synthesis of the diterpene alkaloids, Masamune made monumental contributions in a number of other areas of organic chemistry. Sat had a lifelong fascination with unusual cyclic structures possessing arrays of conjugated pi bonds. The origins of this passion no doubt has its roots in his undergraduate studies on the chemistry of tropones and tropolones, research he performed as an undergraduate at Tohoku University in the laboratory of Professor Tetsuo Nozoe. The theoretically interesting molecules which attracted Sat's attention, and the attention of many, many other scientists in the 1960s, 70s, and 80s, included a variety of fascinating structures predicted to possess imposing instability, instability derived from the fashion in which their structures were expected to be twisted, warped, distorted, stretched, and compressed, or destabilized by effects such as anti-aromaticity. Sat's contributions in this area are too many to enumerate, but include, most notably, the synthesis and investigation of members of the [4] and [10] annulene families. Thus, Masamune and his students were the first to prepare the simple [10] annulenes, which they achieved by an inspired choice of a synthetic route involving the photochemical generation of these exquisitely delicate and unstable structures at low temperature in the absence of other potentially reactive species. This work was a true tour de force, both intellectually and experimentally.

Similar tools were brought to bear as Masamune undertook pursuit of the Holy Grail of the field of theoretically interesting molecules, cyclobutadiene. This simple structure had defied the best efforts of numerous prominent chemists since the molecule was first discussed 100 years earlier. Masamune's keen theoretical analysis of the problem, and again, his extraordinary experimental acumen, enabled him to perform what is regarded as the most impressive and definitive work that was carried out in this very prominent area. Sat's research on the cyclobutadiene problem climaxed with his instantly classic 1978 paper which he elegantly entitled "Cyclobutadiene Is Not Square", a paper in which he revealed unequivocal evidence for the correct ground state structure of this most famous of elusive Huckel ring systems. Space does not permit a full account of his many other triumphs in the field of strained and unusual "unnatural products", but let me call a roll of some of his many conquests in this area: basketene, the trishomocyclo-propenium and pyramidal $(CH)_5$ cations, the aza and oxo [9] annulenes, the bridged 1,5-methano [10] annulene. In addition, in the 1980s and 90s the Masamune group also contributed to the chemistry of strained molecules composed of the Group 14 elements silicon, germanium, and tin. This is yet another case where Sat entered a totally new area and in remarkably short order performed definitive work, solving problems that had eluded the best efforts of the specialists of the field over a period of years.

Another area of significant contributions involves the chemical synthesis of the macrolide antibiotics, a field that first attracted Sat's interest in the late 1960s during his tenure at the University of Alberta. Why was Sat attracted to this problem? No doubt because it was universally regarded as the most formidable problem then confronting the science of organic synthesis. Just several years earlier, referring to one of the most well known and prototypical macrolides, R. B. Woodward had written that in spite of all the recent fabulous advances of the field of synthesis, the synthesis of these molecules "with all our advantages, looks at present quite hopelessly complex". Sat and his group are credited with what is now recognized as the first successful, total synthesis of a stereochemically complex, medicinally important polyoxo macrolide, specifically, the antibiotic methymycin. This monumental achievement was described in a series of three back-to-back-to-back JACS communications that appeared in 1975. I was a graduate student at Harvard at the time and I vividly remember the excitement, and envy, that the appearance of these publications engendered. The solution to this synthetic problem, from today's vantage point, would be described as "classical". However, we are only able to characterize it thus because of the revolution that was about to take place in synthetic organic chemistry, a revolution in which Sat was to be a key architect.

The term "revolution" is appropriate here, because it is no exaggeration to say that the strategy of "double asymmetric synthesis" has revolutionized the way chemists go about the synthesis of a very broad range of molecules, namely stereochemically complex, acyclic compounds. Landmark papers from the Masamune laboratory in the period 1979–1980 established the utility of this important concept in the context of the boron aldol reaction, a process for which Sat was one of the pioneers, and the power of this strategy was demonstrated convincingly by the Masamune group through the total synthesis of 6-deoxyerythronolide B in 1981. The conquest of this molecule was followed in rapid succession by successful ascents of a series of other synthetic Everests, including narbonolide and tylonolide in 1982, key components of rifamycin S and streptovaricin A also in 1982, amphotericin B in the period 1984 to 1988, bryostatin and pimaricin in 1990, and calyculin A in 1994.

Sat's contributions were recognized over the years with numerous honors and awards. In 1978 he received the ACS Award for Creative Work in Synthetic Organic Chemistry, and in 1987 he was the recipient of an Arthur C. Cope Scholar Award. He was a fellow of the Royal Society of Canada and of the American Academy of Arts and Sciences, and was named a Centenary Scholar of the Chemical Society of London in 1980. Sat received the prestigious Fujihara award in Japan in 1997.

Sat had a number of passions outside chemistry, most notably classical music, baseball, and sumo wrestling. A longtime resident of Newton, Massachusetts, he is survived by his wife, Takako (Nozoe), his daughter Hiroko, his son Tohoru, a sister, Michiko Hiyama of Hirosaki, Japan; and four brothers, Tadashi of Sapporo, Japan, Osamu of Akita, Japan, and Shinobu and Tsutomu, both of Tokyo. Sat Masamune is dearly missed by his family, former coworkers, and his colleagues.

RICK L. DANHEISER

May 30, 2004   Cambridge, MA

CONTENTS

SYNTHESIS OF 1,3-DIENES FROM ALKYNES AND ETHYLENE: 1
ACETIC ACID 2-METHYLENE-3-PHENETHYLBUT-3-ENYL ESTER
Miwako Mori, Keisuke Tonogaki, and Atsushi Kinoshita

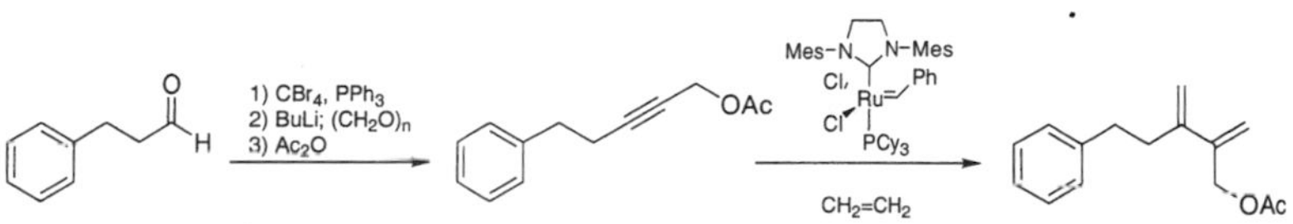

FACILE SYNTHESES OF AMINOCYCLOPROPANES: 14
N,N-DIBENZYL-*N*-(2-ETHENYLCYCLOPROPYL)AMINE
Armin de Meijere, Harald Winsel, and Björn Stecker

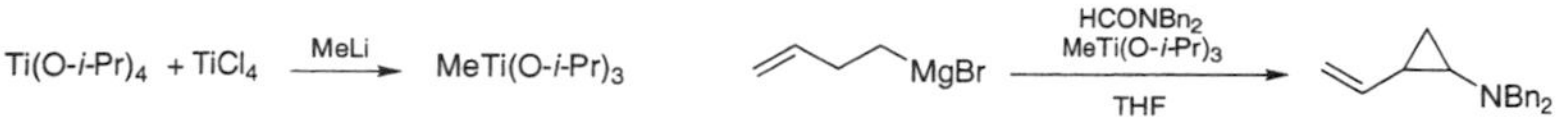

***dl*-SELECTIVE PINACOL-TYPE COUPLING USING ZINC, CHLOROSILANE,**   26
AND CATALYTIC AMOUNTS OF Cp$_2$VCl$_2$:
***dl*-1,2-DICYCLOHEXYLETHANEDIOL**
Toshikazu Hirao, Akiya Ogawa, Motoki Asahara, Yasuaki Muguruma, and Hidehiro Sakurai

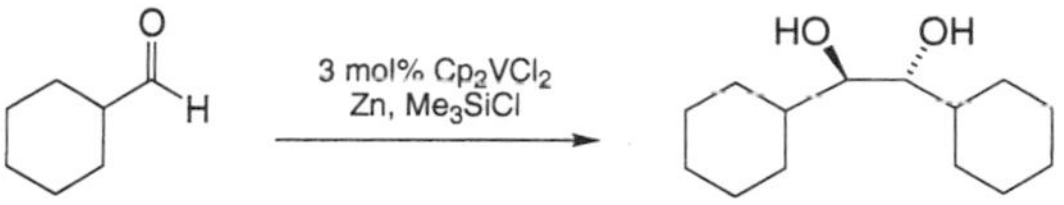

4-NONYLBENZOIC ACID   33
Alois Fürstner, Andreas Leitner, and Günter Seidel

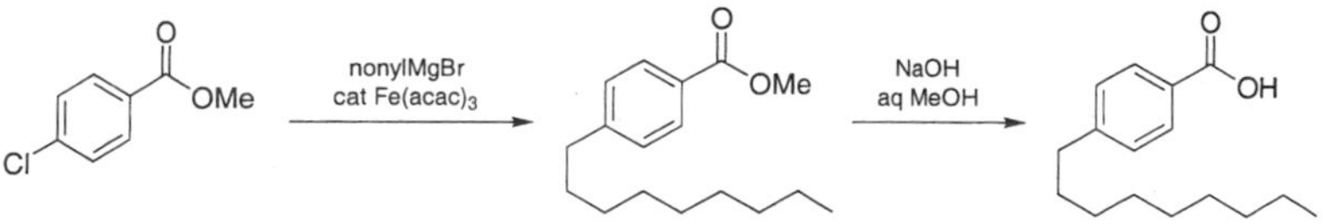

PALLADIUM CATALYZED CROSS-COUPLING OF (Z)-1-HEPTENYLDIMETHYLSILANOL WITH 4-IODOANISOLE: (Z)-1-(HEPTENYL)-4-METHOXYBENZENE

Scott E. Denmark and Zhigang Wang

42

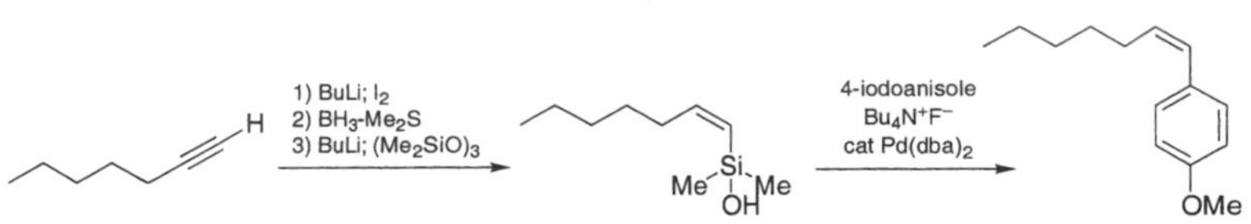

PLATINUM CATALYZED HYDROSILYLATION AND PALLADIUM CATALYZED CROSS-COUPLING: ONE-POT HYDROARYLATION OF 1-HEPTYNE

Scott E. Denmark and Zhigang Wang

54

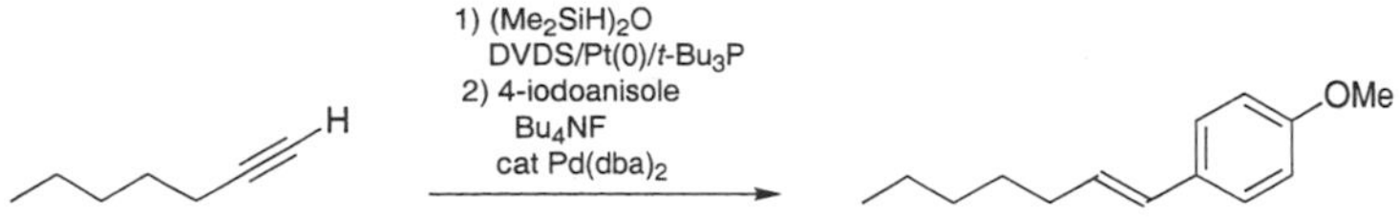

HECK REACTIONS OF ARYL CHLORIDES CATALYZED BY PALLADIUM/TRI-tert-BUTYLPHOSPHINE: (E)-2-METHYL-3-PHENYLACRYLIC ACID BUTYL ESTER AND (E)-4-(2-PHENYLETHENYL)BENZONITRILE

Adam F. Littke and Gregory C. Fu

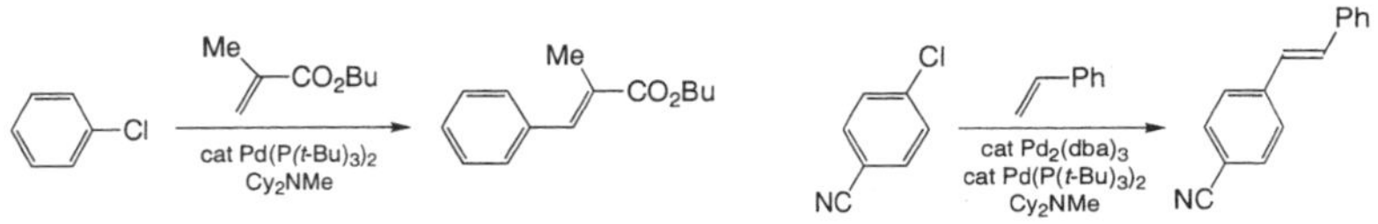

63

SYNTHESIS OF N-(tert-BUTOXYCARBONYL)-β-IODOALANINE METHYL ESTER: A USEFUL BUILDING BLOCK IN THE SYNTHESIS OF NONNATURAL α-AMINO ACIDS VIA PALLADIUM CATALYZED CROSS COUPLING REACTIONS

Richard F. W. Jackson and Manuel Perez-Gonzalez

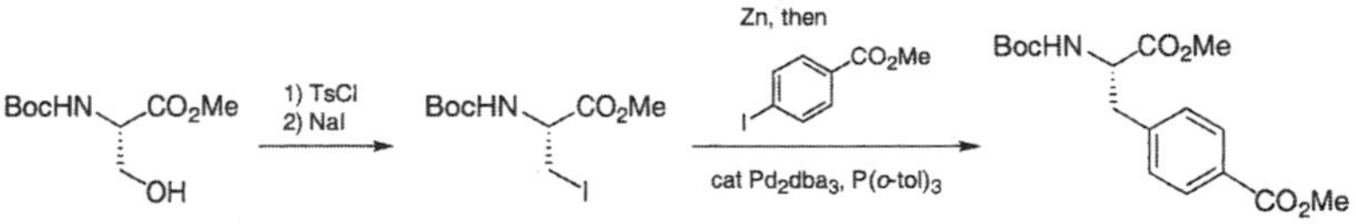

77

SYNTHESIS OF 3-PYRIDYLBORONIC ACID AND ITS PINACOL ESTER. APPLICATION OF 3-PYRIDYLBORONIC ACID IN SUZUKI COUPLING TO PREPARE 3-PYRIDIN-3-YLQUINOLINE

Wenjie Li, Dorian P. Nelson, Mark S. Jensen, R. Scott Hoerrner, Dongwei Cai, and Robert D. Larsen

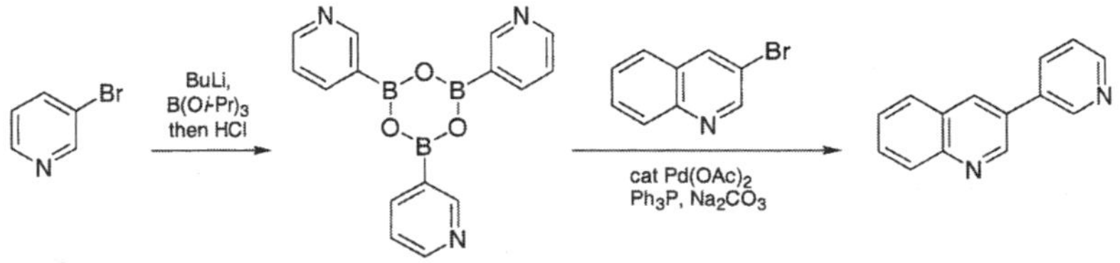

89

SYNTHESIS OF 5-BROMOISOQUINOLINE AND 5-BROMO-8-NITROISOQUINOLINE   98
William Dalby Brown and Alex Haahr Gouliaev

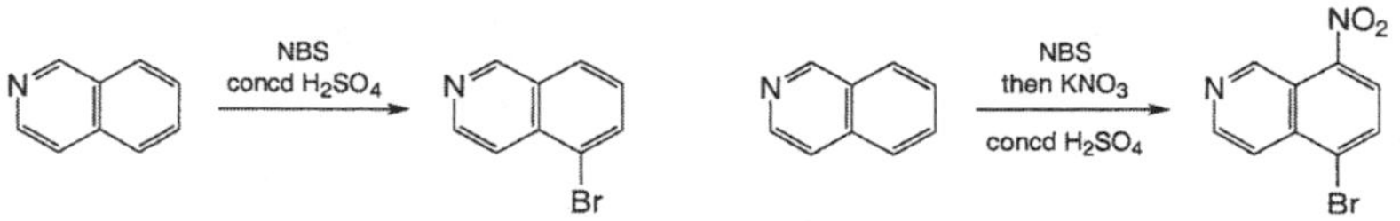

PREPARATION OF 2,4-DISUBSTITUTED IMIDAZOLES: 105
4-(4-METHOXYPHENYL)-2-PHENYL-1H-IMIDAZOLE
Bryan Li, Charles K-F Chiu, Richard F. Hank, Jerry Murry, Joshua Roth, and Harry
Tobiassen

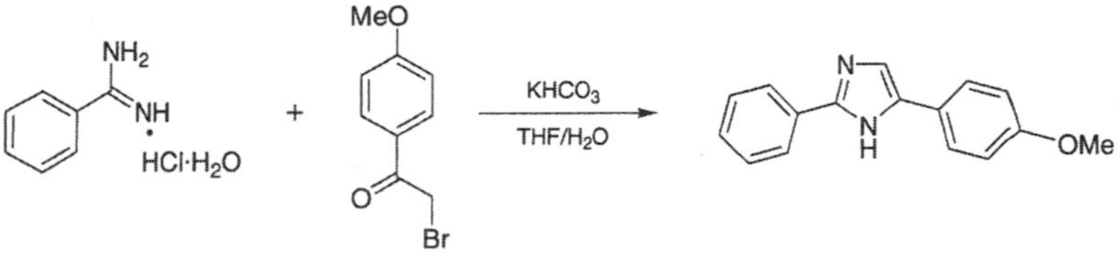

PREPARATION OF N-ARYL-(5R)-HYDROXYMETHYL-2-OXAZOLIDINONES FROM N-   112
ARYL CARBAMATES: N-PHENYL-(5R)-HYDROXYMETHYL-2-OXAZOLIDINONE
Peter R. Manninen and Steven J. Brickner

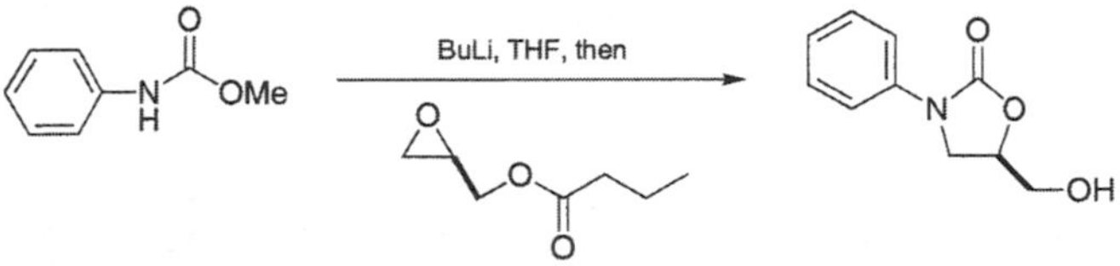

GENERATION AND CYCLIZATION OF 5-HEXENYLLITHIUM: 121
2-CYCLOPENTYLACETOPHENONE
William F. Bailey, Matthew R. Luderer, Michael J. Mealy, and Eric R. Punzalan

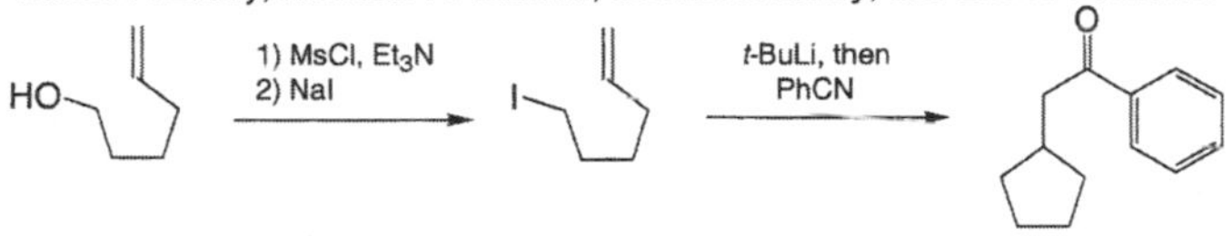

SYNTHESIS OF ORTHO-SUBSTITUTED ARYLBORONIC ESTERS BY IN SITU   134
TRAPPING OF UNSTABLE LITHIO INTERMEDIATES: 2-(5,5-DIMETHYL-1,3,2-
DIOXABORINAN-2-YL) BENZOIC ACID ETHYL ESTER
Jesper Langgaard Kristensen, Morten Lysén, Per Vedsø, and Mikael Begtrup

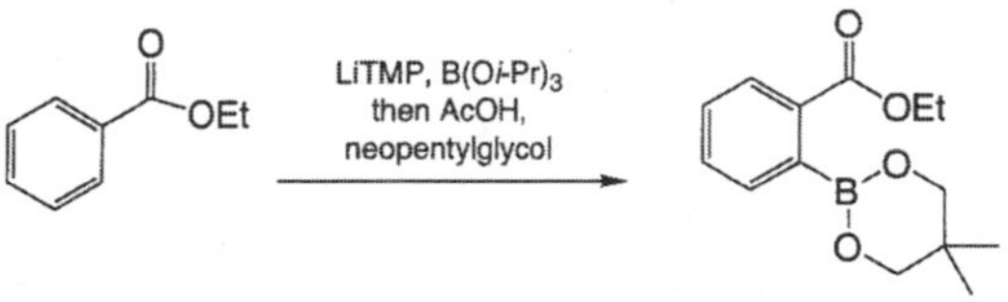

SYNTHESIS OF 1,2:5,6-DIANHYDRO-3,4-O-ISOPROPYLIDENE-L-MANNITOL   140
David A. Nugiel, Kim Jacobs, A. Christine Tabaka, and Chris A. Teleha

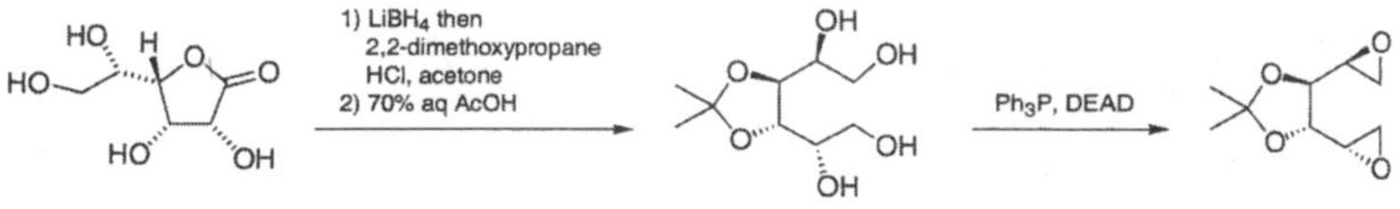

PRACTICAL SYNTHESIS OF NOVEL CHIRAL ALLENAMIDES: 147
(R)-4-PHENYL-3-(1,2-PROPADIENYL)OXAZOLIDIN-2-ONE
H. Xiong, M. R. Tracey, T. Grebe, J. A. Mulder, and R. P. Hsung

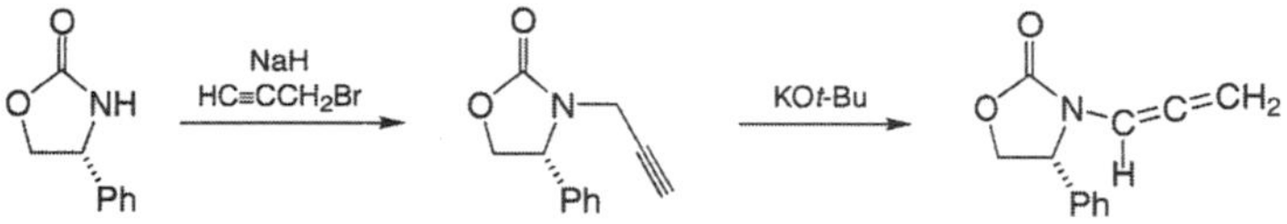

GENERATION OF NONRACEMIC 2-(t-BUTYLDIMETHYLSILYLOXY)-3-   157
BUTYNYLLITHIUM FROM (S)-ETHYL LACTATE:
(S)-4-(t-BUTYLDIMETHYLSILYLOXY)-2-PENTYN-1-OL
James A. Marshall, Mathew M. Yanik, Nicholas D. Adams, Keith C. Ellis, and Harry R.
Chobanian

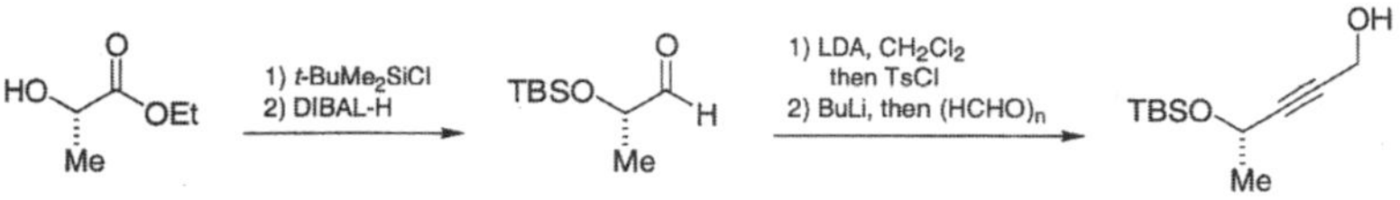

SYNTHESIS OF 9-SPIROEPOXY-endo-TRICYCLO[5.2.2.0²,⁶]UNDECA-4,10-DIEN-8-ONE   171
Vishwakarma Singh, Mini Porinchu, Punitha Vedantham, and Pramod K. Sahu

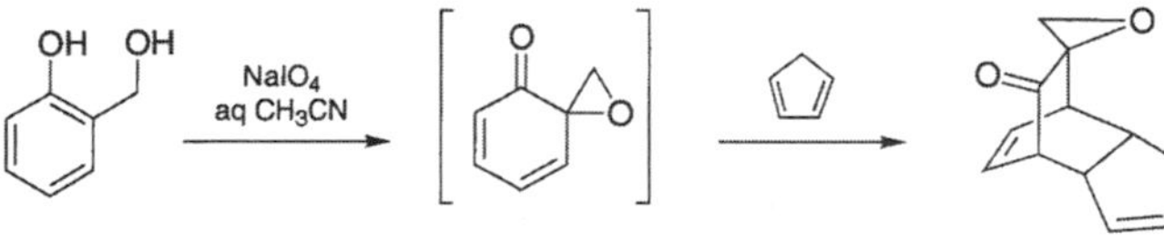

SYNTHESIS AND Ru(II)-BINAP REDUCTION OF A KETOESTER   178
DERIVED FROM HYDROXYPROLINE: 2(S)-(β-tert-BUTOXYCARBONYL-α-(R)-and α-
(S)-HYDROXYETHYL-4(R)-HYDROXYPYRROLIDINE-1-CARBOXYLIC ACID,
tert-BUTYL ESTER
Steven A. King, Joseph Armstrong, and Jennifer Keller

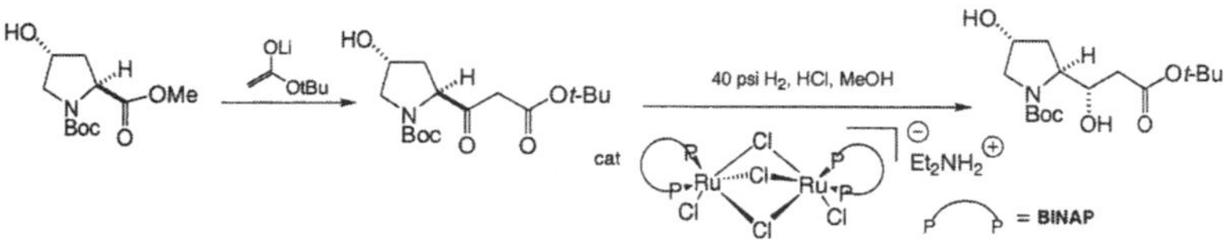

INDIUM/AMMONIUM CHLORIDE-MEDIATED SELECTIVE REDUCTION OF AROMATIC NITRO COMPOUNDS: ETHYL 4-AMINOBENZOATE

188

Bimal K. Banik[1], Indrani Banik, and Frederick F. Becker.

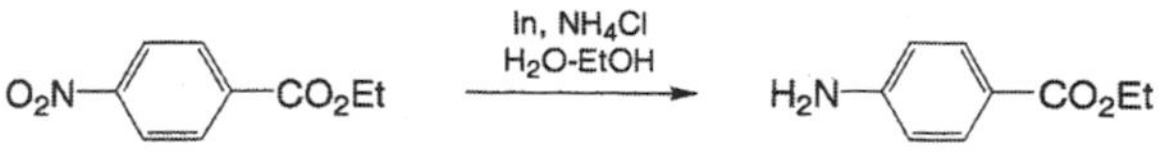

OXIDATION OF PRIMARY ALCOHOLS TO CARBOXYLIC ACIDS WITH SODIUM CHLORITE CATALYZED BY TEMPO AND BLEACH: 4-METHOXYPHENYLACETIC ACID

195

Matthew M. Zhao, Jing Li, Eiichi Mano, Zhiguo J. Song, and David M. Tschaen

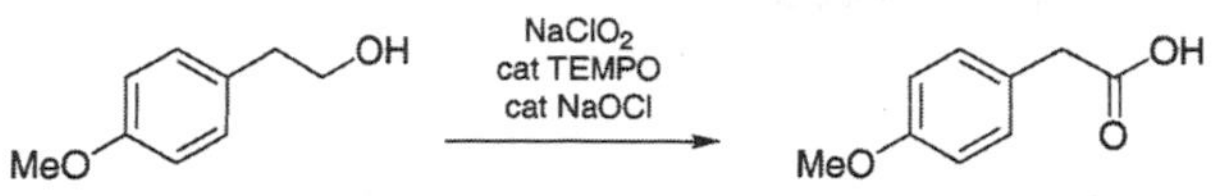

METHYLTRIOXORHENIUM CATALYZED OXIDATION OF SECONDARY AMINES TO NITRONES: N-BENZYLIDENE-BENZYLAMINE N-OXIDE

204

Andrea Goti, Francesca Cardona, and Gianluca Soldaini

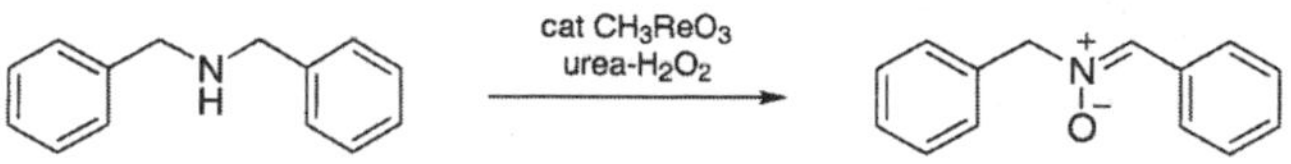

A CONVENIENT PREPARATION OF ORTHOGONALLY PROTECTED Cα,Cα-DISUBSTITUTED AMINO ACID ANALOG OF LYSINE 1-tert-BUTYLCARBONYL-4-((9-FLUORENYLMETHYLOXYCARBONYL)AMINO)-PIPERIDINE-4-CARBOXYLIC ACID

213

Lars G.J. Hammarström, Yanwen Fu, Sid Vail,
Robert P. Hammer, and Mark L. McLaughlin

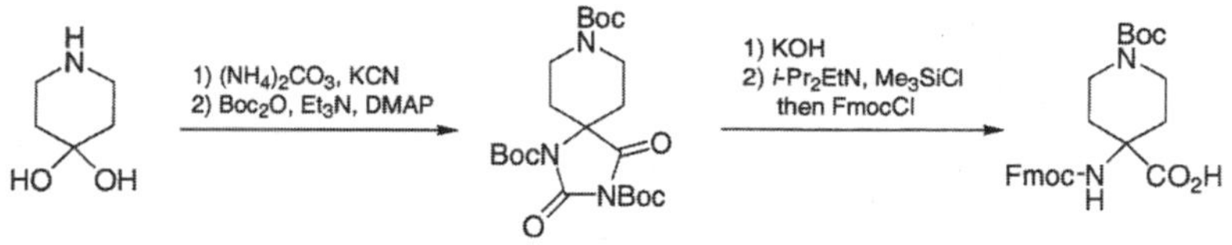

SYNTHESIS AND USE OF GLYCOSYL PHOSPHATES AS GLYCOSYL DONORS

225

Kerry R. Love and Peter H. Seeberger

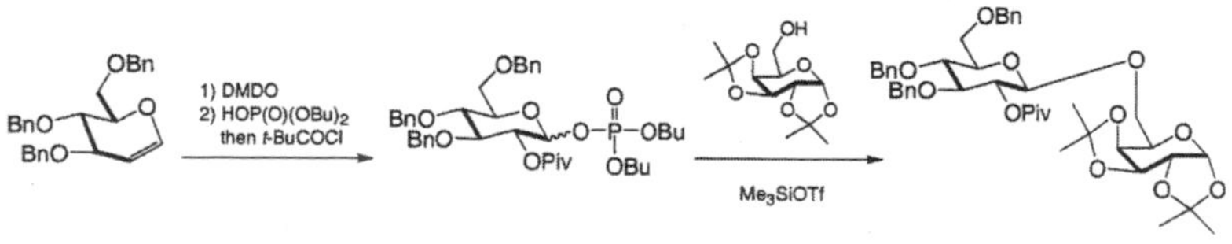

THE USE OF POLYSTYRYLSULFONYL CHLORIDE RESIN AS A SOLID SUPPORTED CONDENSATION REAGENT FOR THE FORMATION OF ESTERS: SYNTHESIS OF N-[(9-FLUORENYLMETHOXY)CARBONYL]-L-ASPARTIC ACID; α-tert-BUTYL ESTER, β-(2-ETHYL[(1E)-(4-NITROPHENYL)AZO]PHENYL]ETHYL ESTER

235

Norbert Zander and Ronald Frank

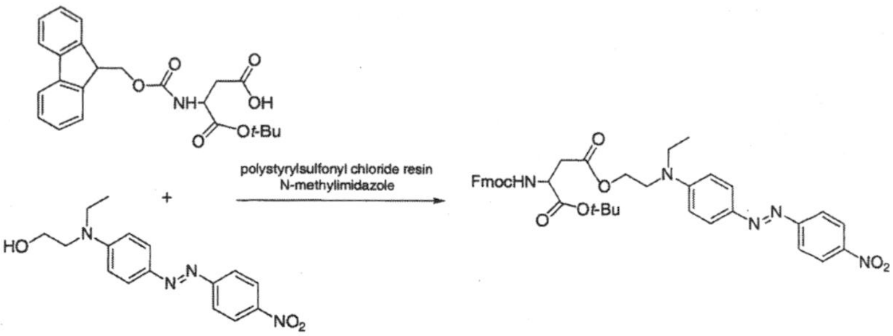

PHENYLSULFENYLATION OF NONACTIVATED δ-CARBON ATOM BY PHOTOLYSIS OF ALKYL BENZENESULFENATES: PREPARATION OF 2-PHENYLTHIO-5-HEPTANOL

244

Goran Petrović, Radomir N. Saičic, and Živorad Cekovic

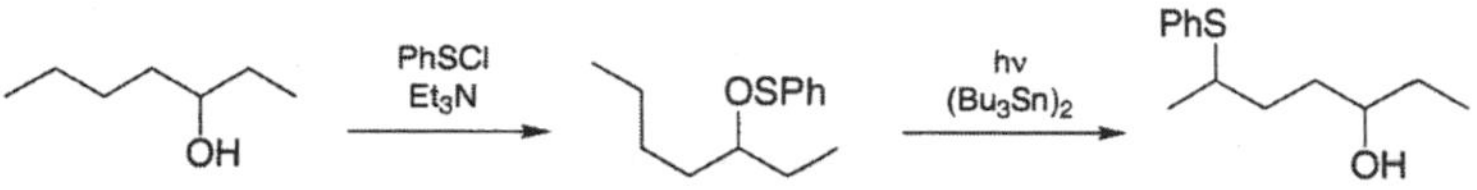

PREPARATION OF PIVALOYL HYDRAZIDE IN WATER

254

Bryan Li, Raymond J. Bemish, David R. Bill, Steven Brenek, Richard A. Buzon, Charles K-F Chiu, and Lisa Newell

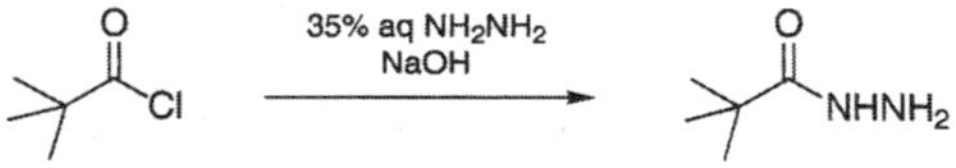

BORIC ACID CATALYZED AMIDE FORMATION FROM CARBOXYLIC ACIDS AND AMINES: N-BENZYL-4-PHENYLBUTYRAMIDE

262

Pingwah Tang

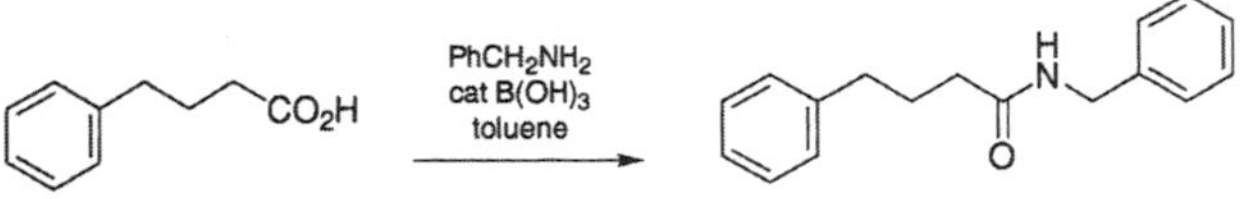

ORGANIC SYNTHESES

SYNTHESIS OF 1,3-DIENES FROM ALKYNES AND ETHYLENE:

ACETIC ACID 2-METHYLENE-3-PHENETHYLBUT-3-ENYL ESTER

[Benzenepentanol, β,γ-bis(methylene)-, acetate]

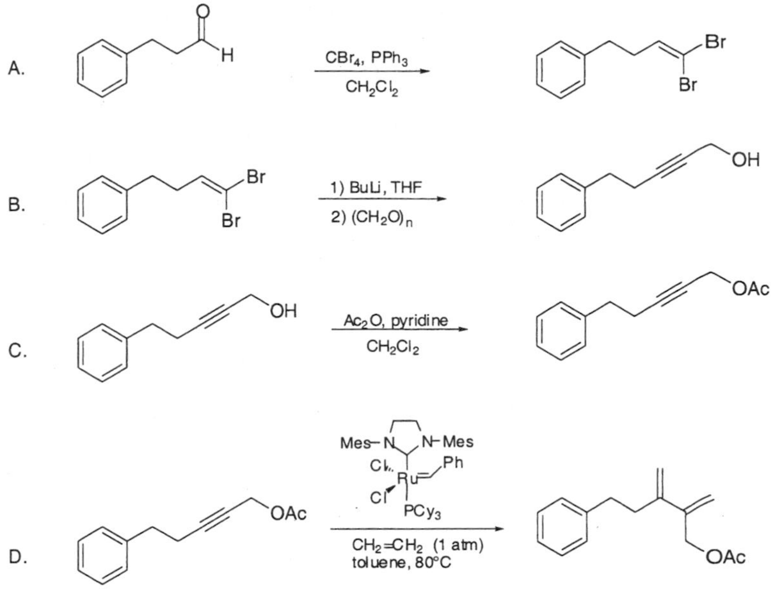

A.

B.

C.

D.

Submitted by Miwako Mori, Keisuke Tonogaki, and Atsushi Kinoshita.[1]

Checked by Scott E. Denmark and Tetsuya Kobayashi.

1. Procedure

A. *(4,4-Dibromobut-3-enyl)benzene.* A flame-dried, 2-L, three-necked flask, equipped with a magnetic stirbar, a rubber septum, an addition funnel fitted with a rubber septum, and an argon inlet is charged with 82.46 g (248.64 mmol) of carbon tetrabromide (Note 1) and then 400 mL of anhydrous dichloromethane (CH_2Cl_2, Note 2)

1

is added by syringe. The solution is cooled in an ice-water bath and a solution of 130.43 g (497.28 mmol) of triphenylphosphine (Note 3) in 400 mL of anhydrous CH_2Cl_2 is then added dropwise via the addition funnel over 30 min. The reaction mixture is stirred at 0°C for 10 min, and then a solution of 16.4 mL (16.7 g, 124.3 mmol) of 3-phenylpropionaldehyde (Note 4) in 266 mL of anhydrous CH_2Cl_2 is added over 10 min via cannula. The solution is stirred at 0°C for 1 hr, and then 500 mL of H_2O is added. The resulting mixture is transferred to a 4-L separatory funnel and the aqueous layer is separated and extracted with three 200-mL portions of CH_2Cl_2. The combined organic layers are washed with 600 mL of brine, dried over Na_2SO_4, filtered, and concentrated under reduced pressure using a rotary evaporator. To the residue is added 500 mL of diethyl ether, and the resulting suspension is filtered to remove triphenylphosphine oxide. The solid that is collected is washed with three 300-mL portions of diethyl ether, and the combined ethereal filtrate is concentrated under reduced pressure using a rotary evaporator. The residue is purified by flash column chromatography on 500 g of silica gel (elution with hexane/ether, 10:1). The fractions containing the product are collected and concentrated under reduced pressure using a rotary evaporator to yield 34.75 g (96%) of (4,4-dibromobut-3-enyl)benzene as a colorless oil (Note 5).

B. *5-Phenylpent-2-yn-1-ol.* A flame-dried, two-necked, 1-L, round-bottomed flask equipped with an argon inlet, rubber septum, and a magnetic stirbar is charged with (4,4-dibromobut-3-enyl)benzene (30.00 g, 103.45 mmol) and then anhydrous tetrahydrofuran (345 mL) (Note 6) is added to the flask by syringe. The reaction mixture is cooled in a −78°C dry ice-acetone bath while a solution of *n*-butyllithium in hexane (1.6M, 148 mL, 237 mmol) (Note 7) is added via syringe over 30 min. The solution is stirred at −78°C for 30 min, and then paraformaldehyde (9.32 g, 310.35 mmol) (Note 8) is added in one portion. The solution is stirred at −78°C for 30 min, and then allowed to warm to room temperature and stirred for 1 hr. Saturated aqueous NH₄Cl (400 mL) is added and the aqueous layer is separated and extracted with three 200-mL portions of ethyl acetate. The combined organic layers are washed with 400 mL of brine, dried over Na₂SO₄, filtered, and concentrated under reduced pressure using a rotary evaporator. The residue is purified by flash column chromatography on 500 g of silica gel (elution with hexane/ethyl acetate, from 7:1 to 1:1). The fractions containing the product are collected and concentrated under reduced pressure using a rotary evaporator to yield 14.32 g (86%) of 5-phenylpent-2-yn-1-ol (Note 9) as a colorless oil.

C. *Acetic acid 5-phenylpent-2-ynyl ester.* A flame-dried, 500-mL, three-necked, round-bottomed flask equipped with an argon inlet, two rubber septa, and a magnetic stirbar is charged with 5-phenylpent-2-yn-1-ol (13.51 g, 84.33 mmol) and then 280 mL of anhydrous CH₂Cl₂ is added by syringe. The solution is cooled in an ice-water bath and pyridine (68 mL, 843 mmol) (Note 10) is added in one portion by syringe, followed by the addition of acetic anhydride (32 mL, 337 mmol) (Note 11) over 10 min. The reaction mixture is stirred at room temperature overnight. The mixture is cooled to 0°C, 10.3 mL of MeOH is added, and the resulting solution is stirred at 0°C for 10 min. 4N Aqueous HCl (300 mL) is added, and the aqueous layer is separated and extracted with three 200-mL portions of ethyl acetate. The combined organic layers are washed with 400 mL of saturated aqueous NaHCO₃ solution, 400 mL of brine, dried over Na₂SO₄, filtered, and concentrated under reduced pressure using a rotary evaporator. The residue is purified by flash column chromatography on 500 g of silica gel (elution

3

with hexane/ethyl acetate, from 7:1 to 3:1). The fractions containing the product are collected and concentrated under reduced pressure using a rotary evaporator to yield 16.77 g (98%) of acetic acid 5-phenylpent-2-ynyl ester (Note 12) as a colorless oil.

D. *Acetic acid 2-methylene-3-phenethylbut-3-enyl ester.*[3]   A flame-dried, 250-mL, two-necked flask equipped with a magnetic stirbar, a rubber septum, and a reflux condenser capped with a three-way stopcock (used as the ethylene inlet adapter) is charged with acetic acid 5-phenylpent-2-ynyl ester (5.16 g, 25.52 mmol) and then 51 mL of anhydrous toluene (Note 13) is added to the flask by syringe. A solution of the second-generation ruthenium carbene complex[4e] (0.217 g, 0.26 mmol, 1 mol %, Note 14) in 5 mL of toluene is then added, and the resulting solution is degassed by repeated freeze-pump-thaw cycles. A 1-gallon (3.8 L) gasbag (Note 15) filled with purified ethylene gas (Notes 16, 17) is connected to the ethylene inlet adapter and the solution is stirred at 80°C for 2 hr under an atmosphere of ethylene gas. After allowing the solution to cool, 6 mL of ethyl vinyl ether is added (Note 18). The solution is concentrated under reduced pressure using a rotary evaporator, and the residue is purified by flash column chromatography on 250 g of silica gel (elution with hexane/ethyl acetate, 7:1). The fractions containing the product are collected and concentrated under reduced pressure using a rotary evaporator to yield 5.68 g (97%) of acetic acid 2-methylene-3-phenethylbut-3-enyl ester (Note 19) as a light yellow oil. Colorless material can be obtained by subjecting this product to Kugelrohr distillation to afford 5.55 g (94%) of the product as a colorless oil (Note 20).

2. Notes

1.   Carbon tetrabromide (99%) was purchased from the Aldrich Chemical Company Inc. and was used without further purification.

2.   Dichloromethane (unstabilized HPLC grade) was purchased from Fisher Scientific and was dried by percolation through two columns packed with neutral alumina under a positive pressure of argon.

3.   Triphenylphosphine (99%) was purchased from the Aldrich Chemical Company Inc. and was used without further purification.

4.   3-Phenylpropionaldehyde (>90%) was purchased from Fluka and was purified by distillation.

5.   The product displayed the following properties: R_f 0.73 (10:1 hexane-ether); IR (CHCl$_3$): cm^{-1} 3087, 3066, 3030, 3015, 2928, 2861, 1604, 1496, 1454, 1083, 1030, 907; ^{1}H NMR (500 MHz, CDCl$_3$): δ 2.42 (q, J = 7.7 Hz, 2 H), 2.74 (t, J = 7.7 Hz, 2 H), 6.42 (t, J = 7.7 Hz, 2 H), 7.19 (br d, J = 7.6 Hz, 2 H), 7.22 (br t, J = 7.6 Hz, 1 H), 7.31 (br t, J = 7.6 Hz, 2 H); ^{13}C NMR (125.7 MHz, CDCl$_3$): δ 33.8, 34.6, 89.5, 126.3, 128.4 (2C), 128.5 (2C), 137.6, 140.6. LRMS EI (relative intensity) m/z: 292 (M$^+$+2, 1.0), 290 (M$^+$, 1.9), 288 (M$^+$-2, 1.2), 227 (0.9), 225 (1.0), 221 (49), 209 (52), 129 (23), 128 (20), 92 (43), 91 (100), 77 (13). Anal. Calcd for C$_{10}$H$_{10}$ Br$_2$: C; 41.42, H; 3.48. Found: C; 41.33, H; 3.40. The spectroscopic data were identical with literature values for material previously prepared by this procedure.[2]

6.   Dry HPLC grade tetrahydrofuran was purchased from Fisher and was dried by percolation through two columns packed with neutral alumina under a positive

pressure of argon.

7.   A hexane solution of *n*-butyllithium was purchased from the Aldrich Chemical Company Inc. and was titrated prior to use.

8.   Paraformaldehyde was purchased from Fisher and was used without further purification.

9.   5-Phenylpent-2-yn-1-ol exhibits the following properties: R_f 0.25 (3:1 hexane-ethyl acetate); IR (CHCl$_3$): cm^{-1} 3610 (OH), 3088, 3066, 3029, 3013, 2932, 2874, 2360, 2223, 1604, 1496, 1455, 1430, 1385, 1342, 1233, 1133, 1078, 1006, 968; ^{1}H NMR (500 MHz, CDCl$_3$): δ 1.60 (t, *J* = 6.0 Hz, 1 H), 2.52 (tt, *J* = 7.5, 2.1 Hz, 2 H), 2.84 (t, *J* = 7.5 Hz, 2 H), 4.23 (dt, *J* = 6.0, 2.1 Hz, 2 H), 7.18-7.26 (m, 3 H), 7.31 (br t, *J* = 7.3 Hz, 2 H); ^{13}C NMR (125.7 MHz, CDCl$_3$): δ 20.9, 34.9, 51.3, 79.0, 85.7, 126.3, 128.38 (2C), 128.40 (2C), 140.5; LRMS EI (relative intensity) *m/z*: 160 (M$^+$, 1.2), 159 (5), 142 (32), 129 (15), 92 (10), 91 (100), 65 (14). Anal. Calcd for C$_{11}$H$_{12}$O: C, 82.46; H, 7.55. Found: C, 82.27; H, 7.68.

10.   Pyridine (99.9%) was purchased from Fisher and was purified by distillation over CaH$_2$.

11.   Acetic anhydride (99.7%) was purchased from Fisher and was purified by distillation.

12. Acetic acid 5-phenylpent-2-ynyl ester has the following properties: R_f 0.5 (3:1 hexane-ethyl acetate); IR (CHCl$_3$): cm^{-1} 3066, 3028, 2948, 2865, 2237 (alkyne), 1737 (CO), 1604, 1496, 1454, 1438, 1380, 1360, 1224, 1210, 1147, 1078, 1026, 967; ^{1}H NMR (500 MHz, CDCl$_3$): δ 2.09 (s, 3 H), 2.52 (tt, *J* = 7.7, 2.1 Hz, 2 H), 2.84 (t, *J* = 7.7 Hz, 2 H), 4.65 (t, *J* = 2.1 Hz, 2 H), 7.18-7.24 (m, 3 H), 7.30 (br t, *J* = 7.1 Hz, 2 H);

^{13}C NMR (125.7 MHz, CDCl$_3$): δ 20.8, 20.9, 34.8, 52.7, 74.7, 86.8, 126.4, 128.4 (4C), 140.4, 170.4; LRMS EI (relative intensity) *m/z*: 202 (M$^+$, 0.14), 201 (0.85), 160 (14), 142 (33), 141 (18), 92 (10), 91 (100), 65 (13). Anal. Calcd for C$_{13}$H$_{14}$O$_2$: C, 77.20; H, 6.98. Found: C, 77.37; H, 7.20.

13. Toluene (99.9%) is purchased from Fisher and dried by distillation under argon from sodium benzophenone ketyl.

14. Second-generation ruthenium-carbene complex, (tricyclohexylphosphine-[1,3-bis(2,4,6-trimethylphenyl)-4,5-dihydroimidazol-2-ylidene][benzylidene]ruthenium (IV) dichloride),[4e] was purchased from Strem Chemicals.

15. The ethylene gasbag was purchased from the Aldrich Chemical Company, Inc., catalog number Z 18,674-0.

16. Ethylene was purchased from Matheson (>99.5%) and was purified by bubbling it sequentially through an aqueous solution of CuCl (2 g of CuCl in 200 mL of saturated aqueous NH$_4$Cl), concentrated H$_2$SO$_4$, and then a KOH tube as shown in Figure 1. The reaction apparatus for the synthesis of the 1,3-diene is shown in Figure 2.

17. The checkers found that this purification of ethylene was not necessary. Without the purification, the checkers obtained the product quantitatively after purification by chromatography and in 97% yield after distillation. Ethyl vinyl ether is added to quench the reaction.[5]

Figure 1. Purification of Ethylene Gas

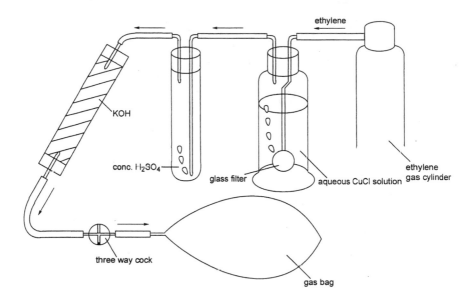

ethylene

KOH

conc. H₂SO₄

glass filter

aqueous CuCl solution

ethylene
gas cylinder

three way cock

gas bag

Figure 2. Reaction Apparatus for the Synthesis of 1,3-Diene

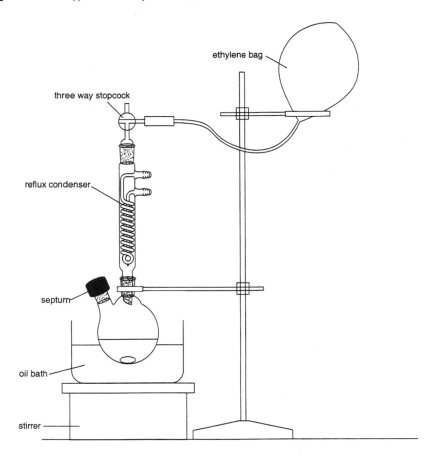

18. Ethyl vinyl ether was purchased from the Aldrich Chemical Company Inc. and was used without purification.

19.   2-Methylene-3-phenethylbut-3-enyl acetate has the following properties: R_f 0.55 (3:1 hexane-ethyl acetate); IR (CHCl$_3$): cm^{-1} 3690, 3087, 3065, 3020, 2945,

2865, 2360, 2342, 1736, 1635, 1601, 1496, 1455, 1370, 1222, 1046, 1030, 905; ^{1}H NMR (500 MHz, CDCl$_3$): δ 2.11 (s, 3 H), 2.58 (br t, J = 8.1 Hz, 2 H), 2.80 (br t, J = 8.1 Hz, 2 H), 4.80 (s, 2 H), 5.03 (br s, 1 H), 5.10 (s, 1 H), 5.31 (br s, 1 H), 5.37 (s, 1 H), 7.16-7.23 (m, 3 H), 7.29 (br t, J = 7.3 Hz, 2 H); ^{13}C NMR (125.7 MHz, CDCl$_3$): δ 21.0, 34.7, 35.9, 65.2, 113.1, 114.8, 125.9, 128.34 (2C), 128.38 (2C), 141.4, 141.9, 144.2, 170.8; LRMS EI (relative intensity) m/z: 230 (M$^+$, 0.04), 170 (30), 155 (28), 142 (37), 141 (17), 130 (21), 129 (20), 128 (14), 104 (10), 92 (18), 91 (100), 65 (15). Anal. Calcd for C$_{13}$H$_{18}$O$_2$: C, 78.23; H, 7.88. Found: C, 78.26; H, 7.91.

20. The checkers found that although the chromatographed product was a light yellow oil, it was analytically pure. Kugelrohr distillation was required to obtain a colorless oil.

Waste Disposal Information

All toxic materials were disposed of in accordance with "Prudent Practices in the Laboratory"; National Academy Press; Washington, DC, 1995.

3. Discussion

This procedure describes the synthesis of 1,3-dienes from alkynes and ethylene (at atmospheric pressure) by ruthenium-catalyzed, cross-enyne metathesis.[3,6] Although this reaction was initially developed using the first-generation ruthenium-carbene complex,[7] the second-generation ruthenium-carbene complex[4] is more effective and the reaction proceeded in the presence of even 1 mol % of the ruthenium catalyst. A number of different functional groups tolerate this reaction. Various 1,3-dienes are obtained in high yields following the procedure described in the text (Table).[3]

(1)   Graduate School of Pharmaceutical Sciences, Hokkaido University, Sapporo 060-0812, Japan.

(2)   (a) Renaud, J.-L; Aubert, C.; Malacria, M. *Tetrahedron* **1999**, *55*, 5113. (b) Harada, T.; Katsuhira, T.; Hara, D.; Kotani, Y.; Maejima, K.; Kaji, R.; Oku, A. *J. Org. Chem.* **1993**, *58*, 4897.

(3)   Tonogaki, K.; Mori, M. *Tetrahedron Lett.* **2002**, *43*, 2235.

(4)   (a) Weskamp, T.; Schattenmann, W. C.; Spiegler, M.; Herrmann, W. A. *Angew. Chem. Int. Ed.* **1998**, *37*, 2490. (b) Huang, J.; Stevens, E. D.; Nolan, S. P.; Peterson, J. L. *J. Am. Chem. Soc.* **1999**, *121*, 2674. (c) Scholl, M.; Trnka, T. M.; Morgan, J. P.; Grubbs, R. H. *Tetrahedron Lett.* **1999**, *40*, 2247. (d) Jafarpour, L.; Nolan, S. P. *Organometallics* **2000**, *19*, 2055. (e) Scholl, M.; Ding, S.; Lee, C. W.; Grubbs, R. H. *Org. Lett.* **1999**, *1*, 953.

(5)   Louie, J.; Grubbs, R. H. *Organometallics* **2002**, *21*, 2153.

(6)   (a) Kinoshita, A.; Sakakibara, N.; Mori, M. *J. Am. Chem. Soc.* **1997**, *119*, 12388. (b) Kinoshita, A.; Sakakibara, N.; Mori, M. *Tetrahedron* **1999**, *55*, 8155. (c) Mori, M.; Tonogaki, K.; Nishiguchi, N. *J. Org. Chem.* **2002**, *67*, 224.

(7)   Schwab, P.; France, M. B.; Ziller, J. W.; Grubbs, R. H. *Angew. Chem., Int. Ed. Engl.* **1995**, *34*, 2039.

TABLE
SYNTHESIS OF VARIOUS 1,3-DIENES[a]

run	alkyne	diene	time (hr)	yield of diene (%)[b]
1	(structure: 4-MeO-phenyl-CH(OAc)-C≡CH)	(structure: 4-MeO-phenyl-CH(OAc)-C(=CH₂)-CH=CH₂)	0.5	100
2	(structure: 4-MeO-phenyl-CH₂-C≡CH)	(structure: 4-MeO-phenyl-CH₂-C(=CH₂)-CH=CH₂)	0.5	83
3	(structure: 4-MeO-phenyl-C≡CH)	(structure: 4-MeO-phenyl-C(=CH₂)-CH=CH₂)	0.5	88
4	(structure: phenyl-CH₂-CH₂-C≡CH)	(structure: phenyl-CH₂-CH₂-C(=CH₂)-CH=CH₂)	0.5	71
5	(structure: phenyl-CH₂-CH₂-C≡C-CH₃)	(structure: phenyl-CH₂-CH₂-C(=CH₂)-C(CH₃)=CH₂)	0.5	85
6[c]	(structure: CH₃CH₂CH₂-C≡C-CH₂CH₂CH₃)	(structure: bis-methylene diene)	24	80
7	(structure: 4-MeO-phenyl-CH₂-C≡C-TMS)	(structure: 4-MeO-phenyl-CH₂-C(=CH₂)-C(TMS)=CH₂)	16	87 (10)
8[d]	(structure: 4-MeO-phenyl-CH₂-C≡C-CO₂Me)	(structure: 4-MeO-phenyl-CH₂-C(=CH₂)-C(CO₂Me)=CH₂)	16	43 (34)

a) All reactions were carried out in toluene at 80°C using 5 mol % of ruthenium carbene complex under an atmosphere of ethylene gas by the representative procedure described in the text.
b) The number in parentheses shows the yield of alkyne recovered.
c) This reaction was carried out in CH_2Cl_2 under reflux condition.
d) This reaction was carried out using 10 mol % of ruthenium carbene complex.

12

Appendix

Chemical Abstracts Nomenclature (Registry Number)

3-Phenylpropionaldehyde: Benzenepropanal; (104-53-0)

(4,4-Dibromobut-3-enyl)benene: Benzene, (4,4-dibromo-3-butenyl)-; (119405-97-9)

5-Phenylpent-2-yn-1-ol: 2-Pentyn-1-ol, 5-phenyl-; (16900-77-9)

Acetic acid 5-phenylpent-2-ynyl ester: 2-Pentyn-1-ol, 5-phenyl-, acetate; (445234-71-9)

Acetic acid 2-methylene-3-phenethylbut-3-enyl ester: Benzenepentanol,

β,γ-bis(methylene)-, acetate; (445234-76-4)

Carbon tetrabromide: Methane, tetrabromo-; (558-13-4)

Triphenylphosphine: Phosphine, triphenyl-; (603-35-0)

Butyllithium: Lithium, butyl-; (109-72-8)

Paraformaldehyde; (30525-89-4)

Acetic anhydride: Acetic acid, anhydride; (108-24-7)

Pyridine; (108-24-7)

Tricyclohexylphosphine[1,3-bis(2,4,6-trimethylphenyl)-4,5-dihydroimidazol-
2-ylidene][benzylidene]ruthenium(IV)dichloride: Ruthenium, [1,3-bis(2,4,6-
trimethylphenyl)-2-
imidazolidinylidene]dichloro(phenylmethylene)(tricyclohexylphosphine)-,
(SP-5-41)-; (246047-72-3)

Ethyl vinyl ether; Ethene, ethoxy-; (109-92-2)

Ethylene: Ethene; (74-85-1)

FACILE SYNTHESES OF AMINOCYCLOPROPANES:

N,N-DIBENZYL-N-(2-ETHENYLCYCLOPROPYL)AMINE

(Benzenemethanamine, N-(2-ethenylcyclopropyl)-N-(phenylmethyl)-

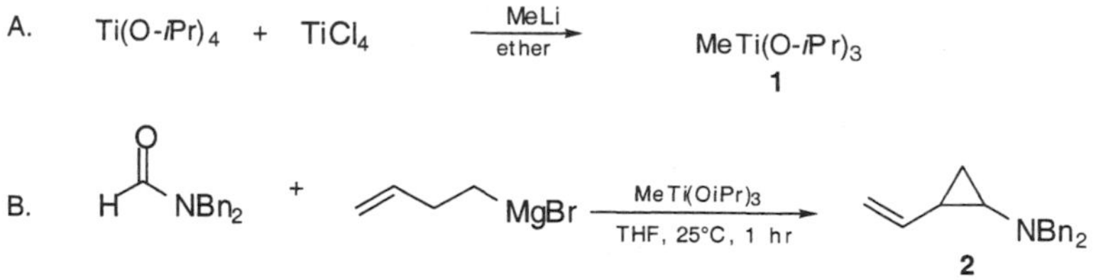

A. $Ti(O\text{-}iPr)_4$ + $TiCl_4$ $\xrightarrow[\text{ether}]{\text{MeLi}}$ $MeTi(O\text{-}iPr)_3$

1

B. H—NBn$_2$ + ⟍⟍MgBr $\xrightarrow[\text{THF, 25°C, 1 hr}]{\text{MeTi(OiPr)}_3}$ ⟍△NBn$_2$

2

Submitted by Armin de Meijere, Harald Winsel, and Björn Stecker.[1]

Checked by Fanglong Yang, Sivaraman Dandapani, and Dennis P. Curran.

A. Procedure

A. *Methyl tris(isopropoxy)titanium*[2] A 500-mL, three-necked, round-bottomed

flask equipped with two rubber septa and a dropping funnel fitted with an argon inlet is

charged with a mixture of 44 mL (150 mmol) of titanium tetraisopropoxide (Note 1) in 50

mL of anhydrous ether and cooled to 0°C in an ice bath. Titanium tetrachloride (6.48

mL, 50 mmol) (Note 1) is added dropwise via syringe pump over 30 min. The resulting

mixture is allowed to warm to room temperature and stirred for 2 hr. The reaction

mixture is then cooled to 0°C and 123 mL (123 mmol) of a 1.0M solution of

methyllithium in ether (Note 1) is added via the dropping funnel over 40 min. The

mixture is allowed to warm to room temperature and stirred for 1 hr. The dropping

funnel is replaced with a short path distillation head, and the ether is removed by distillation at ambient pressure. Distillation of the crude product at reduced pressure (bp 56-57°C, 60 mm) gives 38.5 g (80%) of MeTi(OiPr)$_3$ as a yellow oil (Note 2).

B. *N,N-Dibenzyl-N-(2-ethenylcyclopropyl)amine.* An oven-dried, 250-mL, three-necked, round-bottomed flask containing a magnetic stirbar and equipped with two rubber septa and an addition funnel fitted with a nitrogen inlet is charged with a solution of 9.0 g (40 mmol) of *N,N*-dibenzylformamide (Notes 3 and 4) in 10 mL of anhydrous tetrahydrofuran (Note 5). MeTi(O*i*Pr)$_3$ (11.5 mL, 48 mmol, 1.2 equiv) is added in one portion by syringe, and then 3-butenylmagnesium bromide (1.48M in THF, 40.5 mL, 60 mmol, 1.5 equiv) (Notes 6, 7) is added dropwise via the addition funnel over 30 min. The resulting black suspension is stirred at ambient temperature for an additional hour (Note 8), and the reaction is then quenched by slow addition of 50 mL of diethyl ether followed by 5 mL of water. The mixture is vigorously stirred for 1 hr (Note 9), filtered, and the colorless precipitate is then washed with two 20-mL portions of ether. The combined organic phases are washed with 30 mL of brine and dried over anhydrous MgSO$_4$, filtered, and concentrated at reduced pressure (Note 10) to afford 9.4 g of **2** (1:7 ratio of *cis-* and *trans*-isomers) as a deep yellow oil (Notes 11, 12). This crude product is purified by flash chromatography on 350 g of silica gel (200-400 mesh, elution with 100:1 pentane/ether) (Notes 13, 14) to yield 0.73 g (7%) of the *cis*-diastereomer, 0.51 g (5%) of a mixed fraction, and 6.2 g (59%) of the *trans*-diastereomer of **2**, each as a pale yellow oil (Notes 15, 16).

15

2. Notes

1.   [MeTi(OiPr)$_3$] must be used in pure form rather than as a solution in hexane. Titanium tetraisopropoxide (95%) was obtained from ABCR, Karlsruhe by the submitters and from Aldrich Chemical Company by the checkers and was distilled under nitrogen before use.   Titanium tetrachloride (>99%) was purchased from VWR by the submitters and from Aldrich Chemical Company by the checkers and was used without further purification.   The submitters obtained a solution of methyllithium in ether (1.6M, 5 wt%) from Fluka.

2.   The product is air and moisture sensitive, and the checkers used it immediately after preparation. The submitters report that MeTi(OiPr)$_3$ can be stored for several months in a freezer under an inert atmosphere.

3.   Other N,N-dialkylformamides including commercially available N-methylformanilide, N-formylmorpholine, N-formylpiperidine, N,N-diethylformamide and N,N-dimethylformamide (DMF) can be employed with essentially the same procedure. With N,N-dimethylformamide, care has to be exercised because of the volatility of the product, which can be purified by distillation or chromatography of its hydrochloride salt.

4.   N,N-Dibenzylformamide was prepared as follows. A 250-mL, round-bottomed flask equipped with reflux condenser fitted with an argon inlet was charged with 100 mL of DMF, 40.4 mL (210 mmol) of dibenzylamine, and 12.8 mL (300 mmol) of formic acid. The solution was heated at reflux overnight, allowed to cool to room temperature, and then quenched by carefully adding 40 mL of saturated Na$_2$CO$_3$ solution. The aqueous phase was separated and extracted with four 50-mL portions of

16

ether, and the combined organic layers were washed with two 20-mL portions of brine, dried over $MgSO_4$, and concentrated by rotary evaporation at reduced pressure. The crude product was crystallized from 100 mL of 70:30 ether/pentane to yield 40.2 g (85%) of *N,N*-dibenzylformamide, mp 50-51°C; it is hygroscopic and must be dried before use under reduced pressure (0.007 mm) overnight.

5. The submitters used THF that was freshly distilled from sodium, while the checkers used THF that was dried by pressure filtration through activated alumina.

6. 3-Butenylmagnesium bromide is prepared as follows. A 250-mL, three-necked, round-bottomed flask containing a magnetic stirbar and equipped with two rubber septa and an addition funnel fitted with an argon inlet was charged with 4.8 g (200 mmol) of magnesium turnings (Aldrich) suspended in 70 mL of anhydrous THF. A solution of 18.3 mL (180 mmol) of 4-bromo-1-butene (Aldrich) in 60 mL of anhydrous THF was added by dropping funnel at a rate adjusted to maintain a gentle reflux (total addition time, ca. 3 hr). The reaction mixture was then stirred for an additional hour. To determine the concentration of product, a 1-mL aliquot of the mixture was quenched with 0.1N HCl and then back-titrated with 0.1N NaOH (phenolphthalein indicator) which indicated a concentration of 1.5M.

The concentration of the Grignard reagent should not be higher than 1.5M. The use of more concentrated solutions leads to lower yields. Because of the rather high concentration of the reagents, it is important to stir the reaction mixture vigorously during and after the addition of the Grignard reagent. If the reaction is carried out on a larger scale than that described here, then the use of a mechanical stirrer is recommended.

7.   The submitters used a syringe pump to add the Grignard solution. However, the checkers experienced problems with precipitation of the Grignard reagent and recommend the use of an addition funnel. If the reagent begins to precipitate in the funnel during the addition period, then *gentle* warming with a heat gun accomplishes redissolution.

8.   The reductive cyclopropanation reaction proceeds very rapidly. For example, the submitters report that hydrolysis of the reaction mixture after only 1 min gives the product in 75% yield. However, stirring for an additional 1 hr is recommended, and this should be sufficient for analogous transformations with other Grignard reagents and formamides as well. Higher *N,N*-dialkylcarboxamides may require longer reaction times (up to 5 hr).

9.   The hydrolysis should be carried out with access of air so that the black titanium(II) derivatives are oxidized rapidly and completely to colorless titanium(IV) compounds.

10.   The solvents are removed within 1 hr at a bath temperature of 50°C at 15 mm.

11.   The *cis/trans* diastereomeric ratio is determined by integration of resonances in the ^{1}H NMR spectrum of the crude mixture and is usually found in the range of 1/6.0 to 1/7.5.

12.   TLC analysis of the crude product (elution with 50:1 pentane:ether, visualization with iodine) showed three non-baseline spots: R_f 0.65 (*cis* isomer), R_f 0.52 (unknown impurity), and R_f 0.32 (*trans* isomer). The unknown impurity is intensely

sensitive to iodine and largely coelutes with the *cis*-isomer in the subsequent column chromatography. However, the ^{1}H NMR spectrum of this isomer shows excellent purity despite the presence of this spot on TLC. In 100:1 pentane:ether, R_f values of the *cis* and *trans* isomers are about 0.50 and 0.15, respectively.

13. It is important to wet-pack the column with the eluting solvent. Dry packing results in strong retention of the sample and poor separation, possibly because the small amount of ether in the eluting solvent is adsorbed by the dry silica gel.

14. If separation of the diastereomers is not required, then filtration of the crude product through a pad of 50 g of silica eluting with dichloromethane followed by concentration affords a mixture of diastereomers separated from baseline materials.

15. The submitters report obtaining 1.29 (12%) of the *cis*-diastereoisomer and 8.39 g (80%) of the *trans*-diastereoisomer. The submitters recommend that the purified diastereomers be stored in the freezer (−18°C). Both pure isomers slowly equilibrate to a 21:79 mixture of *cis/trans* isomers upon standing at ambient temperature for several weeks.

16. The products exhibit the following spectroscopic properties: *trans* isomer: ^{1}H NMR (300 MHz, CDCl$_3$) δ), 0.60 (ddd, J = 6.7, 5.5, 4.9 Hz, 1 H), 0.76 (ddd, J = 9.0, 4.6, 4.6 Hz, 1 H), 1.35-1.26 (m, 1 H), 1.79 (ddd, J = 7.1, 4.2, 3.0 Hz, 1 H), 3.60 (d, J = 13.5 Hz, 2 H), 3.70 (d, J = 13.5 Hz, 2 H), 4.83-4.76 (m, 2 H), 5.33 (ddd, J = 17.0, 10.3, 8.5 Hz, 1H), 7.34-7.20 (m, 10H); ^{13}C NMR (75 MHz, CDCl$_3$) 15.9, 25.5, 45.1, 58.2, 112.2, 126.8, 128.0, 129.4, 138.4, 139.6; IR cm^{-1}: 3062, 3027, 2921, 2809, 1635, 1493, 1453, 1364. *cis* isomer: ^{1}H NMR (300 MHz, CDCl$_3$) δ 0.45 (ddd, J = 4.8, 4.8, 4.8 Hz, 1

H), 0.87 (ddd, J = 8.6, 6.8, 4.9 Hz, 1 H), 1.57-1.51 (m, 1 H), 2.06 (ddd, J = 6.8, 6.8, 4.6

Hz, 1H),7.34-7.22 (m, 10 H), 3.51 (d, J = 13.7 Hz, 2 H), 3.74 (d, J = 13.7 Hz, 2 H), 4.97

(dd, J = 10.3, 2.0 Hz, 1 H), 5.14 (dd, J = 17.3, 2.0 Hz, 1 H), 5.88 (ddd, J = 17.3, 10.3,

9.5 Hz, 1 H),; ^{13}C NMR (75 MHz, CDCl$_3$) 15.6, 23.3, 42.4, 57.2, 113.60, 126.8, 128.0,

129.5, 138.3, 138.4; IR cm^{-1}: 3062, 3026, 2922, 2807, 1634, 1493, 1453, 1364.

3. Discussion

The transformation of N,N-dialkylcarboxamides with low valent titanium

compounds formed in situ from Grignard reagents and titanium alkoxides of the type

XTi(O/Pr)$_3$ (X = O/Pr,   Cl,   Me)   to   correspondingly   substituted   N,N-

dialkylcyclopropylamines has been thoroughly examined in recent years since it was

first discovered.[3] Related to the conversion of esters to cyclopropanols,[5] this new and

reasonably general access to this important class of compounds[4] is now easily carried

out and generally furnishes high yields.[6] For example, unsubstituted N,N-dialkylcyclo-

propylamines are obtained from N,N-dialkylformamides and ethylmagnesium bromide

following this protocol in up to 98% yield.

Scheme 1. Generality of the titanium mediated reductive cyclopropanation of
N,N-diakylcarboxamides. More than 60 compounds have been prepared by this route.

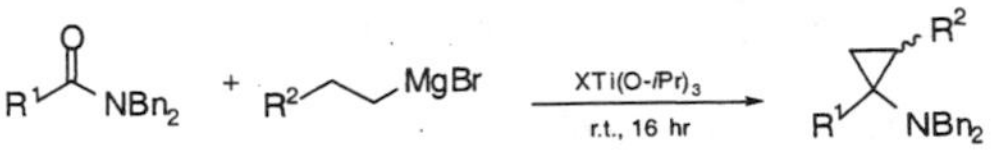

entry	R^1	R^2	yield (%)	*cis/trans*
1	H	H	98	—
2	H	*n*Bu	90	1 : 2.0
3	H	 ⟨cyclopropylmethyl⟩	85	1 : 2.2
4	H	Ph	92	1 : 2.0
5	H	 ⟨alkenyl⟩	92	1 : 7.0
6	Me	H	77	—
7	Et	H	70	—

$R^1 = R^2 = H$, Alkyl, Alkenyl, Aryl, functionally substituted Alkyl, Alkenyl, Aryl
X = Me, Cl, O*i*Pr

The products can be purified by distillation or crystallization as hydrochloride salts, but they can often be used for subsequent transformations without purification.

The best titanium mediator appears to be methyltitanium triisopropoxide, yet good yields are also obtained with titanium tetraisopropoxide and chlorotitanium triisopropoxide. The methyl group on titanium serves as a dummy alkyl ligand which is eliminated as methane after hydride transfer from the other alkyl group introduced from the Grignard reagent so that essentially only one equivalent of the latter would be required. However, yields do increase from 80–85% to over 90% in most cases by using an excess of the Grignard reagent (up to 2.5 equiv.). In the presence of titanium tetraisopropoxide and with the use of more than two equivalents of the Grignard reagent, the desired product is usually obtained from an *N,N*-dialkylformamide in over 80% yield along with 10-15% of the corresponding dialkylamine.

21

The reaction is best performed in THF. In less polar solvents such as diethyl ether or benzene, the chemical yields are significantly lower.

In the presence of 1 equiv. of chlorotrimethylsilane, the transformation can be performed with substoichiometric amounts of titanium tetraisopropoxide (25 mol%) to yield 73% of the diastereomeric mixture of the corresponding N,N-dialkylaminocyclopropane.[7]

Ethenylcyclopropylamines such as the described one are essentially functionally substituted cyclopropylamines, since the ethenyl group can easily be converted to functionally substituted ethyl groups, e. g., by hydroboration/oxidation or hydroboration/amination, and eventually the benzyl groups can be removed by catalytic hydrogenation to yield the corresponding primary cyclopropylamines. Ethenylcyclopropylamines can also easily be transformed to cyclopenten-4-ylamines by thermal rearrangement.[3c] Functionality can also directly be introduced with the carboxamide,[8a] the Grignard reagent,[8b] or the use of functionalized organozinc reagents.[8c]

A wide range of alkenyl-, aryl-, alkyl-, and dialkyl-substituted cyclopropylamines can be made by reaction of an N,N-dialkylformamide (or higher N,N-dialkylcarboxamide) with cyclohexylmagnesium halide in the presence of methyltitanium triisopropoxide (or titanium tetraisopropoxide) and a 1,3-diene, a styrene, a terminal or an internal alkene following this protocol with some minor modifications.[9] In these cases the reacting low valent titanium intermediate is generated by ligand exchange,[10] and the overall transformation corresponds to an aminocyclopropanation of the added

alkene. For example, a solution of N-Boc-protected pyrroline (4.36 g, 25.8 mmol) and

Ti(OiPr)$_4$ (9.06 mL, 30.9 mmol) in THF (100 mL) is treated with MeMgCl (10.3 mL of a

3M solution in THF, 30.9 mmol, added within 10 min) at 0°C and the mixture is warmed

to ambient temperature. N,N-Dibenzylformamide (6.95 g, 30.9 mmol) is added in one

portion, then dropwise with stirring a solution of cyclohexylmagnesium bromide

(28.1 mL of a 2.2M solution in diethyl ether, 61.9 mmol) within 2 hr. The mixture is

heated under reflux for an additional 1 hr, then cooled to ambient temperature. Addition

of 100 mL of water and 100 mL of pentane, filtration and further workup with filtration

through a pad of 10 g of basic aluminum oxide and subsequent crystallization from

pentane yields 8.75 g (90%) of crystalline N-Boc-protected exo-6-N,N-dibenzylamino-3-

azabicyclo[3.1.0]hexane.[9b]

1. Institut für Organische und Biomoleckulare Chemie der Georg-August-Universität Göttingen, Tammannstrasse 2, D-37077 Göttingen, Germany; Fax: + 49 (0)551 399475; E-mail: armin.deMeijere@chemie.uni-goettingen.de

2. Reetz, M. T.; J. Westermann, J.; Steinbach, R.; Wenderoth, B.; Peter, R.; Ostarek, R.; Maus, S. Chem. Ber. **1985**, *118*, 1421-1440.

3. (a) Chaplinski, V.; de Meijere, A. Angew. Chem. Int. Ed. Engl. **1996**, *35*, 413-414. (b) Chaplinski, V.; Winsel, H.; Kordes, M.; de Meijere, A. Synlett **1997**, 111-114. (c) Williams, C. M.; de Meijere, A. J. Chem. Soc., Perkin

4. *Trans.***1998**, 3699-3702. (d) Kordes, M.; Winsel, H.; de Meijere, A. *Eur. J. Org. Chem.* **2000**, 3235-3245.

5. Salaün, J. *Topics in Current Chemistry* **2000**, *207*, 1-67.

6. Kulinkovich, O. G.; Sviridov, S. V.; Vasilevski, D. A.; Pritytskaya, T. S. *Zh. Org. Khim.* **1989**, *25*, 2244-2245; *J. Org. Chem. USSR (Engl. Transl.)* **1989**, *25*, 2027-2028.

7. Reviews: (a) Kulinkovich, O. G.; de Meijere, A. *Chem. Rev.* **2000**, *100*, 2789-2834. (b) de Meijere, A.; Kozhushkov, S. I.; Savchenko, A. I. in *Titanium and Zirconium in Organic Synthesis*, (Ed.: I. Marek), Wiley-VGH, Weinheim, **2002**, 390-434.

8. de Meijere, A.; Stecker, B., to be published. Stecker, B., Dissertation, *Universität Göttingen*, **2002**.

9. (a) Winsel, H.; Gazizova, V.; Kulinkovich, O.; Pavlov, V.; de Meijere, A. *Synlett* **1999**, 1999-2003. (b) Winsel, H. Dissertation, Universität Göttingen, **2000**. (c) Wiedemann, S.; Marek, I.; de Meijere, A. *Synlett.* **2002**, 879-882.

10. (a) Williams, C. M.; Chaplinski, V.; Schreiner, P. R.; de Meijere, A. *Tetrahedron Lett.* **1998**, *39*, 7695-7698. (b) de Meijere, A.; Williams, C. M.; Chaplinski, V.; Kourdioukov, A.; Sviridov, S. V.; Savtchenko, A.; Kordes, M.; Stratmann, C. *Chem. Eur. J.* **2002**, *8*, 3789-3801.

11. (a) Lee, J. C.; Cha, J. K. *J. Org. Chem.* **1997**, *62*, 1584-1585. (b) Lee, J. C.; U, J. S.; Blackstock, S. C.; Cha, J. K. *J. Am. Chem. Soc.* **1997**, *119*, 10241-10242. (c) Lee, J. C.; Sung, M. J.; Cha, J. K. *Tetrahedron Lett.* **2001**, *42*, 2059-2061.

Appendix

Chemical Abstracts Nomenclature (Registry Number)

N,*N*-Dibenzyl-*N*-(2-ethenylcyclopropyl)amine: Benzenemethanamine, *N*-(2-ethenylcyclopropyl)-*N*-(phenylmethyl)-; (220247-75-5).

N,*N*-Dibenzylformamide: Formamide, *N*,*N*-bis(phenylmethyl)-; (5464-77-7).

Dibenzylamine: Benzenemethanamine, *N*-(phenylmethyl)-; (103-49-1).

Formic acid ; (64-18-6).

Methyl tris(isopropoxy)titanium: Titanium, methyltris(2-propanolato)-, (T-4)-; (18006-13-8).

Titanium tetraisoproproxide: 2-Propanol, titanium (4+)salt ; (546-68-9)

Titanium tetrachloride: Titanium chloride ($TiCl_4$)(T-4) (9); (7550-45-0).

Methyllithium: Lithium, methyl- ; (917-54-4)

4-Bromo-1-butene: 1-Butene, 4-bromo-; (5162-44-7).

3-Butenylmagnesium bromide: Magnesium, bromo-3-butenyl-; (7013-09-5).

dl-SELECTIVE PINACOL-TYPE COUPLING USING ZINC, CHLOROSILANE, AND CATALYTIC AMOUNTS OF Cp₂VCl₂;

dl-1,2-DICYCLOHEXYLETHANEDIOL

(1,2-Ethanediol,1,2-dicyclohexyl-)

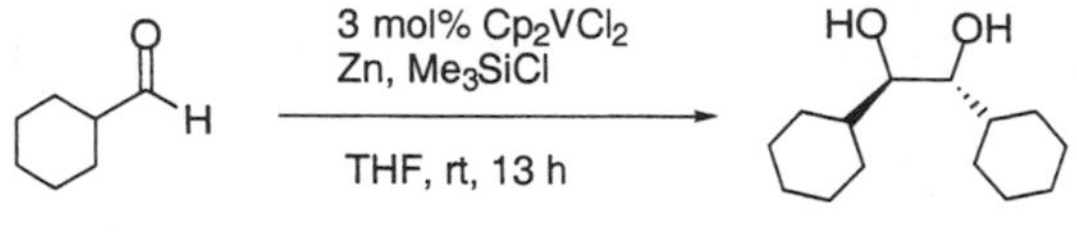

Submitted by Toshikazu Hirao, Akiya Ogawa, Motoki Asahara, Yasuaki Muguruma, and Hidehiro Sakurai.[1]

Checked by Helga Krause and Alois Fürstner.

1. Procedure

A 500-mL, two-necked, round-bottomed flask equipped with a rubber septum fitted with an argon inlet needle, a 200-mL, pressure-equalizing addition funnel capped with a rubber septum, and a magnetic stirbar is charged with zinc powder (13 g, 200 mmol) (Note 1) and dichlorodicyclopentadienylvanadium (0.756 g, 3.0 mmol) (Note 2). The flask is flame-dried and purged with argon, allowed to cool to room temperature, and then a solution of chlorotrimethylsilane (21.7 g, 200 mmol) (Note 3) in 200 mL of tetrahydrofuran (THF) (Note 4) is added dropwise over ca. 15 min via the addition funnel. The reaction mixture is stirred at room temperature for 1 hr during which time the color of the solution changes from red purple to light blue. Cyclohexanecarboxaldehyde (11.2 g, 100 mmol) (Note 5) is added via syringe over 5 min and the reaction mixture is stirred at room temperature for 13 hr. Diethyl ether (100 mL) and 100 mL of 2M aqueous HCl solution are added to the resulting mixture. The organic phase is separated and washed with 50 mL of saturated aqueous sodium bicarbonate solution, two 50-mL portions of water, and 50 mL of brine, dried over sodium sulfate, and concentrated by rotary evaporation to give 9.86-10.79 g (87-95%) of 1,2-dicyclohexylethanediol as a mixture of diastereomers (*dl*/meso = 87/13, Note 6). Three to four consecutive recrystallizations from ethanol (10 mL/g) affords 2.3-3.4 g (20-30%) of the *dl*-isomer in pure form (Notes 7,8).

2. Notes

1. Zinc powder was purchased from Wako Pure Chemical Industries and used as received.

2. Dichlorodicyclopentadienylvanadium was purchased from Aldrich Chemical Company, Inc. and used as received.

3. Chlorotrimethylsilane was purchased from Wako Pure Chemical Industries and freshly distilled from CaH_2 before use.

4. THF was obtained from Kanto Chemical Co., Inc. as dehydrated stabilizer-free grade.

5. Cyclohexanecarboxaldehyde was purchased from Tokyo Chemical Industry Co., Ltd. and used as received.

6. The diastereomeric ratio was determined by the NMR integral ratio of the methine protons adjacent to the hydroxyl group (3.33 ppm for the *dl*-isomer; 3.44 ppm for the *meso*-isomer).

7. The checkers obtained 6.7 g (60%) after the first recrystallization; this sample

showed a *dl*/meso ratio of 95:5. The submitters report obtaining 5.5-7.0 g (49-62%) of the pure *dl* isomer after three recrystallizations.

8. The physical properties of the *dl*-isomer are as follows: mp 136-138°C; IR (KBr) 3320, 2915, 2850, 1450, 1410, 1270, 1105, 1020, 890, 735 cm^{-1}; ^{1}HNMR (400 MHz, CDCl$_3$) δ 1.0-1.3 (m, 10H), 1.5-1.9 (m, 14H), 3.33 (br, 2H); ^{13}CNMR (100 MHz, CDCl$_3$) δ 26.1, 26.3, 26.5, 28.3, 29.7, 40.5, 75.2; MS (EI) *m/z* (rel. intensity): 226 (0.7, [M$^+$]), 143 (17), 113 (50), 112 (40), 96 (13), 95 (100), 81 (10), 67 (10), 55 (16), 41 (11); Anal. Found: C, 74.18; H, 11.52%. Calcd for C$_{14}$H$_{26}$O$_2$: C, 74.29; H, 11.58%.

Waste Disposal Information

All toxic materials were disposed of in accordance with "Prudent Practices in the Laboratory"; National Academy Press; Washington, DC, 1995.

3. Discussion

Metal-induced reductive dimerization of carbonyl compounds is a useful synthetic method for the formation of vicinally functionalized carbon-carbon bonds. For stoichiometric reductive dimerizations, low-valent metals such as aluminum amalgam, titanium, vanadium, zinc, and samarium have been employed. Alternatively, ternary systems consisting of *catalytic* amounts of a metal salt or metal complex, a chlorosilane, and a stoichiometric co-reductant provide a catalytic method for the formation of pinacols based on reversible redox couples.[2] The homocoupling of aldehydes is effected by vanadium or titanium catalysts in the presence of Me$_3$SiCl and Zn or Al to give the 1,2-diol derivatives; high selectivity for the *dl*-isomer is observed in the case of secondary aliphatic or aromatic aldehydes.

A variety of such ternary catalytic systems has been developed for diastereoselective carbon-carbon bond formations (Table). A Cp-substituted vanadium catalyst is superior to the unsubstituted one,[3] whereas a reduced species generated from $VOCl_3$ and a co-reductant is an excellent catalyst for the reductive coupling of aromatic aldehydes.[4] A trinuclear complex derived from Cp_2TiCl_2 and $MgBr_2$ is similarly effective for *dl*-selective pinacol coupling.[5] The observed *dl*-selectivity may be explained by minimization of steric effects through *anti*-orientation of the bulky substituents in the intermediate.

Chlorosilanes appear to contribute to the catalytic reactions in various ways. Importantly, they are necessary to liberate the catalyst from the primary product formed and they are also thought to facilitate the electron transfer to the carbonyl moiety, generating the stabilized silyloxyalkyl radicals that subsequently undergo dimerization. Moreover, the observed diastereoselectivity partly depends on the substituents of the chlorosilanes. Similar ligand and additive effects are observed in diastereoselective titanium-catalyzed coupling reactions of aromatic aldehydes.[6,7]

The cat. $Cp_2VCl_2/R_3SiCl/Zn$ system outlined above can also be used for the reductive coupling of aldimines with *meso*-diastereoselectivity.[8] The observed selectivity depends on the substituents on the nitrogen as well as the silicon atoms. 1,5- and 1,6-dialdehydes undergo intramolecular pinacol coupling to give cyclic *vic*-diols with excellent selectivity.[3,6] The reductive coupling has also been applied to the catalytic diastereoselective cyclization of arylidene malononitriles and ketonitriles to give the corresponding cyclopentene and cyclopentenol derivatives, respectively.[9]

1. Department of Applied Chemistry, Graduate School of Engineering, Osaka University, Yamada-oka, Suita, Osaka 565-0871, Japan.

2. Hirao, T. *Chem. Rev.* **1997**, *97*, 2707; Hirao, T. *Synlett* **1999**, 175; A. Fürstner, *Chem. Eur. J.* **1998**, *4*, 567.

3. Hirao, T.; Hasegawa, T.; Muguruma, Y.; Ikeda, I. *Abstracts for the 6th International Conference on New Aspects of Organic Chemistry*, 1994; p 175; Hirao, T.; Asahara, M.; Muguruma, Y.; Ogawa, A. *J. Org. Chem.* **1998**, *63*, 2812; Hirao, T.; Hasegawa, T.; Muguruma, Y.; Ikeda, I. *J. Org. Chem.* **1996**, *61*, 366; Hirao, T.; Hatano, B.; Asahara, M.; Muguruma, Y.; Ogawa, A. *Tetrahedron Lett.* **1998**, *39*, 5247.

4. Hirao, T.; Hatano, B.; Imamoto, Y.; Ogawa, A. *J. Org. Chem.* **1999**, *64*, 7665.

5. Gansäuer, A. *Chem. Commun.* **1997**, 457.

6. Lipski, T. A.; Hilfiker, M. A.; Nelson, S. G. *J. Org. Chem.* **1997**, *62*, 4566.

7. Gansäuer, A.; Bauer, D. *J. Org. Chem.* **1998**, *63*, 2070.

8. Hatano, B.; Ogawa, A.; Hirao, T. *J. Org. Chem.* **1998**, *63*, 9421.

9. Zhou, L.; Hirao, T. *Tetrahedron Lett.* **2000**, *41*, 8517; Zhou, L.; Hirao, T. *Tetrahedron* **2001**, *57*, 6927.

Table. Cat. Cp$_2$VCl$_2$/Me$_3$SiCl/Zn-Induced Pinacol Coupling of Aliphatic Aldehydes[3]

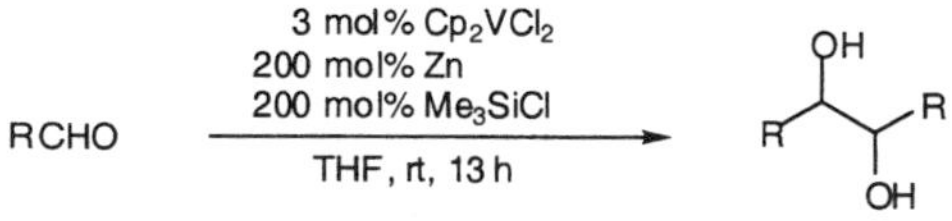

RCHO $\xrightarrow[\text{THF, rt, 13 h}]{\begin{array}{c}\text{3 mol\% Cp}_2\text{VCl}_2\\\text{200 mol\% Zn}\\\text{200 mol\% Me}_3\text{SiCl}\end{array}}$

aldehyde	yield / % [*dl*/ *meso*]	
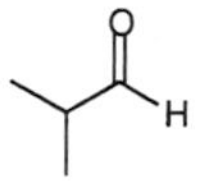	89	[91/9]
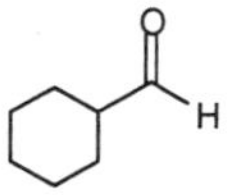	86	[87/13]
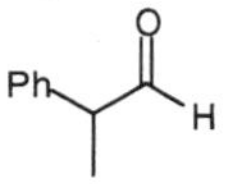	66	[94/6]
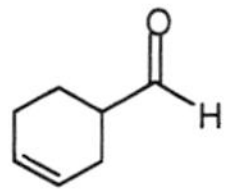	97	[85/15]
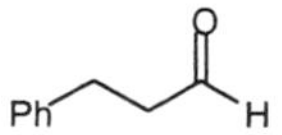	97	[64/36]

31

Appendix

Chemical Abstracts Nomenclature (Registry Number)

Dichlorodicyclopentadienylvanadium; Vanadium, dichlorobis (η^5-2,4-cyclopentadien-1-yl)-; (12083-48-6)

Chlorotrimethylsilane: Silane, chlorotrimethyl-; (75-77-4)

Cyclohexanecarboxaldehyde; (2043-61-0)

Zinc; (7440-66-6)

dl-1,2-Cyclohexylethanediol: 1,2-Ethanediol, 1,2-dicyclohexyl-; (92319-61-4)

4-NONYLBENZOIC ACID

(Benzoic acid, 4-nonyl-)

A.

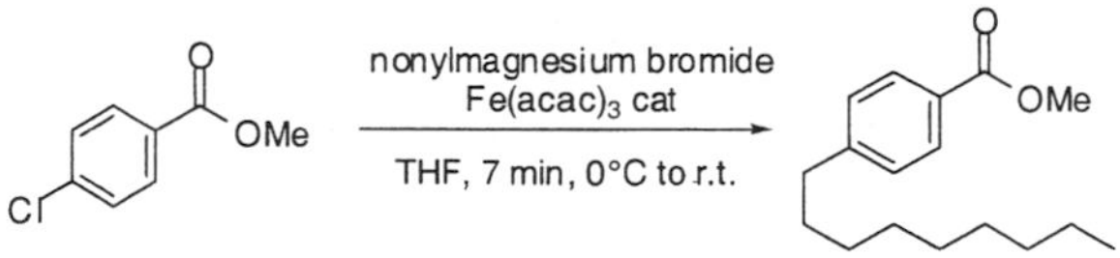

nonylmagnesium bromide
Fe(acac)₃ cat

THF, 7 min, 0°C to r.t.

B.

NaOH

H₂O/MeOH

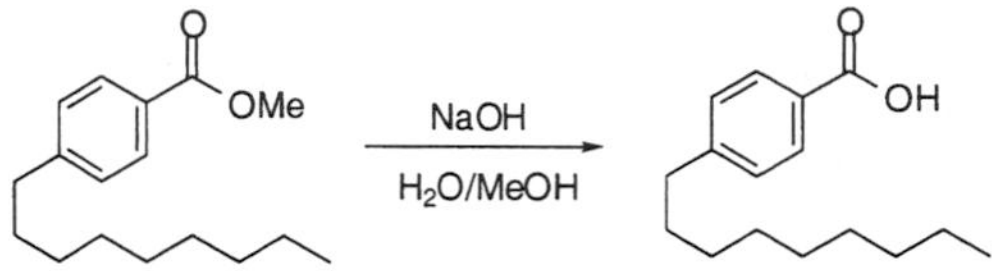

Submitted by Alois Fürstner, Andreas Leitner, and Günter Seidel.

Checked by Renee Kontnik and Steven Wolff.

1. Procedure

A.   *4-Nonylbenzoic acid methyl ester.*   A 250-ml, three-necked, round-bottomed flask equipped with a Teflon-coated magnetic stirbar, glass stopper, a reflux condenser fitted with an argon stopcock inlet, and a pressure-equalizing dropping funnel is charged with magnesium turnings (2.95 g, 121.0 mmol). The flask is evacuated and flame-dried while the magnesium turnings are gently stirred. After reaching ambient temperature, the apparatus is flushed with argon, the magnesium turnings are suspended in 20 mL of tetrahydrofuan (Note 1), and 1,2-dibromoethane (0.3 ml, 3.6 mmol) (Note 2) is introduced. A solution of 1-bromononane (20.52 g, 97.0 mmol) (Note 2) in 100 mL of THF (Note 1) is then added via the addition funnel to the suspension over a period of ca. 45 min at such a rate as to maintain gentle reflux. After the addition is complete, the mixture is refluxed for another 20 min and then the resulting solution of nonylmagnesium bromide is allowed to cool to ambient temperature. In the meantime, an oven-dried, 2-L, two-necked, round-bottomed flask equipped with a magnetic stirbar, rubber septum, and an argon inlet is flushed with argon and charged with 4-chlorobenzoic acid methyl ester (13.0 g, 76.2 mmol) (Note 2), ferric acetylacetonate [Fe(acac)$_3$] (1.35 g, 3.82 mmol) (Note 2), 450 ml of THF (Note 1), and 25 mL of N-methylpyrrolidinone (NMP) (Note 1). The flask is immersed in an ice bath and the solution of nonylmagnesium bromide prepared above is immediately added within one min via a polyethylene cannula. This causes an immediate color change from red to black-violet. The ice bath is removed and the resulting dark mixture is stirred for 7-10 min at ambient temperature, diluted with 200 mL of diethyl ether, and then carefully quenched by addition of 300 mL of 1M HCl with stirring. The mixture is transferred into a separatory funnel (the flask is rinsed with 200 mL of Et$_2$O), and the aqueous phase is separated and extracted with 200 mL of Et$_2$O. The combined organic phases are washed with 300 mL of saturated aq NaHCO$_3$, dried over Na$_2$SO$_4$, and concentrated by rotary evaporation at reduced pressure (Note 3). The resulting crude orange-red residue is purified by short path distillation under high vacuum (1 x 10^{-4} torr) to give 15.81-16.85 g (79-84%) of 4-nonylbenzoic acid methyl ester as a colorless syrup, bp 103-105°C (Notes 4, 5).

B.   *4-Nonylbenzoic acid.* A 500-mL, round-bottomed flask equipped with a Teflon-coated magnetic stirbar and a reflux condenser is charged with 4-nonylbenzoic acid methyl ester (10.07 g, 38.37 mmol), 100 mL of methanol (Note 1), and 96 mL of 1M aqueous NaOH. The resulting mixture is heated at reflux for 18 hr and then allowed to cool to room temperature. The reaction mixture is carefully acidified by addition of 200 mL of 1M aqueous HCl, and the resulting solution is transferred to a separatory funnel and extracted with four 250-mL portions of ethyl acetate. The combined organic layers are dried over Na_2SO_4, filtered, and concentrated by rotary evaporation at reduced pressure. The residue (ca. 9.5 g) is recrystallized from 70 mL of hexanes to give 8.32-8.35 g (87-88%) of 4-nonylbenzoic acid as a white solid (Notes 6, 7).

2. Notes

1. The submitters used THF that was freshly distilled over Na/K alloy; NMP was distilled over CaH_2. The checkers used commercially available anhydrous THF and NMP obtained from Aldrich Chemical Co. There appears to be a slight increase in yield when THF distilled from sodium/benzophenone ketyl is used instead of the commercial THF. All other solvents used were of reagent grade quality and were used without further purification.

2. 1,2-Dibromoethane, 1-bromononane (98%), 4-chlorobenzoic acid methyl ester (99%), and $Fe(acac)_3$ (99.9%) were purchased from Aldrich Chemical Co. and used as received.

3. The checkers found that a dark red oil separated upon concentration of the ethereal solution of the crude ester. In this instance, the ester was diluted with 200 mL of ether, washed with 200 mL of brine, and then again concentrated.

4. The submitters obtained the product in 87-90% yield.

5. The product is ≥ 95% pure by GC (Agilent 6890 Series, column: HP-5MS; 5% phenyl methyl siloxane, 30 m x 250 µm 0.25 µm); the remainder is octadecane formed by oxidative coupling of the Grignard reagent; T-program: 70°C (3.5 min)→280°C (20°C/min); 12.62 min retention time; ^{1}H NMR (300 MHz, CDCl$_3$): δ 0.86 (t, J = 6.7 Hz, 3H), 1.2-1.36 (m, 12H), 1.60 (m, 2H), 2.63 (t, J = 7.7 Hz, 2H), 3.88 (s, 3H), 7.22 (d, J = 8.4 Hz, 2H), 7.93 (d, J = 8.3 Hz, 2H); ^{13}C NMR (75 MHz, CDCl$_3$): δ 14.1, 22.7, 29.2, 29.3, 29.4, 29.7, 31.1, 31.9, 36.0, 51.9, 127.6, 128.4, 129.6, 148.5, 167.2,; IR (film): 3032, 2953, 2926, 2855, 1725, 1611, 1465, 1435, 1278, 1178, 1109, 1021, 854 cm $^{-1}$; MS (EI): m/z (rel. Intensity) 262 (63, [M$^+$], 231 (24), 163 (13), 150 (100), 121, (8), 105, (9), 91 (41), 57 (8), 43 (13). Anal. Calcd for C$_{17}$H$_{26}$O$_2$: C, 77.82; H, 9.99. Found C, 78.30; H, 10.45.

6. The submitters obtained the product in 94% yield.

7. The product has the following physical properties: mp 92.5-94.3°C; ^{1}H NMR (400 MHz, CDCl$_3$): δ 0.86 (t, J = 6.9 Hz, 3H), 1.2-1.38 (m, 12H), 1.62 (m, 2H), 2.66 (t, J = 7.7 Hz, 2H), 7.26 (d, J = 8.2 Hz, 2H), 8.01 (d, J = 8.2 Hz, 2H); ^{13}C NMR (100 MHz, CDCl$_3$): δ 14.1, 22.7, 29.2, 29.3, 29.4, 29.5, 31.1, 31.8, 36.1, 126.7, 128.6, 130.4, 149.6, 171.9; IR (film): 3072, 2924, 2852, 2669, 2554, 1683, 1609, 1575, 1469, 1424, 1321, 1290, 945, 859, 758 cm^{-1}; MS (EI): m/z (rel. intensity): 248 (56, [M$^+$]), 177 (7), 149 (9), 136 (100), 107 (9), 92 (38), 57 (14); 29 Anal. Calcd for C$_{16}$H$_{24}$O$_2$-H$_2$O: C, 72.14; H, 9.84. Found C, 72.53; H, 9.76.

Waste Disposal Information

All toxic materials were disposed in accordance with "Prudent Practices in the Laboratory"; National Academy Press; Washington, DC 1995.

3. Discussion

Modern cross coupling chemistry is heavily dominated by the use of palladium and nickel complexes as the catalysts, which show an impressively wide scope and an excellent compatibility with many functional groups.[2] This favorable application profile usually overcompensates the disadvantages resulting from the high price of the palladium precursors, the concerns about the toxicity of nickel salts, the need for ancillary ligands to render the complexes sufficiently active and stable, and the extended reaction times that are necessary in certain cases.

Our group has recently developed an alternative method for alkyl-(hetero)aryl- as well as aryl-heteroaryl cross coupling reactions catalyzed by iron salts.[3,4] This methodology was inspired by early reports of Kochi et al.[5,6] on iron-catalyzed cross coupling of vinyl halides and is distinguished by several notable advantages.

(1) The expensive noble metal catalysts are replaced by cheap, air stable, commercially available and toxicologically benign iron salts without any loss in efficiency. The reactions are usually carried out under "ligand free" conditions using inexpensive Grignard reagents as the preferred coupling partners.

(2) Aryl chlorides, which are the most attractive type of substrates due to their low cost and ready availability, perform inherently better than the corresponding aryl

37

bromides or iodides. Moreover, aryl triflates and even electron-deficient aryl tosylates can be used as the starting materials.

(3) Most iron-catalyzed reactions proceed at unprecedentedly high rates and are finished within a few minutes even when carried out at or below ambient temperature.

(4) Due to the efficiency with which the iron catalysts activate the C-Cl bond, several functional groups are tolerated that normally would react with a Grignard reagent.

(5) When applied to polyfunctional substrates, iron catalyzed reactions allow either for selective, exhaustive, or consecutive cross-coupling processes to be carried out in "one pot."

Most of these chemical features are evident from the preparation of 4-nonylbenzoic acid described above which requires less than 10 min reaction time at 0°C→ r.t. even when performed on a > 15 g scale. No competing attack of the Grignard reagent onto the methyl ester of the substrate can be detected under the reaction conditions. It should be noticed that 4-nonylbenzoic acid, like other alkyl benzoic acid derivatives that are equally available by this iron-catalyzed cross coupling method, is of some practical relevance as liquid crystalline material or a component thereof.[7,8] The additional examples compiled in the Table illustrate the scope and performance of this new protocol in more detail. [3,4]

1. Max-Planck-Institut für Kohlenforschung, Kaiser-Wilhelm-Platz 1, D-45470 Mülheim/Ruhr, Germany.

2. *Metal Catalyzed Cross-coupling Reactions;* Diederich, F.; Stang, P. J., Eds.; Wiley-VCH: Weinheim, **1998**.

3. Fürstner, A.; Leitner, A. *Angew. Chem., Int. Ed.* **2002**, *41*, 609.

4. Fürstner, A.; Leitner, A.; Méndez, M.; Krause, H. *J. Am. Chem. Soc.* **2002**, *124*, 13856.

5. (a) Tamara, M. Kochi, J. K. *J. Am. Chem. Soc.* **1971**, *93*, 1487. (b) Tamura, M.; Kochi J. *Synthesis* **1971**, 303. (c) Neumann, S. M.; Kochi, J. K. *J. Org. Chem.* **1976**, *41*, 502.

6. See also: Cahiez, G; Avedissian, H. *Synthesis* **1998**, 1199.

7. Sage, I. In *Ullmann's Encyclopedia of Industrial Chemistry*, VCH, Weinheim, **1990**, Vol. A15, pp. 359.

8. See also: (a) Kato, T.; Fukumasa, M.; Frechet, J. M. J. *Chem. Mater.* **1995**, *7*, 368. (b) Kato, T.; Frechet, J. M. J.; Wilson, P. G.; Saito, T.; Uryu, T.; Fujishima, A.; Jin, C.; Kaneuchi, F. *Chem. Mater.* **1993**, *5*, 1094. (c) Ojha, D. P.; Kumar, D.; Pisipati, V. G. K. M. *Z. Naturforsch.* **2002**, *57a*, 189.

TABLE. IRON CATALYZED CROSS COUPLING REACTIONS OF GRIGNARD
REAGENTS WITH (HETERO)ARYL CHLORIDES OR -TRIFLATES

Substrate	Product	Yield
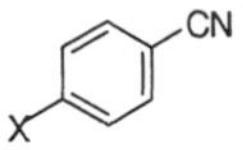	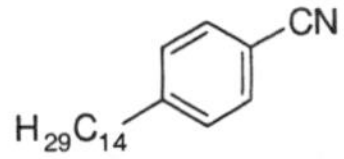	91% (X = Cl) 80% (X = OTf)
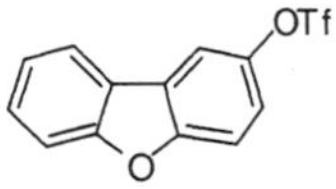	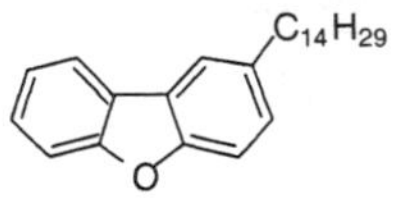	81%
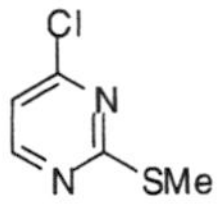	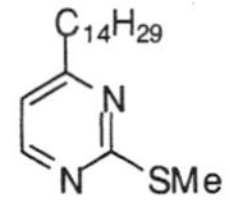	89%
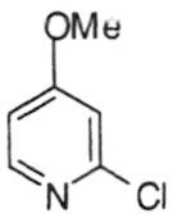	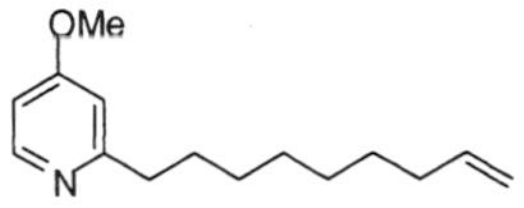	81%
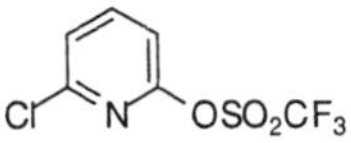	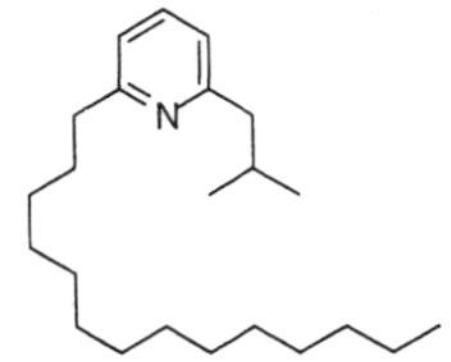	71%[a]
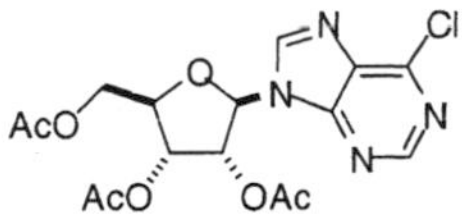	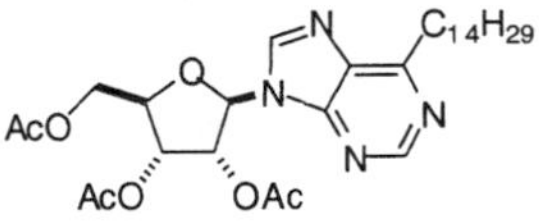	72%
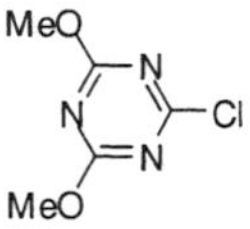	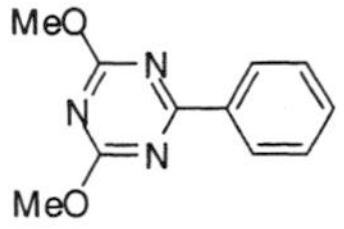	63%
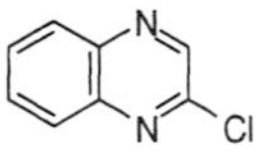	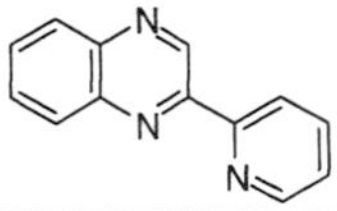	82%

a) Consecutive addition of isobutylmagnesium bromide and tetradecylmagnesium bromide within
an overall reactioin time of 8 min.

40

Appendix

Chemical Abstracts Nomenclature (Registry Number)

4-Nonylbenzoic Acid: Benzoic acid, 4-nonyl-; (38289-46-2)

Magnesium; (7439-95-4)

1-Bromononane: Nonane, 1-bromo-; (693-58-3)

Nonylmagnesium bromide: Magnesium, bromononyl-; (39691-62-8)

Ferric acetylacetonate: Tris (2,4-pentanedionato)iron (III); (14024-18-1)

N-Methylpyrrolidinone: 2-Pyrrolidinone, 1-methyl-; (872-50-4)

4-Chlorobenzoic acid methyl ester: Benzoic acid, 4-chloro-, methyl ester; (1126-46-1).

PALLADIUM CATALYZED CROSS-COUPLING OF

(Z)-1-HEPTENYLDIMETHYLSILANOL WITH 4-IODOANISOLE:

(Z)-1-HEPTENYL-4-METHOXYBENZENE

[Benzene,1-(1Z)-1-heptenyl-4-methoxy-]

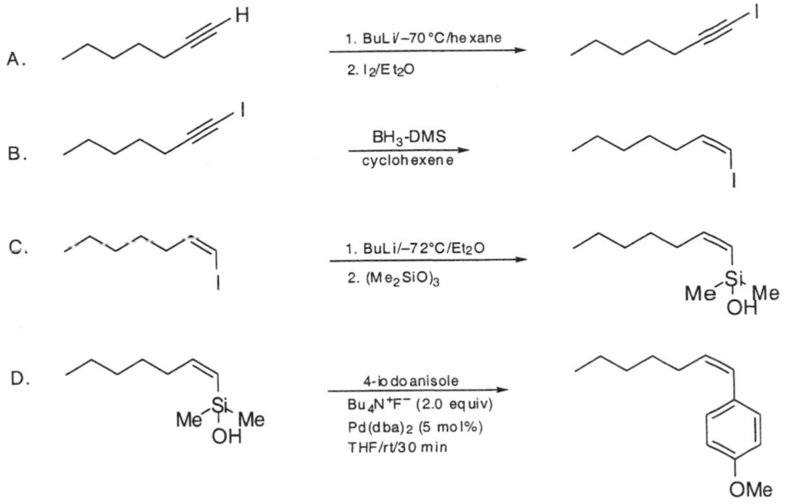

Submitted by Scott E. Denmark and Zhigang Wang.[1]

Checked by Matthew Campbell and Dennis P. Curran.

1. Procedure

A.   *1-Iodo-1-heptyne*.[2] A solution of 9.62 g (100 mmol) of 1-heptyne (Note 1) in 50 mL of hexane (Note 2) is placed in a flame-dried, three-necked, 500-mL, round-bottomed flask equipped with a large stirbar, a pressure equalizing addition funnel, and an N_2 inlet. The solution is cooled to –50 to –55°C (Note 3) and a solution of *n*-BuLi in hexane (61.0 mL, 1.65M, 100 mmol) (Note 4) is then added dropwise over 30 min. The resulting thick suspension is stirred at –70°C for 1.5 hr. A solution of 25.4 g (100 mmol) of Iodine (Note 5) in 130 mL of diethyl ether (Note 6) is added over 45 min. The cooling bath is removed and the reaction solution is allowed to warm to room temperature. Water (100 mL) is added and the resulting mixture is stirred at room temperature for 10 min. The aqueous layer is separated and extracted with 100 mL of hexane, and the combined organic phases are washed with two 200-mL portions of water, 50 mL of 20% aqueous sodium thiosulfate ($Na_2S_2O_3$) solution, and then dried over anhydrous $MgSO_4$ (Note 7). The solvents are removed by rotary evaporation (Note 8) and the residue is dissolved in 30 mL of pentane (Note 9) and the solution is then passed through a short column of silica gel (Note 10) followed by further elution with 250 mL of pentane. The solvent is removed by rotary evaporation and the residue is purified by Kugelrohr distillation to afford 17.9-18.5 g (81-83%) of 1-iodo-1-heptyne as a colorless liquid, bp 60-65°C at 1.2 mm (lit.[2] 55-60°C at 5 mm) (Notes 11, 12).

B. *(Z)-1-Iodo-1-heptene.*[2] A solution of 8.52 g of (112 mmol) of borane-dimethylsulfide complex (Note 13) in 100 mL of ether is added to a flame-dried, three-necked, 300-mL, round-bottomed flask equipped with stirbar, temperature probe, and N_2 inlet. The solution is cooled to 5°C with an ice-bath. Cyclohexene (18.4 g, 224 mmol) (Note 14) is then added by syringe over 10 min while keeping the temperature below 15°C. The mixture is stirred at 5°C for 15 min. A white solid precipitates either towards the end of the addition or during the subsequent stirring period. The reaction mixture is

43

allowed to warm to room temperature and is stirred for 1 hr. The non-homogeneous solution is cooled to 2-3°C and 22.7 g (102 mmol) of 1-iodo-1-heptyne is added by syringe over 10 min. The reaction mixture is stirred at 2-3°C for 30 min, and then the cooling bath is removed and the mixture is stirred at room temperature for 1 hr. The solution is cooled to 2-3°C and 51 mL of acetic acid (AcOH) (Note 15) is slowly added. The ice-bath is removed after the addition of AcOH is completed and the mixture is stirred at room temperature for 2 hr. Ether (100 mL) is added and the solution is washed with four 75-mL portions of water, dried over anhydrous $MgSO_4$, filtered, and concentrated by rotary evaporation to give a pale yellow liquid which is dissolved in 50 mL of pentane. The solution is passed through a short column of silica gel (Note 10) followed by further elution with 250 mL of pentane. The solvent is removed by rotary evaporation to give the crude product which is purified by Kugelrohr distillation to afford 15.7 g (69%) of (Z)-1-iodo-1-heptene as a colorless liquid, bp 75-80°C at 1.5 mm (lit.[2] 60-63°C at 5.5 mm) (Notes 11, 16).

C. *(Z)-1-(1-Heptenyl)-1,1-dimethylsilanol.* A solution of 13.0 g (58 mmol) of (Z)-1-iodo-1-heptene in 50 mL of ether is placed in a flame-dried, three-necked, 300-mL, round-bottomed flask equipped with stirbar, septum, temperature probe, and N_2 inlet. The solution is cooled to –72°C and a solution of n-BuLi in hexane (Note 4) (35.6 mL, 1.63M, 58 mmol) is then added by syringe over 15 min. The mixture is stirred at –72°C for 30 min and then a solution of 4.30 g (19.3 mmol) of hexamethylcyclotrisiloxane (Note 17) in 50 mL of ether is added over 15 min. The cooling bath is removed, and the reaction solution is allowed to warm to room

44

temperature and stirred for 14 hr. The solution is then cooled to 0°C and 50 mL of water is slowly added to quench the reaction. The aqueous layer is separated and extracted with three 50-mL portions of diethyl ether. The combined organic phases are washed with 30 mL of water and two 30-mL portions of brine, dried over anhydrous $MgSO_4$, filtered, and concentrated by rotary evaporator and vacuum drying to give a crude product which is distilled to afford 7.34 g (73%) of (Z)-1-(1-heptenyl)-1,1-dimethylsilanol as a colorless liquid, bp 54-55°C at 0.15 mm (lit.[4] 120°C at 0.9 mm) (Notes 18, 19).

D. (Z)-1-(1-Heptenyl)-4-methoxybenzene. A solution of 4.14 g (24 mmol) of (Z)-1-(1-heptenyl)-1,1-dimethylsilanol in 2.0 mL of THF is added to a flame-dried, 250 mL, three-necked, round-bottomed flask equipped with a stirbar, addition funnel, temperature probe, and argon inlet. A solution of tetrabutylammonium fluoride in THF (40 mL, 1.0M, 40 mmol) (Note 20) is then added over 5 min. The mixture is stirred at room temperature for 10 min and then 0.575 g (1.0 mmol) of $Pd(dba)_2$ (Note 21) is added. A solution of 4-iodoanisole in THF (20 mL, 1.0M, 20 mmol) (Note 22) is added slowly through the addition funnel (Note 23) such that the temperature of the solution is kept between 30-33°C (addition time 30 min). The mixture is stirred for 30 min after complete addition of 4-iodoanisole and then 70 mL of diethyl ether is added and the mixture is stirred for an additional 10 min. The mixture is passed through a short column of silica gel (Note 10) followed by further elution with 200 mL of diethyl ether. The combined eluate is concentrated by rotary evaporation and vacuum drying to give a residue which is purified by Kugelrohr distillation to afford a colorless or light yellow liquid. This liquid is further

45

purified by chromatography (Note 24) followed by Kugelrohr distillation to afford 3.29 g (80%) of (Z)-1-(1-heptenyl)-4-methoxybenzene as a colorless oil, bp 90-95°C at 0.1 mm (lit.[4] 80-85°C at 1.3 mm) (Notes 11, 25).

8. Notes

1.   1-Heptyne was purchased from GFS Chemicals Inc. and was used without further purification.

2.   Hexane was purchased from Fisher Scientific Company and was freshly distilled from sodium.

3.   Unless otherwise noted, all temperatures refer to internal temperatures measured with Teflon-coated thermocouples.

4.   Butyllithium was purchased from FMC Corporation Lithium Division and its concentration was determined by double titration by the Gilman method.[3]

5. I_2 was purchased from Mallinckrodt Inc. and was used as received.

6.   Ether was freshly distilled from sodium/benzophenone.

7.   Anhydrous $MgSO_4$ was purchased from Fisher Scientific Company.

8.   The vacuum for the rotary evaporator is about 15 mm.

9.   Pentane was purchased from Fisher Scientific Company and was distilled from $CaCl_2$.

10.   Silica gel was purchased from VWR Scientific (230-400 mesh). A 45-mm diameter column was employed and 50 g silica gel was loaded as a pentane slurry.

11. Boiling points (bp) correspond to uncorrected air-bath temperatures in the Buchi GKR-50 Kugelrohr.

12. The analytical data are as follows: ^{1}H NMR (500 MHz, CDCl$_3$) δ: 0.90 (t, J = 7.1, 3 H), 1.34 (m, 4 H), 1.51 (m, 2 H), 2.35 (t, J = 7.1, 2 H); ^{13}C NMR (125.6 MHz, CDCl$_3$) δ: −7.2, 13.9, 20.7, 22.1, 28.2, 30.9, 94.6; IR (NaCl) cm^{-1}: 2956, 2932, 2859, 1462; Anal. Calcd. for C$_7$H$_{11}$I: C, 37.86; H, 4.99; I, 57.15; Found: C, 37.90; H, 5.08; I, 57.50.

13. Borane-methyl sulfide complex (neat) was purchased from Aldrich Chemical Company, Inc. and was used as received.

14. Cyclohexene was purchased from Aldrich Chemical Company, Inc. and was used as received.

15. Acetic acid (100%) was purchased from J. T. Baker Company and was used as received.

16. The checkers obtained the product in 65-74% yield. The analytical data are as follows: ^{1}H NMR (500 MHz, CDCl$_3$) δ: 0.90 (t, J = 7.1, 3 H), 1.32 (m, 4 H), 1.34 (m, 2 H), 2.13 (m, 2 H), 6.17 (m, 2 H); ^{13}C NMR (125.6 MHz, CDCl$_3$) δ: 14.0, 22.5, 27.6, 31.3, 34.6, 82.1, 141.5; IR (NaCl) cm^{-1}: 2956, 2927, 2857, 1608, 1463, 1276; GC analysis: t_R (Z) 6.34 min (99.3%), t_R(E) 6.57 min (0.7%) (HP-5, 150 °C, 15 psi); Anal. Calcd. for C$_7$H$_{13}$I: C, 37.52; H, 5.85; I, 56.63; Found: C, 37.72; H, 5.96; I, 56.81.

17. Hexamethylcyclotrisiloxane was purchased from Gelest Inc. and was used as received.

18. The (Z)-1,1-dimethyl-1-heptenylsilanol easily forms 1,1,1',1'-tetramethyl-1,1'-diheptenyldisiloxane. The crude product was distilled (60-62°C at 0.5 mm) to afford 8.02 g (80%) of (Z)-1,1-dimethyl-1-heptenylsilanol containing 2–5% of 1,1,1',1'-tetramethyl-1,1'-diheptenyldisiloxane. To obtain analytically pure product, the silanol was distilled twice again (bp 54-55°C, 0.15 mm).

19. The checkers obtained the product in 67-89% yield. The analytical data are as follows: ^{1}H NMR (500 MHz, CDCl$_3$) δ: 0.24 (s, 6 H), 0.89 (t, J = 7.1, 3 H), 1.30 (m, 4 H), 1.39 (m, 2 H), 1.69 (br, 1 H), 2.19 (m, 2 H), 5.47 (dt, J = 14.1, 1.3, 1 H), 6.36 (dt, J = 14.1, 7.5, 1 H); ^{13}C NMR (125.6 MHz, CDCl$_3$) δ: 1.7, 14.0, 22.6, 29.3, 31.5, 33.6, 127.8, 151.6; IR (NaCl) cm^{-1}: 3274 (br), 2959, 2928, 2858, 1607, 1465, 1253, 1067, 864, 785; GC analysis: t_R 7.15 min (HP-1, 100 °C, 10 psi); Anal. Calcd. for C$_9$H$_{20}$OSi: C, 62.72; H, 11.70; Found: C, 62.56; H, 12.04.

20. A solution of tetrabutylammonium fluoride (TBAF) in THF (1.0M) is prepared from colorless, crystalline tetrabutylammonium fluoride trihydrate purchased from Fluka Chemical Corp.

21. Pd(dba)$_2$ was prepared by the literature procedure without recrystallization from CHCl$_3$;[5] See Note 8, *Org. Synth.*, **2004**, *81*, 57.

22. 4-Iodoanisole was purchased from Aldrich Chemical Company, Inc., and was purified by chromatography on silica gel (hexane/EtOAc, 50/1) prior to use.

23. The coupling reaction is very exothermic. The reaction temperature can be controlled by the rate of addition of 4-iodoanisole.

24. A 55 mm diameter column is employed and 200 g silica gel is used (pentane/EtOAc, 50/1).

25. The checkers obtained the product in 90-97% yield. The analytical data are as follows: ^{1}H NMR (400 MHz, CDCl$_3$) δ: 0.89 (t, J = 7.1, 3 H), 1.31 (m, 4 H), 1.45 (qn, J = 7.3, 2 H), 2.31 (qd, J = 7.3, 2 H), 3.81 (s, 3 H), 5.57 (dt, J = 11.7, 7.1, 1 H), 6.33 (d, J = 11.7, 1 H), 6.87 (d, J = 8.8, 2 H), 7.22 (d, J = 8.8, 2 H); ^{13}C NMR (100.6 MHz, CDCl$_3$) δ: 13.9, 22.6, 28.6, 29.7, 31.6, 55.0, 113.5, 128.0, 129.9, 130.5, 131.5, 158.1; IR (NaCl) cm^{-1}: 3005, 2956, 2927, 2856, 1608, 1511, 1463, 1301, 1249, 1175, 1038, 837; GC analysis: t_R (Z) 6.05 min (99.1%), t_R (E) 6.53 min (0.9%) (HP-5, 250 °C, 15psi); Anal. Calcd. for C$_{14}$H$_{20}$O: C, 82.30; H, 9.87; Found: C, 82.25; H, 9.96.

Waste Disposal Information

All toxic materials were disposed of in accordance with "Prudent Practices in the Laboratory", National Academy Press; Washington, DC, 1995.

3. Discussion

Palladium-catalyzed, cross-coupling reactions of organosilicon reagents with aryl heteroaryl and alkenyl halides have emerged in recent years as a synthetically viable alternative to the traditional organotin (Stille) and organoboron (Suzuki) cross-coupling reactions. The most extensively investigated of these processes involves the use of mono-, di-, and trifluorosilanes as the coupling partners activated by a fluoride source.[6]

Recently, aryl trialkoxysiliconates have also shown promise for carbon-carbon bond formation under similar conditions.[7]

Our recent demonstration that a silicon-oxygen bond is the key structural feature that lends high reactivity of organosilicon compounds toward cross-coupling has led to the development of a wide range of silicon-based substrates for this process.[4,8] The most thoroughly investigated of these derivatives are the dialkylsilanols. These compounds are easily prepared by a number of different methods, the most common of which illustrated here is the reaction of organolithium compounds with hexamethylcyclotrisiloxane (D_3).[9,10] The addition is general for alkenyl-, aryl-, heteroaryl- and alkynyllithium agents and all three silicon groups in D_3 are available.

The palladium-catalyzed cross-coupling of alkenylsilanols has been extensively studied with respect to the structure of both the silicon component and the acceptor halide. The preferred catalyst for coupling of aryl iodides is $Pd(dba)_2$ and for aryl bromides it is $[allylPdCl]_2$. The most effective promoter is tetrabutylammonium fluoride used as a 1.0M solution in THF. In general the coupling reactions occur under mild conditions (room temperature, in 10 min to 12 hr) and some are even exothermic.

Varying substitution patterns have been studied and while the rate of the coupling is attenuated by steric crowding of the dimethylsilanol, the yields and stereospecificities are still high. In all cases studied to date, the coupling reactions proceed with retention of configuration. The examples compiled in Figure 1 represent successful couplings of a number of classes of alkenylsilanols in combination with both aryl and alkenyl halides.[8]

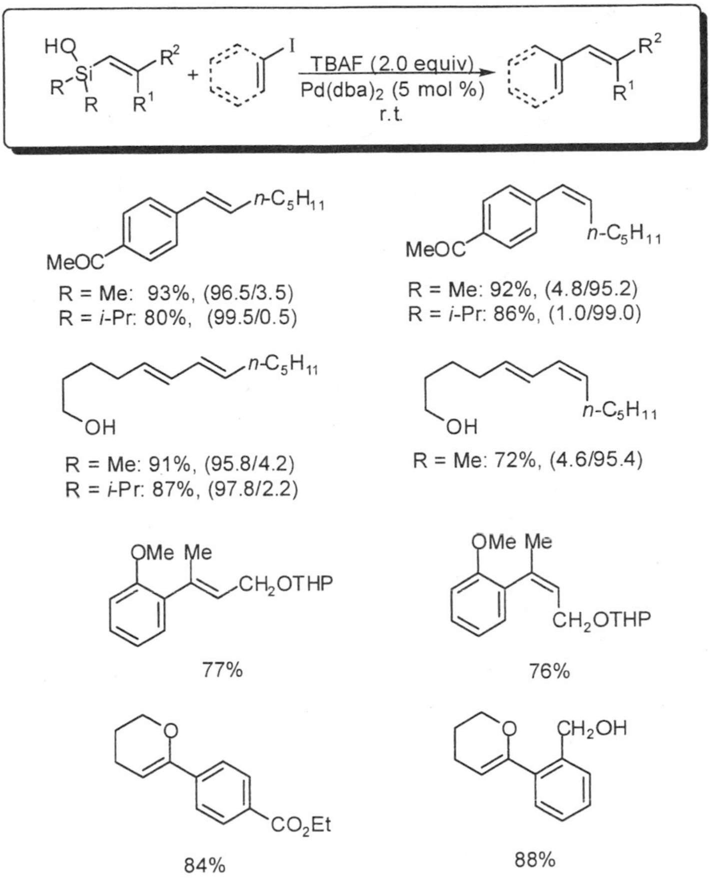

Figure 1

1. Department of Chemistry, Roger Adams Lab., University of Illinois at Urbana-Champaign. Urbana, IL, 61801, E-mail: denmark@scs.uiuc.edu.

2. Ravid, U.; Silverstein, R. M.; Smith, L. R. *Tetrahedron* **1978**, *34*, 1449-1452.

3. Gilman, H.; Cartledge, F. K.; Sin, S.-Y. *J. Organomet. Chem.* **1963**, *1*, 8-14.

4. Denmark, S. E.; Wehrli, D. *Org. Lett.* **2000**, *2*, 565-568.

5. Ukai, T.; Kawazura, H.; Ishii Y. *J. Organomet. Chem.* **1974**, *65*, 253-266.

6. Hiyama, T. Organosilicon Compounds in Cross-Coupling Reactions. In *Metal Catalyzed Cross-Coupling Reactions*; Diederich, F., Stang,   P. J., Eds.; Wiley-VCH: Weinhein, 1998; Chapter 10.

7. Tamao, K.; Kobayashi, K.; Ito, Y. *Tetrahedron Lett.* **1989**, *30*, 6051-6054. (b) Mowery, M. E.; DeShong, P. *J. Org. Chem.* **1999**, *64*, 1684-1688.   (c) Mowery, M. E.; DeShong, P. *J. Org. Chem.* **1999**, *64*, 3266-3270.   (d) Mowery, M. E.; DeShong, P. *Org. Lett.* **1999**, *1*, 2137-2140. (e) Lee, H. M.; Nolan, S. P. *Org. Lett.* **2000**, *2*, 2053-2055.

8. Denmark, S. E.; Sweis, R. F. *Acc. Chem. Res.* **2002**, *35*, 835-846.

9. Hirabayashi, K.; Takahisa, E.; Nishihara, Y.; Mori, A.; Hiyama, T. *Bull. Chem. Soc. Jpn.* **1998**, *71*, 2409-2417.

10. Sieburth, S. McN.; Fensterbank, L. *J. Org. Chem.* **1993**, *58*, 6314-6318. (b) Sieburth, S. McN.; Mu, W. *J. Org. Chem.* **1993**, *58*, 7584-7586.

Appendix

Chemical Abstracts Nomenclature (Registry Number)

1-Heptyne: 1-Heptyne; (628-71-7)

Cyclohexene; (110-83-8)

1-Iodo-1-heptyne: 1-Heptyne, 1-iodo-; (54573—13-6)

Butyllithium: Lithium, butyl; (109-72-8)

(Z)-1-Iodo-1-Heptene: 1-Heptene, 1-iodo-, (1Z)-; (63318-29-6)

Borane-dimethylsulfide complex: Boron, trihydro[thiobis[methane]]-(T-4)-; (13292-87-0)

(Z)-1,1-Dimethyl-1-heptenylsilanol: Silanol, (1Z)-1-heptenyldimethyl-; (261717-40-2)

(Z)-1-Heptenyl-4-methoxybenzene: Benzene, 1-(1Z)-1-heptenyl-4-methoxy-; (80638-85-3)

Iodine; (7553-56-2)

4-Iodoanisole: Benzene, 1-iodo-4-methoxy-; (696-62-8)

Hexamethylcyclotrisiloxane: Cyclotrisiloxane, hexamethyl-; (54-05-9)

Tetrabutylammonium fluoride trihydrate: 1-Butanaminium, N, N, N-tributyl, fluoride, trihydrate; (87749-50-6)

Pd(dba)$_2$: Palladium, bis[(1,2,4,5-η)-1,5-diphenyl-1,4-pentadien-3-one]-; (32005-36-0)

PLATINUM CATALYZED HYDROSILYLATION AND PALLADIUM CATALYZED CROSS-COUPLING: ONE-POT HYDROARYLATION OF 1-HEPTYNE TO (E)-1-(1-HEPTENYL)-4-METHOXYBENZENE

[Benzene, 1-(1E)-1-heptenyl-4-methoxy-]

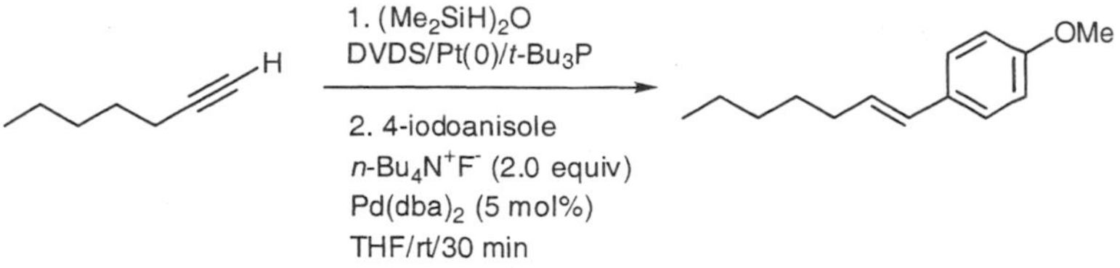

Reaction scheme:
1. (Me$_2$SiH)$_2$O
DVDS/Pt(0)/t-Bu$_3$P

2. 4-iodoanisole
n-Bu$_4$N$^+$F$^-$ (2.0 equiv)
Pd(dba)$_2$ (5 mol%)
THF/rt/30 min

Submitted by Scott E. Denmark and Zhigang Wang.[1]

Checked by Adam I. Keller and Dennis P. Curran.

1. Procedure

(E)-1-(1-Heptenyl)-4-methoxybenzene. A solution of 2.79 mL (2.12 g, 15.6 mmol) of 1,1,3,3-tetramethyldisiloxane in 3.0 mL of tetrahydrofuran (THF) is added to a flame-dried, three-necked, 250-mL, round-bottomed flask equipped with a temperature probe, addition funnel, argon inlet, and stirbar (Note 1). *t*-Bu$_3$P-Pt(0) complex (200 μL) (Notes 2 and 3) is added by syringe. 1-Heptyne (3.43 mL, 2.51 g, 26.0 mmol) (Note 4) is then slowly added by syringe with external cooling in a water bath so that the temperature of the reaction solution is not allowed to exceed 30°C (Note 5). The mixture is stirred at room temperature for 60 min after complete addition of 1-heptyne. A solution of tetrabutylammonium fluoride in THF (40 mL, 1.0M, 40 mmol) is then added over 10 min

(Notes 6 and 7). After complete addition of TBAF, the solution is stirred for 10 min and then 0.575 g (1.0 mmol) of Pd(dba)$_2$ is added in one portion (Note 8). A solution of 4-iodoanisole (Note 9) in THF (20 mL, 1.0M, 20 mmol) is then slowly added via the addition funnel such that the temperature of the reaction solution is kept between 30-33°C (addition time 45 min). The mixture is stirred at room temperature for 30 min after the complete addition of 4-iodoanisole. Diethyl ether (50 mL) is added and the mixture is stirred for an additional 5 min. The mixture is filtered through a short column of silica gel (Note 10), eluting with 200 mL of diethyl ether. The filtrate is concentrated by rotary evaporation and vacuum drying to give a residue that is purified by Kugelrohr distillation, chromatography, and Kugelrohr distillation to afford 3.38-3.42 g (83-84%) of (E)-1-(1-heptenyl)-4-methoxybenzene as a colorless oil, bp 90-95°C/0.1 mm (lit.[4] 85-90°C/ 1.2 mm) (Notes 11, 12, and 13).

2. Notes

1.   1,1,3,3-Tetramethydisiloxane was purchased from Lancaster Synthesis Inc. and used as received. THF was obtained from Mallinckrodt Inc. and freshly distilled from sodium/benzophenone ketyl.

2. t-Bu$_3$P-Pt(0) complex was prepared by the literature procedure:[2] t-Bu$_3$P (32 mg, 0.158 mmol) (Strem Chemicals) was dissolved in platinum(0)-1,3-divinyl-1,1,3,3-tetramethyldisiloxane complex (1.5 mL of xylene solution obtained from Aldrich Chemical Company, Inc., catalog number 47-951-9). The mixture was stirred at 65°C

(oil bath) for 5 min and then was slowly cooled to room temperature. This solution could be stored under N_2 in the freezer indefinitely.

3. t-Bu$_3$P is pyrophoric and should be handled and stored under an inert atmosphere of nitrogen or argon.

4.   1-Heptyne was purchased from GFS Chemicals Inc. and was used without further purification.

5.   The hydrosilylation of 1-heptyne with 1,1,3,3-tetramethyldisiloxane is exothermic. If the temperature of the reaction exceeds 30°C, the amount of the undesired α-isomer will increase. However, the solution should not be cooled to below 10°C as the reaction becomes very sluggish.

6.   A solution of tetrabutylammonium fluoride (TBAF) in THF (1.0M) was prepared from colorless, crystalline tetrabutylammonium fluoride trihydrate purchased from Fluka Chemical Corp.

7.   Caution: TBAF reacts exothermally with the disiloxane mixture and the first few drops must be added very slowly and carefully.

8.   Pd(dba)$_2$ was prepared by the literature procedure:[3] palladium chloride (1.05 g, 5.92 mmol) was added to hot (ca. 50°C) methanol (150 mL) containing dibenzylideneacetone (DBA) (4.60 g, 19.6 mmol) and sodium acetate (3.90 g, 47.5 mmol). The mixture was stirred for 4 hr at 40°C to give a reddish-purple precipitate and allowed to cool to complete the precipitation. The precipitate was isolated by filtration, washed successively with water and acetone, and dried in vacuo.

9. 4-Iodoanisole was purchased from Aldrich Chemical Company, Inc., and was purified by chromatography on silica gel (elution with hexane/EtOAc, 50/1) prior to use.

10. Silica gel was purchased from VWR Scientific (230-400 mesh). A 45-mm diameter column was employed and 60 g of silica gel was loaded as a pentane slurry.

11. The boiling point (bp) corresponds to uncorrected air-bath temperature of a Büchi GKR-50 Kugelrohr.

12. The crude product was purified by Kugelrohr distillation (98-103°/0.2 mm) followed by column chromatography (55-mm diameter column, 200 g of silica gel, elution with pentane/EtOAc, 50/1), and another Kugelrohr distillation.

13. The analytical data are as follows: ^{1}H NMR (500 MHz, CDCl$_3$) δ: 0.91 (t, J = 7.1, 3 H), 1.33 (m, 4 H), 1.46 (m, 2 H), 2.18 (qd, J = 7.1, 1.5, 2 H), 3.80 (s, 3 H), 6.09 (dt, J = 15.9, 6.9, 1 H), 6.33 (d, J = 15.9, 1 H), 6.84 (d, J = 8.8, 2 H), 7.28 (d, J = 8.8, 2 H); ^{13}C NMR (100.6 MHz, CDCl$_3$) δ: 14.1, 22.6, 29.2, 31.4, 33.0, 55.2, 113.8, 126.9, 128.9, 129.1, 130.8, 158.5; IR (NaCl) cm^{-1}: 2956, 2927, 2855, 1608, 1511, 1464, 1291, 1248, 1175, 1037, 964, 833; Anal. calcd for C$_{14}$H$_{20}$O: C, 82.30; H, 9.87; Found: C, 82.10; H, 10.07; GC analysis: t$_R$(Z) = 6.07 min (1.6%), t$_R$(E) = 6.56 min (95.3%), tR(α) = 5.91 min (3.1%) (HP-5, 250 °C, 15 psi).

Waste Disposal Information

All toxic materials were disposed of in accordance with "Prudent Practices in the Laboratory", National Academy Press: Washington, DC, 1995

3. Discussion

One-pot coupling reactions of organoboron- and organotin reagents with aryl or alkenyl halides and their applications in organic synthesis have been well documented.[5,6] There were only a few examples of hydrosilylation/cross-coupling applied to specific synthetic target molecules.[7,8] The procedures presented here provide a general route to (E)-olefins by palladium-catalyzed cross-coupling of alkenyl disiloxanes with aryl or alkenyl iodides.[9]

Tetramethyldisiloxane is an inexpensive, commercially available reagent. It reacts with alkynes to generate the alkenyldisiloxanes in situ, which can then react with a variety of iodides bearing different substituent patterns. The processes display very good functional group compatibility and afford products in high yield and with excellent regio- and stereoselectivity (Table 1). A notable characteristic of this protocol is the extremely mild reaction conditions employed. The hydrosilylation was complete at room temperature after 30 min. Most coupling reactions proceeded in 10 min to a few hours at room temperature.

The generality and functional group compatibility were also excellent with regard to the alkynes (Table 2). Free hydroxyl groups and even a remote double bond were compatible and give very good yields and purities.

The couplings of reactive iodides (with electron-withdrawing groups) were exothermic. However, the reaction temperature can be modulated by the slow addition of iodides last. These protocols are suitable for larger scale applications as well.

Table 1. One-Pot Hydrosilylation/Cross-Coupling of 1-Heptyne with Aryl Iodides

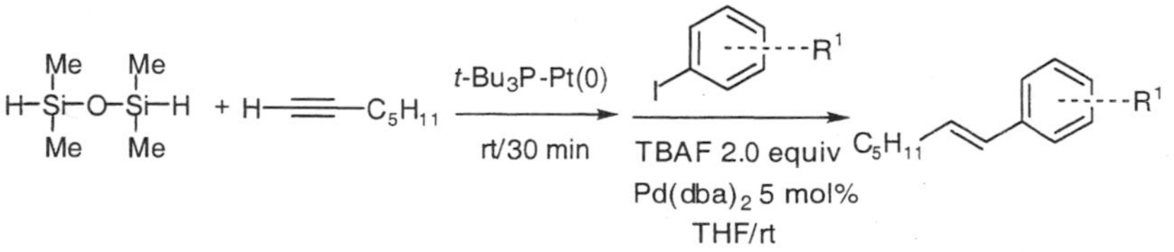

Entry	Iodide, R^1	Time, min	Product, R^1	Yield, %
1	1-napthyl	10	1-napthyl	82
2	$4\text{-}(CH_3CO)C_6H_4$	10	$4\text{-}(CH_3CO)C_6H_4$	94
3	$4\text{-}(CH_3O)C_6H_4$	10	$4\text{-}(CH_3O)C_6H_4$	89
4	$3\text{-}(NO_2)C_6H_4$	10	$3\text{-}(NO_2)C_6H_4$	85
5	$3\text{-}(NO_2)C_6H_4$	60 (1 mol% Pd)	$3\text{-}(NO_2)C_6H_4$	89
6	$3\text{-}(CH_3)C_6H_4$	10	$3\text{-}(CH_3)C_6H_4$	78
7	$2\text{-}(CH_3OCO)C_6H_4$	20 h (10 mol% Ph_3As)	$2\text{-}(CH_3OCO)C_6H_4$	88
8	$2\text{-}(CH_3O)C_6H_4$	10	$2\text{-}(CH_3O)C_6H_4$	82

Table 2. One-Pot Hydrosilyation/Cross-Coupling of 1-Alkynes with Aryl Iodides

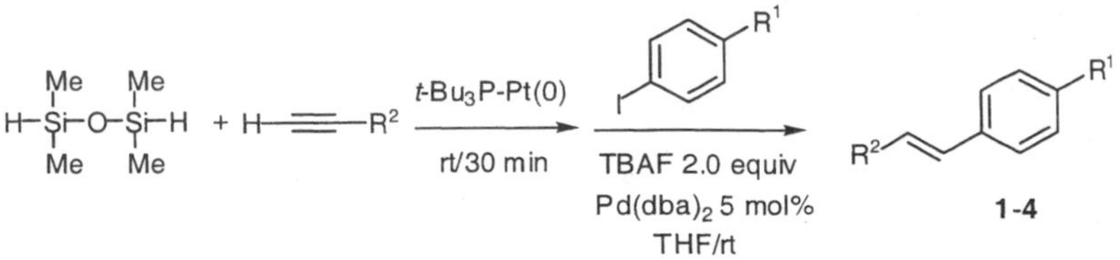

Entry	Alkyne, R^2	Iodide, R^1	Time, min	Product	Yield, %
1	C_6H_5	4-$(CH_3CO)C_6H_4$	10	**1a**	89
2	C_6H_5	4-$(CH_3O)C_6H_4$	10	**1b**	74
3	$HO(CH_2)_3$	4-$(CH_3CO)C_6H_4$	30	**2a**	82
4	$HO(CH_2)_3$	4-$(CH_3O)C_6H_4$	60	**2b**	89
5	$C_6H_5C(OH)(CH_3)$	4-$(CH_3CO)C_6H_4$	24 h	**3a**	72
6	$C_6H_5C(OH)(CH_3)$	4-$(CH_3O)C_6H_4$	24 h	**3b**	79
7	$CH_2=CHCH_2O(CH_2)_3$	4-$(CH_3CO)C_6H_4$	10	**4a**	78
8	$CH_2=CHCH_2O(CH_2)_3$	4-$(CH_3O)C_6H_4$	10	**4b**	76

1. Department of Chemistry, Roger Adams Laboratory, University of Illinois at Urbana-Champaign, Urbana, IL 61801, E-mail: denmark@MITscs.uiuc.edu.

2. Chandra, G.; Lo, P. Y.; Hitchcock, P. B.; Lappert, M. F. *Organometallics* **1987**, *6*, 191-192.

3. Ukai, T.; Kawazura, H.; Ishii Y. *J. Organomet. Chem.* **1974**, *65*, 253-266.

4. Denmark, S. E.; Wehrli, D. *Org. Lett.* **2000**, *2*, 565-568.

5. (a) Miyaura, N.; Yano, T.; Suzuki, A. *Tetrahedron Lett.* **1980**, *21*, 2865-2868. (b) Miyaura, N.; Ishiyama, T.; Sasaki, H.; Ishikawa, M.; Satoh, M.; Suzuki, A. *J. Am. Chem. Soc.* **1989**, *111*, 314-321. (c) Ishiyama, T.; Nishijima, K.-i.; Miyaura, N.; Suzuki, A. *J. Am. Chem. Soc.* **1993**, *115*, 7219-7225. (d) Baudoin, O.; Guenard, D.; Gueritte, F. *J. Org. Chem.* **2000**, *65*, 9268-9271.

6. Maleczka, R. E., Jr.; Gallagher, W. P.; Terstiege, I. *J. Am. Chem. Soc.* **2000**, *122*, 384-385.

7. Tamao, K.; Kobayashi, K.; Ito, Y. *Tetrahedron Lett.* **1989**, *30*, 6051-6054.

8. Takahashi, K.; Minami, T.; Ohara, Y.; Hiyama, T. *Tetrahedron Lett.* **1993**, *34*, 8263-8266.

9. Denmark, S. E.; Wang, Z. *Org. Lett.* **2001**, *3*, 1073-1076.

Appendix

Chemical Abstracts Nomenclature (Registry Number)

1-Heptyne: 1-Heptyne; (628-71-7)

(*E*)-1-Heptenyl-4-methoxybenzene: Benzene, 1-(1*E*)-1-heptenyl-4-methoxy-; (135987-61-0)

4-Iodoanisole: Benzene, 1-iodo-4-methoxy-; (696-62-8)

1,1,3,3-Tetramethyldisiloxane: Disiloxane, 1,1,3,3-tetramethyl-; (3277-26-7)

Tri-(tert-butyl)phosphine: Phosphine, tris(1,1-dimethylethyl)-; (13716-12-6)

Platinum(0)-1,3-divinyl-1,1,3,3-tetramethyldisiloxane complex: Platinum, 1,3-diethenyl-1,1,3,3-tetramethyldisiloxane complex; (68478-92-2)

Tetrabutylammonium fluoride trihydrate: 1-Butanaminium, N, N, N-tributyl, fluoride, trihydrate; (87749-50-6)

Pd(dba)$_2$: Palladium, bis[(1,2,4,5-η)-1,5-diphenyl-1,4-pentadien-3-one]-; (32005-36-0)

Palladium chloride; Palladium chloride (PdCl$_2$); (7647-10-1)

Dibenzylideneacetone (DBA): 1,4-Pentadien-3-one, 1,5-diphenyl-; (538-58-9)

HECK REACTIONS OF ARYL CHLORIDES CATALYZED BY PALLADIUM/TRI-tert-BUTYLPHOSPHINE: (E)-2-METHYL-3-PHENYLACRYLIC ACID BUTYL ESTER AND (E)-4-(2-PHENYLETHENYL)BENZONITRILE

(2-Propenoic acid, 2-methyl-3-phenyl-, butyl ester and Benzonitrile, 4-[(1E)-2-phenylethenyl])

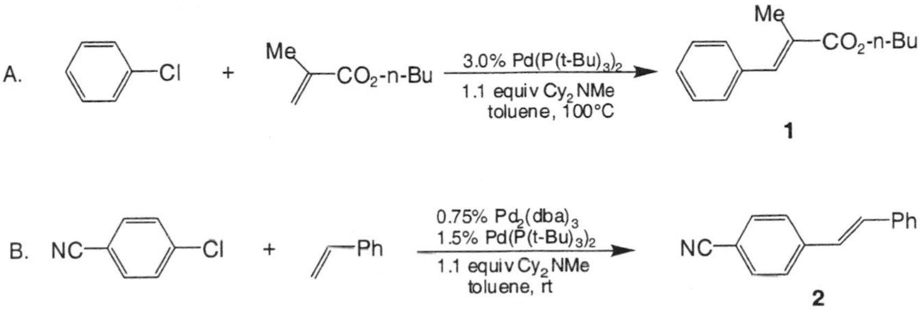

A.

B.

Submitted by Adam F. Littke and Gregory C. Fu.[1]
Checked by Michael H. Ober and Scott E. Denmark.

1. Procedure

A. (E)-2-Methyl-3-phenylacrylic acid butyl ester (**1**). An oven-dried, 250-mL, three-necked, round-bottomed flask equipped with a reflux condenser (fitted with an argon inlet adapter), rubber septum, glass stopper, and a teflon-coated magnetic stir bar is cooled to room temperature under a flow of argon. The flask is charged with bis(tri-tert-butylphosphine)palladium (Pd(P(t-Bu)$_3$)$_2$) (0.482 g, 0.943 mmol, 3.0 mol% Pd) (Notes 1, 2) and again purged with argon. Toluene (32 mL) (Notes 9, 10) is added, and the mixture is stirred at room temperature, resulting in a homogeneous brown-orange solution. Chlorobenzene (3.20 mL, 31.5 mmol) (Note 11), *N*-methyldicyclohexylamine (Cy$_2$NMe) (7.50 mL, 35.0 mmol) (Note 12), and butyl methacrylate (5.50 mL, 34.6 mmol) (Note 12) are then added successively via syringe. The resulting mixture is allowed to stir at room temperature for 5 min, resulting in a homogeneous light-orange solution. The rubber septum is then replaced with a glass stopper, and the flask is heated in a 100°C oil bath under a positive pressure of argon for 22 hr (Note 13). Upon heating, the solution becomes bright canary-yellow in color, and within 10-15 min the formation of a white precipitate (the amine hydrochloride salt) is observed. Upon completion of the reaction, shiny deposits of palladium metal form on the sides of the flask, and a large quantity of white precipitate is present. The reaction mixture is allowed to cool to room temperature and then diluted with 100 mL of diethyl ether. The resulting solution is washed with 100 mL of H$_2$O, and the aqueous layer is extracted with three 50-mL portions of diethyl ether. The combined organic phases are washed with 100 mL of brine and then concentrated by rotary evaporation. Any residual solvent is removed at 0.5 mm. The crude product, a dark-brown oil, is then purified by flash column chromatography (Note 14) to afford 6.67 - 6.72 g (95%) of **1** as a pale red-orange liquid. This liquid appears to be pure by ^{1}H and ^{13}C NMR spectroscopy; however, if desired, the discoloration can be removed by filtering the product through a small column of silica gel (3 cm diameter x 10 cm height), which furnishes 6.49-6.62 g (95-96%) of **1** as a clear, colorless liquid (Notes 15 and 16).

   B.   *(E)-4-(2-Phenylethenyl)benzonitrile* (**2**).   An oven-dried, 250-mL, three-necked, round-bottomed flask equipped with an argon inlet adapter, rubber septum, glass stopper, and a teflon-coated magnetic stir bar is cooled to room temperature under a flow of argon. The flask is charged successively with bis(tri-*tert*-butylphosphine)palladium [(Pd(P(*t*-Bu)$_3$)$_2$] (0.238 g, 0.466 mmol, 1.5 mol% Pd) (Notes 1, 2), tris(dibenzylideneacetone)dipalladium(0) (Pd$_2$(dba)$_3$) (0.213 g, 0.233 mmol, 1.5 mol% Pd) (Note 3), and 4-chlorobenzonitrile (4.25 g, 30.9 mmol) (Note 17). The flask is purged with argon, and 62 mL of toluene is added (Notes 9, 10, 18). The resulting mixture is stirred at room temperature, resulting in a dark, red-purple solution. N-Methyldicyclohexylamine (7.5 mL, 35.0 mmol) (Note 12) and styrene (3.8 mL, 33.2 mmol) (Note 19) are then added via syringe. The reaction mixture is allowed to stir at room temperature under a positive pressure of argon for 72 hr (Note 13). Within the first 1-2 hr, the color changes from deep red-purple to dark brown, and a precipitate (the amine hydrochloride salt) begins to form. Upon completion of the reaction, 100 mL of ethyl acetate is added, and the resulting solution is washed with 100 mL of water. The aqueous layer is separated and extracted with three 50-mL portions of diethyl ether, and the combined organic phases are concentrated by rotary evaporation. Any residual solvent is removed at 0.5 mm. The crude product, a yellow solid, is purified via flash column chromatography (Note 20) to afford 5.30-5.34 g (83-84%) of **2** as white crystalline spheres (Notes 21, 22, 23).

2. Notes

1.   Pd(P(t-Bu)$_3$)$_2$ was prepared according to the following procedure. In a nitrogen-filled Vacuum Atmospheres glovebox, a 100-mL, one-neck, round-bottomed flask equipped with a teflon-coated magnetic stir bar is charged with tris(dibenzylideneacetone)dipalladium(0) (Pd$_2$(dba)$_3$) (2.98 g, 3.25 mmol) (Note 3). A solution of tri-*tert*-butylphosphine (P(t-Bu)$_3$) (2.88 g, 14.2 mmol) (Note 4) in 43 mL of N, N-dimethylformamide (DMF) (Note 5) is then added to the reaction flask via a glass pipette, and the resulting dark green-brown solution is stirred at room temperature for 23 hr. The reaction mixture is then filtered through a 30-mL medium-porosity glass frit to collect the unpurified product, Pd(P(t-Bu)$_3$)$_2$, as a gray solid. The reaction flask is rinsed with three 6-mL portions of DMF and 5 mL of methanol (Note 6), which are passed through the glass frit (the final rinses should be colorless). The reaction product

is then dissolved in 100 mL of toluene (Note 7), and the resulting solution is filtered through a 3-cm diameter 0.45 μm Gelman acrodisc (to remove some insoluble black material) into a 250-mL Schlenk tube, affording a homogeneous orange-yellow solution. The Schlenk tube is removed from the glovebox, and the toluene solution is concentrated under high vacuum (0.5mm) to a volume of approximately 25 mL, at which point a white crystalline solid begins to precipitate. The Schlenk tube is taken into the glovebox, and the toluene solution and crystalline solid are transferred to a 250-mL Erlenmeyer flask via a glass pipette. MeOH (100 mL) is then added slowly via pipette over 10 min, resulting in the precipitation of additional white crystalline solid. The solution is allowed to stand for one hour, and then the mother liquor is separated from the solid via pipette. The solid is washed with two 10-mL portions of MeOH, transferred to a tared 20-mL glass vial, and dried under high vacuum, affording 2.41 g (72%) of $Pd(P(t-Bu)_3)_2$ as a white, crystalline solid (Note 8). The $Pd(P(t-Bu)_3)_2$ can be stored indefinitely under nitrogen in a Vacuum Atmospheres glovebox. Although $Pd[P(t-Bu)_3]_2$ has been reported to be "stable in air in the solid state,"[2] if a glovebox is not available, it is recommended that $Pd[P(t-Bu)_3]_2$ be stored in a tightly capped vial in a desiccator, preferably under argon or nitrogen.

2.   Alternatively, $Pd[P(t-Bu)_3]_2$ may be purchased from Strem Chemicals.

3. $Pd_2(dba)_3$ was purchased from the Aldrich Chemical Company and used as received.

4. $P(t-Bu)_3$ (99%) was purchased from Strem Chemicals and used as received.

5.   DMF (anhydrous, DriSolv) was purchased from EM Science and degassed under high vacuum for 10-15 min prior to use.

6.   Methanol (certified A.C.S., purchased from Fisher Scientific) was distilled from $Mg(OMe)_2$ and was degassed by three freeze-pump-thaw cycles prior to use.

7. Toluene (J. T. Baker; CYCLE-TRAINER solvent delivery kegs) was vigorously purged with argon for 2 hr and then passed through two packed columns of neutral alumina and copper(II) oxide under argon pressure.[3]

8. ^{31}P{^{1}H} NMR (C$_6$D$_6$, 202 MHz): δ 85.3; ^{1}H NMR (C$_6$D$_6$, 500 MHz): δ 1.51 (t, J = 5.4 Hz). The NMR sample must be prepared under inert atmosphere to avoid aerobic oxidation of the catalyst as evidenced by free P(t-Bu)$_3$ at δ 1.27 ppm in the ^{1}H NMR spectrum. Even with these precautions, ca. 1% of P(t-Bu)$_3$ is observed.

9. Toluene (anhydrous, 99.8%, Sure/Seal^{TM} bottle) was purchased from the Aldrich Chemical Company and used as received.

10. 1,4-Dioxane is an equally suitable solvent for these Heck couplings (and is the solvent used in the published procedures); however, due to the lower cost and the lower toxicity of toluene, it was chosen as the solvent for these reactions.

11. Chlorobenzene (anhydrous, 99.8%, Sure/Seal^{TM} bottle) was purchased from the Aldrich Chemical Company and used as received.

12. N-Methyldicyclohexylamine (97%) and butyl methacrylate (99%) were purchased from the Aldrich Chemical Company and gently sparged with argon for 5-10 min prior to use.

13. The progress of the reaction was monitored by GC.

14. Flash column chromatography was performed using silica gel (6 cm diameter x 35 cm height), eluting with 19/1 hexane/diethyl ether.

15. Compound 1 has the following properties: bp 111°C (1 mm) ^{1}H NMR (CDCl$_3$, 500 MHz) δ: 0.98 (t, J = 7.5, 3H), 1.45 (sext, J = 1.0, 2H), 1.71 (qn, J = 1.0, 2H), 2.12 (d, J = 1.5, 3H), 4.22 (t, J = 6.6, 2H), 7.33 (m, 1H), 7.39 (m, 4H), 7.68 (apparent d, J = 1.5, 1H); ^{13}C NMR (CDCl$_3$ 126 126MHz): δ 13.7, 14.0, 19.3, 30.7, 64.8, 128.2, 128.4, 128.7, 129.6, 136.0, 138.8, 168.8; IR (neat, cm^{-1}): 3059, 3026, 2960, 2933, 2873, 1709, 1635, 1492, 1448, 1388, 1356, 1254, 1201, 1115, 1074, 1003, 931, 766, 704; Anal. Calcd for C$_{14}$H$_{18}$O$_2$: C, 77.03; H, 8.31. Found: C, 77.02; H, 8.30.

16. The checkers found that the product could be further purified by distillation (110°C/1 mm).

17. 4-Chlorobenzonitrile (99%) was purchased from Aldrich Chemical Company and used as received.

18. To allow more efficient stirring, it was beneficial to run this reaction at half of the concentration (2 mL solvent per mmol of aryl chloride) of the original published procedure (1 mL solvent per mmol of aryl chloride).

19. Styrene (99+%) was purchased from Aldrich Chemical Company and gently sparged with argon for 5-10 min prior to use.

20. Flash column chromatography was performed using silica gel (10 cm diameter x 27 cm height), eluting with 4/1 toluene/hexane. A small amount of aryl chloride that remained unreacted after 72 hr was recovered mixed with a small quantity (<5%) of the desired product.

21. Compound **2** has the following properties: mp (corr.) 115-117°C (lit.[4] mp 115 °C); ^{1}H NMR (CDCl$_3$, 500 MHz): δ 7.09 (d, J = 16.5, 1H), 7.22 (d, J = 16.5, 1H), 7.33 (d, J = 7.1, 1H), 7.40 (t, J = 7.1, 2H), 7.54 (d, J = 7.5, 2H), 7.59 (d, J = 8.3, 2H), 7.64 (d, J = 8.4, 2H); ^{13}C NMR (CDCl$_3$, 126 MHz): δ 110.8, 119.3, 126.9, 127.0, 127.1, 128.9, 129.1, 132.6, 132.7, 136.5, 142.1; IR (neat, cm^{-1}): 3029, 2964, 2225, 1648, 1604, 1531, 1450, 1278, 1226, 1174, 1095, 966, 821, 769; Anal. Calcd for C$_{15}$H$_{11}$N: C, 87.78; H, 5.40; N, 6.82. Found: C, 87.43; H, 5.26; N, 6.86.

22. The checkers found that the product could be further purified by sublimation (127°C/0.5 mm).

23. The submitters obtained the product in 89% yield.

Waste Disposal Information

All toxic materials were disposed of in accordance with "Prudent Practices in the Laboratory"; National Academy Press; Washington, DC, 1995.

3. Discussion

Since its discovery in the early 1970's, the palladium-catalyzed arylation of olefins (Heck reaction; eq 1 and Figure 1)[5,6] has been applied to a diverse array of fields, ranging from natural products synthesis[7,8] to materials science[9] to bioorganic chemistry.[10] This powerful carbon-carbon bond-forming process has been practiced on an industrial scale for the production of compounds such as naproxen[11] and octyl methoxycinnamate.[12] The Heck reaction is typically performed in the presence of a palladium/tertiary phosphine catalyst and a stoichiometric amount of an inorganic or organic base. High functional-group tolerance and the ready availability and low cost of simple olefins contribute to the exceptional utility of the Heck arylation.

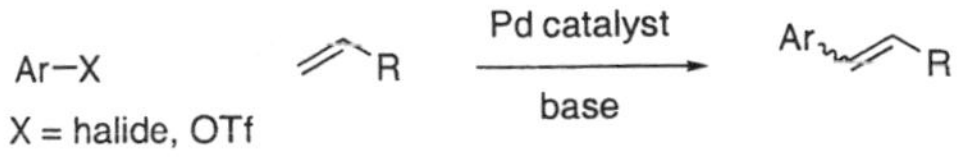

$$Ar-X \qquad \diagup\!\!\!\diagdown R \xrightarrow[\text{base}]{\text{Pd catalyst}} Ar\diagup\!\!\!\searrow R \qquad (1)$$

$$X = \text{halide, OTf}$$

FIGURE 1

OUTLINE OF THE CATALYTIC CYCLE FOR THE HECK COUPLING REACTION

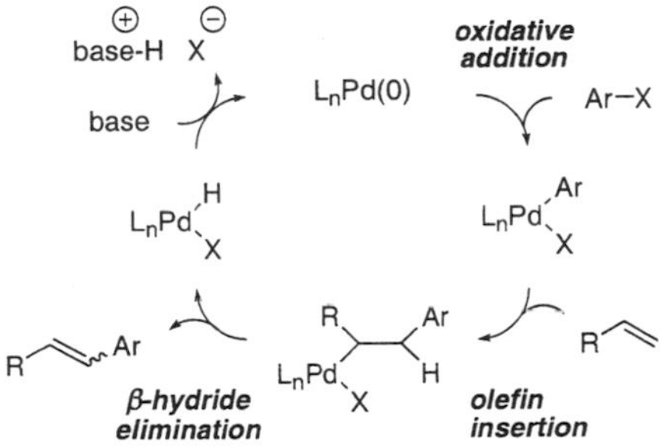

Until recently, one important unsolved problem for the Heck reaction was the poor reactivity of aryl chlorides, which are arguably the most attractive class of aryl halides, due to their lower price and greater availability as compared with the corresponding bromides and iodides.[13,14] For the few catalyst systems that have displayed activity for Heck couplings of aryl chlorides (e.g., those of Milstein (bulky, electron-rich chelating bisphosphines),[15] Herrmann (palladacycles, N-heterocyclic carbenes),[16] Reetz (tetraphenylphosphonium salts),[17] and Beller (phosphites),[18,19,20] the scope of the reactions has been quite narrow and the reaction temperatures have been rather high ($\geq$120 °C). This need for elevated temperatures can be problematic for a variety of reasons, including decomposition of thermally unstable substrates and decreased regio- and stereoselectivities.

Pd/P(t-Bu)$_3$, in the presence of Cy$_2$NMe, is an unusually mild and versatile catalyst for Heck reactions of aryl chlorides (Tables 1 and 2) (as well as for room-temperature reactions of aryl bromides).[21,22,23] Example A, the coupling of chlorobenzene with butyl methacrylate, illustrates the application of this method to the stereoselective synthesis of a trisubstituted olefin; α-methylcinnamic acid derivatives are an important family of compounds that possess biological activity (e.g., hypolipidemic[24] and antibiotic[25]) and serve as intermediates in the synthesis of pharmaceuticals (e.g., Sulindac, a non-steroidal anti-inflammatory drug[26]). Example B, the coupling of 4-chlorobenzonitrile with styrene, demonstrates that Pd/P(t-Bu)$_3$ can catalyze the Heck reaction of activated aryl chlorides *at room temperature*.

From a practical point of view, it is worth noting that Heck reactions catalyzed by Pd/P(t-Bu)$_3$ do not typically require rigorously purified reagents or solvents. In addition, the palladium and phosphine sources, Pd[P(t-Bu)$_3$]$_2$ and Pd$_2$(dba)$_3$, are commercially available and can be handled in air.

Thus, in terms of scope, mildness, and convenience, Pd/P(t-Bu)$_3$ provides an attractive method for achieving Heck couplings of aryl chlorides.

TABLE 1.

HECK COUPLINGS OF ARYL CHLORIDES AT ELEVATED TEMPERATURE

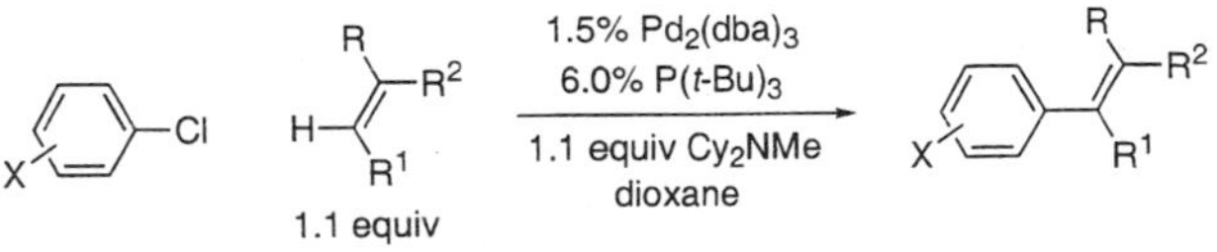

Entry	Aryl Chloride	Olefin	Temperature	Product	Yield[a,b]
1[c]	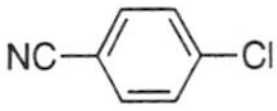 NC—⟨⟩—Cl	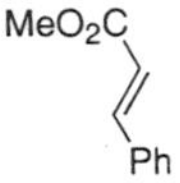 MeO$_2$C / Ph	70 °C	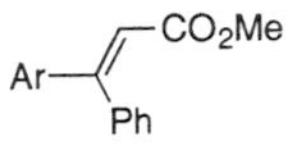 Ar—⟨⟩CO$_2$Me / Ph	67%[d]
2	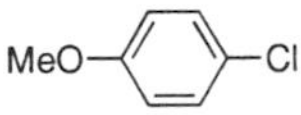 MeO—⟨⟩—Cl	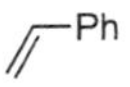 Ph	120 °C	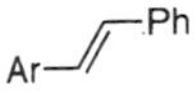 Ar—Ph	72%
3	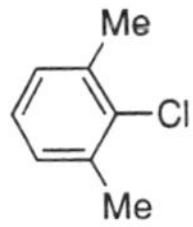 Me / ⟨⟩—Cl / Me	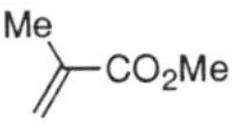 Me—CO$_2$Me	120 °C	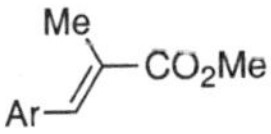 Me / Ar—CO$_2$Me	80%
4	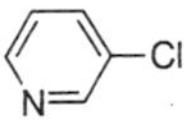 ⟨N⟩—Cl	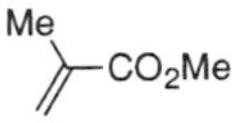 Me—CO$_2$Me	100 °C	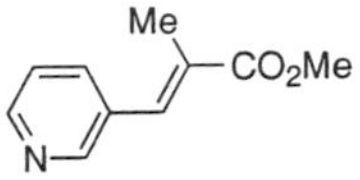 Me—CO$_2$Me	76%
5	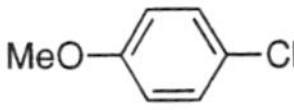 MeO—⟨⟩—Cl	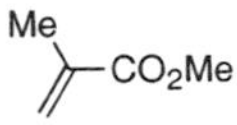 Me—CO$_2$Me	120 °C	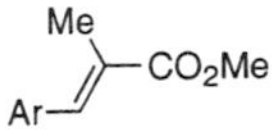 Me / Ar—CO$_2$Me	72%

[a]Isolated yield, average of two runs. [b]E:Z ratio is >20:1, as determined by ^{1}H NMR. [c]3.6% P(t-Bu)$_3$ was used. [d]Product includes 2.5% P(t-Bu)$_3$/OP(t-Bu)$_3$.

TABLE 2.

HECK COUPLINGS OF ACTIVATED ARYL CHLORIDES

AT ROOM TEMPERATURE

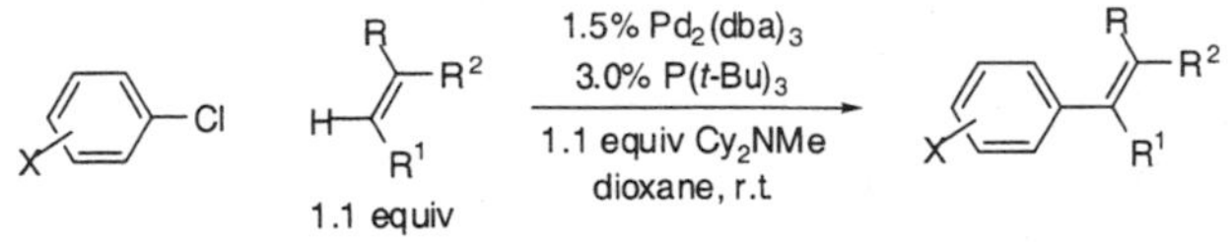

$$\text{Ar–Cl} + \underset{\substack{H}}{\overset{R}{\underset{R^1}{=}}}\!\!R^2 \xrightarrow[\substack{1.1\ \text{equiv Cy}_2\text{NMe} \\ \text{dioxane, r.t.}}]{\substack{1.5\%\ \text{Pd}_2(\text{dba})_3 \\ 3.0\%\ \text{P}(t\text{-Bu})_3}} \text{product}$$

1.1 equiv (aryl chloride), 1.1 equiv (olefin)

Entry	Aryl Chloride	Olefin	Product	Yield[a,b]
1	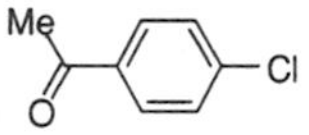 4-acetylphenyl Cl (Me–C(O)–C₆H₄–Cl)	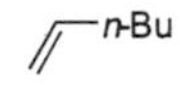 CH₂=CH–n-Bu	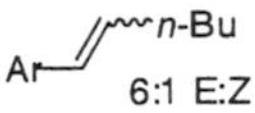 Ar–CH=CH–n-Bu, 6:1 E:Z	70%
2	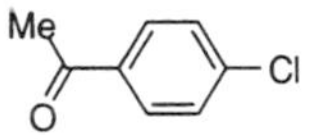 4-acetylphenyl Cl	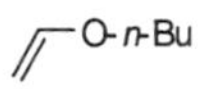 CH₂=CH–O-n-Bu	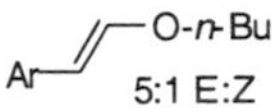 Ar–CH=CH–O-n-Bu, 5:1 E:Z (10); 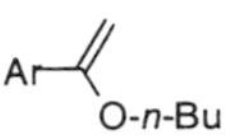 Ar–C(=CH₂)–O-n-Bu (1)	87%[c]
3	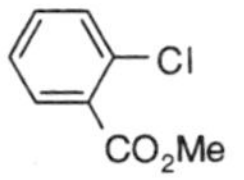 2-(CO₂Me)phenyl Cl	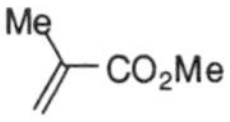 CH₂=C(Me)–CO₂Me	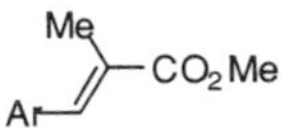 Ar–CH=C(Me)–CO₂Me	90%
4	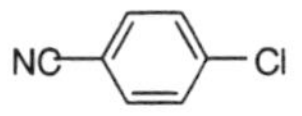 4-NC–C₆H₄–Cl	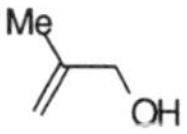 CH₂=C(Me)–CH₂OH	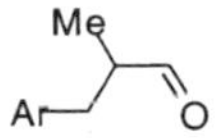 Ar–CH₂–CH(Me)–CHO	79%
5[d]	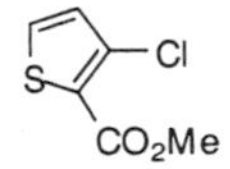 3-Cl-thiophene-2-(CO₂Me)	 2,3-dihydrofuran	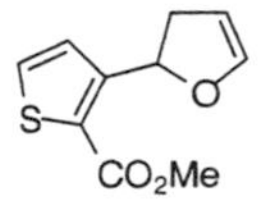 thiophene-dihydrofuran product (CO₂Me)	87%
6[e]	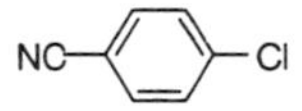 4-NC–C₆H₄–Cl	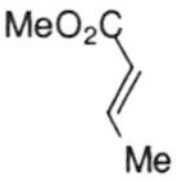 MeO₂C–CH=CH–Me	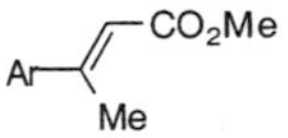 Ar–C(=C(Me)...)–CO₂Me	57%[f]

[a]Isolated yield, average of two runs. [b]Unless otherwise indicated, the E:Z ratio is >20:1, as determined by ^{1}H NMR. [c]Product includes 5% 4'-chloroacetophenone. [d]3 equiv. of olefin was used. [e]2 equiv. of olefin was used. [f]Product includes 2% P(t-Bu)$_3$/OP(t-Bu)$_3$.

1. Department of Chemistry, Massachusetts Institute of Technology, Cambridge, MA 02139.

2. (a) Otsuka, S.; Yoshida, T.; Matsumoto, M.; Nakatsu, K. *J. Am. Chem. Soc.* **1976**, *98*, 5850-5858. (b) Yoshida, T.; Otsuka, S. *J. Am. Chem. Soc.* **1977**, *99*, 2134-2140. (c) Yoshida, T.; Otsuka, S. *Inorg. Synth.* **1990**, *28*, 113-119.

3. (a) Pangborn, A. B.; Giardello, M. A.; Grubbs, R. H.; Rosen, R. K.; Timmers, F. J. *Organometallics* **1996**, *15*, 1518-1520. (b) Alaimo, P. J.; Peters, D. W.; Arnold, J.; Bergman, R. J. *J. Chem. Ed.* **2001**, *78*, 64-64.

4. Gusten, H.; Salzwedel, M. *Tetrahedron* **1967**, *23*, 173-185.

5. (a) Mizoroki, T.; Mori, K.; Ozaki, A. *Bull. Chem. Soc. Jpn.* **1971**, *44*, 581. (b) Heck, R. F.; Nolley, J. P., Jr. *J. Org. Chem.* **1972**, *37*, 2320-2322.

6. For reviews of the Heck reaction, see: (a) Bräse, S.; de Meijere, A. In *Metal-Catalyzed Cross-Coupling Reactions*; Diederich, F., Stang, P. J., Eds.; Wiley: New York, 1998; Chapter 3. (b) Beletskaya, I. P.; Cheprakov, A. V. *Chem. Rev.* **2000**, *100*, 3009-3066. (c) Heck, R. F. In *Comprehensive Organic Synthesis*; Trost, B. M., Ed.; Pergamon: New York, 1991; Vol. 4, Chapter 4.3. (d) Heck R. F. *Org. React.* **1982**, *27*, 345-390. (e) Crisp, G. T. *Chem. Soc. Rev.* **1998**, *27*, 427-436. (f) de Meijere, A.; Meyer, F. E. *Angew. Chem., Int. Ed. Engl.* **1994**, *33*, 2379-2411. (g) Jeffery, T. In *Advances in Metal-Organic Chemistry*; Liebeskind, L. S., Ed.; JAI: London, 1996; Vol. 5, pp. 153-260. (h) Cabri, W.; Candiani, I. *Acc. Chem. Res.* **1995**, *28*, 2-7.

7. For example, see: (a) Taxol: Danishefsky, S. J.; Masters, J. J.; Young, W. B.; Link, J. T.; Snyder, L. B.; Magee, T. V.; Jung, D. K.; Isaacs, R. C. A.; Bornmann, W. G.; Alaimo, C. A.; Coburn, C. A.; Di Grandi, M. J. *J. Am. Chem. Soc.* **1996**, *118*, 2843-2859. (b) Scopadulcic acid: Overman, L. E.; Ricca, D. J.; Tran, V. D. *J. Am. Chem. Soc.* **1993**, *115*, 2042-2044.

73

8. For overviews of applications of the Heck reaction in natural products synthesis, see: (a) Link, J. T.; Overman, L. E. In *Metal-Catalyzed Cross-Coupling Reactions*; Diederich, F., Stang, P. J., Eds.; Wiley-VCH: New York, 1998; Chapter 6. (b) Bräse, S.; de Meijere, A. In *Metal-Catalyzed Cross-Coupling Reactions*; Diederich, F., Stang, P. J., Eds.; Wiley: New York, 1998; Chapter 3.6. (c) Nicolaou, K. C.; Sorensen, E. J. *Classics in Total Synthesis*; VCH: New York, 1996; Chapter 31. These authors refer to the Heck reaction as "one of the true "power tools" of contemporary organic synthesis" (p. 566).

9. For example, see: (a) *Step-Growth Polymers for High-Performance Materials*; Hedrick, J. L., Labadie, J. W., Eds.; ACS Symp. Ser. 624; American Chemical Society: Washington, DC, 1996; Chapters 1, 2, and 4. (b) DeVries, R. A.; Vosejpka, P. C.; Ash, M. L. *Catalysis of Organic Reactions*; Herkes, F. E., Ed.; Marcel Dekker: New York, 1998; Chapter 37. (c) Tietze, L. F.; Kettschau, G.; Heuschert, U.; Nordmann, G. *Chem. Eur. J.* **2001**, *7*, 368-373.

10. For some recent examples, see: (a) Haberli, A.; Leumann, C. J. *Org. Lett.* **2001**, *3*, 489-492. (b) Burke, T. R., Jr.; Liu, D.-G.; Gao, Y. *J. Org. Chem.* **2000**, *65*, 6288-6292.

11. Stinson, S. C. *Chem. Eng. News* January 18, 1999, p. 81.

12. Octyl methoxycinnamate (OMC) is the most common UV-B sunscreen that is on the market: Eisenstadt, A. In *Catalysis of Organic Reactions*; Herkes, F. E., Ed.; Marcel Dekker: New York, 1998; Chapter 33.

13. For discussions of coupling reactions of aryl chlorides, see: (a) Grushin, V. V.; Alper, H. In *Activation of Unreactive Bonds and Organic Synthesis*; Murai, S., Ed.; Springer-Verlag: Berlin, 1999; pp. 193-226. (b) Grushin, V. V.; Alper, H. *Chem. Rev.* **1994**, *94*, 1047-1062.

14. The low reactivity of aryl chlorides is usually attributed to the strength of the C-Cl bond (bond dissociation energies for aryl-X: Cl = 96 kcal/mol; Br = 81 kcal/mol; I = 65 kcal/mol).

15. (a) Ben-David, Y.; Portnoy, M.; Gozin, M.; Milstein, D. *Organometallics* **1992**, *11*, 1995-1996. (b) Portnoy, M.; Ben-David, Y.; Milstein, D. *Organometallics* **1993**, *12*, 4734-4735. (c) Portnoy, M.; Ben-David, Y.; Rousso, I.; Milstein, D. *Organometallics* **1994**, *13*, 3465-3479.

16. (a) Herrmann, W. A.; Brossmer, C.; Ofele, K.; Reisinger, C.-P.; Priermeier, T.; Beller, M.; Fischer, H. *Angew. Chem., Int. Ed. Engl.* **1995**, *34*, 1844-1848. (b) Herrmann, W. A.; Brossmer, C.; Reisinger, C.-P.; Reirmeier, T. H.; Ofele, K.; Beller, M. *Chem. Eur. J.* **1997**, *3*, 1357-1364. (c) Herrmann, W. A.; Elison, M.; Fischer, J.; Köcher, C.; Artus, G. R. J. *Angew. Chem., Int. Ed. Engl.* **1995**, *34*, 2371-2374. (d) See also: Herrmann, W. A.; Brossmer, C.; Ofele, K.; Beller, M.; Fischer, H. *J. Mol. Catal. A* **1995**, *103*, 133-146.

17. Reetz, M. T.; Lohmer, G.; Schwickardi, R. *Angew. Chem. Int. Ed.* **1998**, *37*, 481-483.

18. Beller, M.; Zapf, A. *Synlett* **1998**, 792-793.

19. See also: Kaufmann, D. E.; Nouroozian, M.; Henze, H. *Synlett* **1996**, 1091-1092.

20. For very early work on Heck reactions of aryl chlorides, see: (a) Davison, J. B.; Simon, N. M.; Sojka, S. A. *J. Mol. Cat.* **1984**, *22*, 349-352. (b) Spencer, A. *J. Organomet. Chem.* **1984**, *270*, 115-120.

21. Littke, A. F.; Fu, G. C. *J. Am. Chem. Soc.* **2001**, *123*, 6989-7000. See also: Littke, A. F.; Fu, G. C. *J. Org. Chem.* **1999**, *64*, 10-11.

22. See also: Shaughnessy, K. H.; Kim, P.; Hartwig, J. F. *J. Am. Chem. Soc.* **1999**, *121*, 2123-2132.

23. Generally, for Pd/P(t-Bu)$_3$-catalyzed Heck couplings that proceed at elevated temperature, a 1:2 Pd:phosphine ratio is preferred. For reactions that occur at room temperature, a 1:1 Pd:phosphine ratio is usually desirable (2:1 mixture of Pd(P(t-Bu)$_3$)$_2$:Pd$_2$(dba)$_3$).

24. For example, see: Watanabe, T.; Hayashi, K.; Yoshimatsu, S.; Sakai, K.; Takeyama, S.; Takashima, K. *J. Med. Chem.* **1980**, *23*, 50-59.

25. For example, see: Buchanan, J. G.; Hill, D. G.; Wightman, R. H.; Boddy, I. K.; Hewitt, B. D. *Tetrahedron* **1995**, *51*, 6033-6050.

26. Eisenstadt, A. In *Catalysis of Organic Reactions*; Herkes, F. E., Ed.; Marcel Dekker: New York, 1998; Chapter 33.

Appendix
Chemical Abstracts Nomenclature; (Registry Number)

(E)-2-Methyl-3-phenylacrylic acid butyl ester: 2-Propenoic acid, 2-methyl-3-phenyl-, butyl ester, (2E)-; (21511-00-5).

Bis (tri-tert-butylphosphine)palladium: Palladium, bis[tris(1,1,-dimethylethyl)phosphine]- ; (53199-31-8).

N-Methyldicyclohexylamine: Cyclohexanamine, N-cyclohexyl-N-methyl- ; (7560-83-0)

Chlorobenzene: Benzene, chloro-; (108-90-7)

Butyl methacrylate: 1-Propenoic acid, 2-methyl-, butyl ester ; (97-88-1)

(E)-4-(2-Phenylethenyl)benzonitrile: Benzonitrile, 4-[(1E)-2'phenylethenyl]- ; (13041-79-7)

Tris(dibenzylideneacetone)dipalladium: Palladium,tris[μ-[(1,2-η:4,5-η)-(1E,4E)-1,5,-diphenyl-1,4-pentadien-3-one]]di- ; (51364-51-3]

4-Chlorobenzonitrile: Benzonitrile, 4-chloro- ; (623-03-0)

Tri-*tert*-butylphosphiine: Phosphine, tris(1,1-dimethylethyl)- ; (13716-12-6)

SYNTHESIS OF N-(*tert*-BUTOXYCARBONYL)-β-IODOALANINE METHYL ESTER: A USEFUL BUILDING BLOCK IN THE SYNTHESIS OF NONNATURAL α-AMINO ACIDS VIA PALLADIUM CATALYZED CROSS COUPLING REACTIONS

(L-Alanine, N-[(1,1-dimethylethoxy)carbonyl]-3-iodo-, methyl ester)

Submitted by Richard F. W. Jackson and Manuel Perez-Gonzalez[1].

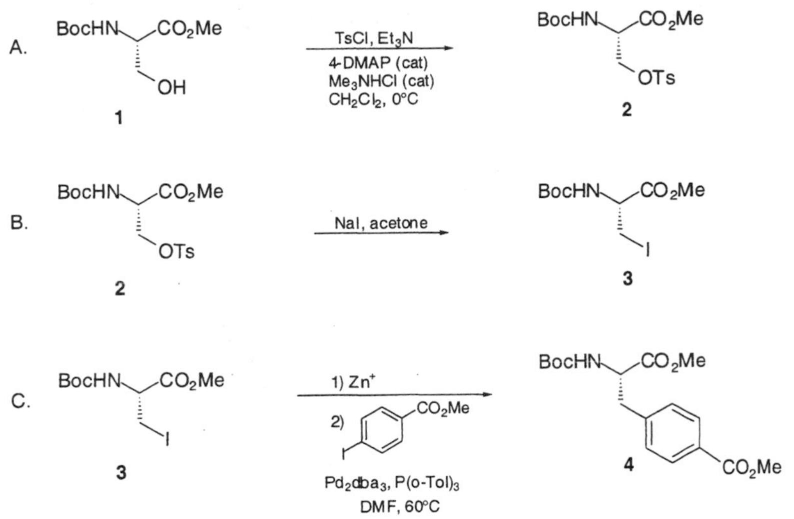

Checked by Rick L. Danheiser and Aimee L. Crombie.

1. Procedure

A.   *N-(tert-Butoxycarbonyl)-O-(p-toluenesulfonyl)-L-serine methyl ester* (**2**). A one-necked, 500-mL, round-bottomed flask equipped with a rubber septum and magnetic stirbar was charged with 26.1 g (119 mmol) of N-(*tert*-butoxycarbonyl)-L-serine methyl ester (**1**) (Note 1) and 200 mL of CH₂Cl₂ (Note 2). The solution is cooled in an ice bath at 0°C while 0.700 g (6.0 mmol) of 4-dimethylaminopyridine (4-DMAP), 1.1 g (12 mmol) of Me$_3$NHCl (Note 3), and 22.7 g (119 mmol) of freshly recrystallized p-toluenesulfonyl chloride (TsCl) (Note 4) are added. The septum is replaced with a dropping funnel charged with 17 mL (119 mmol) of triethylamine (Et$_3$N) in 50 ml of CH$_2$Cl$_2$ which is added dropwise to the reaction mixture at 0°C over 40 min (Note 5). The resulting slurry is stirred at 0°C for 2 hr and then poured into a mixture of 100 mL of ice, 100 mL of water, and 50 mL of 2M HCl solution. The aqueous layer is extracted with 100 mL of CH$_2$Cl$_2$, and the combined organic layers are washed with two 60-mL portions of brine, dried over magnesium sulfate, and concentrated by rotary evaporation to yield 59.4 g of a light yellow solid. This product may contain ca. 15% of starting material and TsCl that can be efficiently removed by crystallization according to the following procedure. The solid is dissolved in 140 mL of hot diethyl ether, filtered, and the filtrate is allowed to cool to room temperature and then to 0°C. Once crystallization begins (Note 6), a total of 250 mL of petroleum ether is added in five portions over 2 hr and then crystallization is allowed to proceed at −20°C overnight. The crystals are collected by suction filtration on a Büchner funnel and air-dried to give 28.3-30.8 g (64-69%) of **2** as a white solid (Note 7).

B. *N-(tert-Butoxycarbonyl)-β-iodoalanine methyl ester* (**3**). A one-necked, 250-mL, round-bottomed flask equipped with a rubber septum and magnetic stirbar is charged with 27.8 g (74.0 mmol) of N-(*tert*-butoxycarbonyl)-O-(*p*-toluenesulfonyl)-L-serine methyl ester (**2**) and 160 mL of acetone (Note 8). The solution is stirred at room temperature and 13.4 g (89.0 mmol) of NaI (Note 9) is added in one portion. The reaction mixture is stirred in the dark for 3 days, after which an additional 3.3 g (22 mmol) of NaI is added and stirring is continued for an additional day (Note 10). The reaction mixture is then suction filtered through a sintered glass funnel and the filtrate is collected in a one-necked, 500-mL, round-bottomed flask. The solid is washed with acetone until it is colorless. The solid is discarded and the filtrate is concentrated by rotary evaporation under reduced pressure (Note 11). The residual yellow oil is partitioned between 150 mL of diethyl ether and 60 mL of 1M sodium thiosulfate ($Na_2S_2O_3$) solution. The organic layer is separated and washed with 40 mL of 1M $Na_2S_2O_3$ 50 mL of brine, dried over magnesium sulfate, filtered, and concentrated by rotary evaporation (Note 11) to afford 23.2 g of a colorless oil that solidifies on standing at 0°C. The solid is dissolved in 30 mL of hot (40°C) petroleum ether (bp 35-45°C), cooled to room temperature and then to 0°C. Once a precipitate appears (Note 6), the mixture is cooled at –20°C for 1 hr and the white solid is collected on a Büchner funnel and washed with cold petroleum ether to yield 19.4-20.0 g (80-82%) of N-(*tert*-butoxycarbonyl)-β-iodoalanine methyl ester **3** as white to pale yellow crystals (Note 12).

C. *N-(tert-Butoxycarbonyl)-β-[4-(methoxycarbonyl)phenyl]alanine methyl ester* (**4**). A three-necked, 100-mL, round-bottomed flask equipped with an argon inlet adapter, reflux condenser, rubber septum, and magnetic stirbar (Note 13) is charged with 2.7 g (42 mmol) of zinc dust and 0.039 g (0.15 mmol) of I_2. The flask is evacuated to 0.03 mm and heated with a heat gun for 10 min, flushed three times with argon, and then allowed to cool to room temperature. Dimethylformamide (1 mL) (Note 14) is added *via* syringe, and then a solution of 10.8 g (33 mmol) of iodide **3** in 14 mL of DMF is added dropwise *via* syringe to the well-stirred suspension of zinc dust. The reaction mixture is stirred for 30 min at 0°C to produce a solution of the zinc reagent and a suspension of excess zinc dust. The ice bath is removed and 7.9 g (30 mmol) of methyl 4-iodobenzoate (Note 15), 0.137 g (0.15 mmol) of tris(dibenzylideneacetone)dipalladium (0), and 0.183 g (0.6 mmol) of tri-*o*-tolylphosphine are added. The reaction mixture is then stirred at 60°C for 5 hr. The resulting mixture is poured into a conical flask containing 300 mL of water. An additional 100 mL of 10% citric acid solution is added in order to break the black emulsion. The aqueous layer is extracted with two 150-mL portions of diethyl ether, and the combined organic layers are washed with two 100-mL portions of water and 100 mL of brine. At this point, the organic extracts are decanted from a black solid [consisting of palladium and zinc(0), together with a bright white solid; (Note 16)], dried over magnesium sulfate, filtered, and concentrated by rotary evaporation to yield 9.1 g of a pale yellow solid. This material is purified by column chromatography (Note 17). The product is charged on a column of 200 g of silica gel (Sorbent 60A, 32-63 μm) and eluted with 1 L of 20% EtOAc-hexane. At that point, fraction collection (25-mL fractions) is begun, and elution is continued with 500 mL of 25% EtOAc-hexane and then 200 mL of 30% EtOAc-hexane. The desired product is collected in fractions 13-59 which are concentrated by rotary evaporation to yield 3.5-3.9 g (35-39%) of **4** as a white solid (Note18).

2. Notes

1. The checkers purchased N-(*tert*-butoxycarbonyl)-L-serine methyl ester from Aldrich Chemical Company, Inc. The submitters prepared this compound, a thick amber oil, according to the procedure described in Dondoni, A.; Perrone, D. *Org. Synth.* Coll. Vol. X, **2004**, 320.

2. The checkers purified dichloromethane by pressure filtration through activated alumina.

3. The checkers purchased 4-DMAP and Me$_3$N·HCl from Aldrich Chemical Company, Inc. Et$_3$N was purchased from Alfa Aesar Chemical Company and was distilled from CaH$_2$. The submitters obtained 4-DMAP and Et$_3$N from Lancaster and Me$_3$N·HCl from Aldrich Chemical Company, Inc.

4. TsCl was obtained from Aldrich Chemical Company, Inc. and purified by recrystallization according to the following procedure. p-Toluenesulfonyl chloride (85 g) is dissolved in 150 mL of hot CHCl$_3$ and 200 mL of petroleum ether (room temperature) is added in one portion to the clear, colorless solution. The resulting cloudy solution is clarified by addition of ca. 5 g of charcoal, stirred for 1 min, and filtered on a Büchner

funnel. The filtrate is concentrated to ca. 1/5th of its original volume by rotary evaporation, and the solid which appears is collected by filtration and dried under reduced pressure (25°C, 0.03 mm) to afford 68 g of TsCl as bright white crystals.

5. Triethylamine addition must be carried out slowly to avoid base-promoted elimination of p-toluenesulfonate in the final product.

6. It may be necessary to scratch the flask in order to start crystallization.

7. The product exhibits the following properties: mp 74-76°C, lit.[7a] 74-75°C; $[\alpha]_D^{20}$ +3.0 (methanol, c 2.0), lit.[7b] +4.6 (methanol, c 2); Rf 0.24 (petroleum ether/EtOAc 3:1); IR (CH_2Cl_2) cm^{-1}: 3384, 2979, 1753, 1714, 1511, 1367, 1191, 1177; ^{1}H NMR (500 MHz, $CDCl_3$) δ: 1.43 (s, 9 H), 2.46 (s, 3 H), 3.71 (s, 3 H), 4.29 (dd, J = 10.1, 3.1 Hz, 1 H), 4.40 (dd, J = 10.1, 3.1 Hz, 1 H), 4.50-4.53 (m, 1 H), 5.30 (d, J = 7.9 Hz, NH), 7.37 (app d, J = 7.9 Hz, 2 H), 7.77 (app d, J = 8.2 Hz, 2 H); ^{13}C NMR (125 MHz, $CDCl_3$) δ: 21.9, 28.4 (3 C), 53.1 (2 C), 69.7, 80.7, 128.2 (2 C), 130.1 (2 C), 132.5, 145.4, 155.1, 169.2; MS (EI) m/z 314 (M$^+$-CO_2Me, 20%), 300 [(M$^+$-O^{t}Bu), 8%], 258 [(M$^+$-BocNH$_2$), 13%], 215 (11%), 155 (TolSO$_2$, 43%), 57 (100%); HRMS (EI) m/z: [M – CO_2Me]$^+$ calcd for $C_{14}H_{20}NO_5S$, 314.1057; found, 314.1059.

8. A.C.S. Reagent grade acetone (99.5%) was purchased from Aldrich Chemical Company, Inc.

9. The checkers purchased sodium iodide from Mallinckrodt Chemical Co. and dried it under vacuum (0.03 mm) at 100°C for 2 days.

10. The reaction can be monitored by TLC: elution with 3:1 petroleum ether/ ethyl acetate, Rf = 0.24 (O-tosylserine), Rf = 0.60 (iodoalanine).

11. Due to the unstable nature of the iodide, the solution should be cooled in a ice-water bath at 0°C during concentration.

12. Iodoalanine derivative 3 is also available from Aldrich Chemical Company, Inc. Compound 3 exhibits the following spectroscopic and physical

properties: mp 45-47°C, lit.[6b] mp = 51°C; $[\alpha]_D^{20}$ -3.7 (MeOH), c 3.0, lit.[6b] $[\alpha]_D^{20}$ = -4.0 (c 3 methanol); R_f 0.60 (petroleum ether/AcOEt 3:1); IR (CH_2Cl_2) cm^{-1}: 1163, 1259, 1422, 1714, 1748, 2986, 3054, 3424; ^1H NMR (500 MHz, $CDCl_3$) δ: 1.47 (s, 9 H), 3.55-3.67 (m, 2 H), 3.81 (s, 3 H), 4.53-4.54 (m, 1 H), 5.36 (d, J = 6.4 Hz, 1 H); C NMR (125 MHz, $CDCl_3$ δ: 8.6, 29.0 (3 C), 53.7, 54.4, 81.2, 155.5, 170.8,; MS (EI) m/z 270 (M^+-CO_2Me 25%), 214 [(M^+-Boc), 8%], 57 (100%); HRMS (ESI) m/z: [M + Na]$^+$ calcd for $C_9H_{16}INO_4$, 352.0016; found, 352.0021; Anal. Calcd for $C_9H_{16}INO_4$: C, 32.8; H 4.9; N, 4.3. Found: C, 33.2; H 4.8; N, 4.1.

13.   The submitters used a two-necked, round-bottomed flask equipped with a magnetic stirbar, three-way stopcock connected to vacuum and a N_2 source, and a solid addition tube. The addition tube was charged with methyl 4-iodobenzoate, tris(dibenzylideneacetone)dipalladium(0), and tri-o-tolylphosphine. After the zinc reagent was formed and the ice bath was removed, the solids contained in the addition tube were added in one portion by inverting the tube.

14.   DMF was distilled from CaH_2 and stored over activated 4Å molecular sieves.

15.   Methyl 4-iodobenzoate was purchased from Avocado Chemical Co. Zinc dust (< 10 microns, 95% purity) was purchased from Aldrich Chemical Ccompany, Inc. The checkers purchased Pd_2dba_3 and P(o-Tol)$_3$ from Strem Chemicals, Inc. and I_2 from Mallinckrodt. The submitters purchased Pd_2dba_3 and P(o-Tol)$_3$ from Aldrich Chemical Company, Inc.

16.   This solid was identified as dimethyl biphenyl-4,4'-dicarboxylate: ^1H NMR (200MHz, $CDCl_3$) δ: 3.96 (s, 6H), 7.67-7.72 (BB', 4H), 8.10-8.16 (AA', 4H); mp 219-220°C, (lit. mp 214-217°C: Catalogue handbook of Fine Chemicals. Aldrich 1999-2000).

17.   The submitters purified the product by the following procedure.   The residual pale yellow solid is dissolved in 50 ml of diethyl ether and the remaining solid is filtered off (Note 16). The filtrate is concentrated to a volume of ca. 25 mL, and the solution is allowed to crystallize at 0°C. Once crystallization begins, 50 mL of petroleum ether is added in two portions over 10 hr, and then crystallization is allowed to proceed overnight at 0°C. The white solid is collected by filtration and washed with a mixture of 3:1 petroleum ether-diethyl ether to afford 3.8 g of **4**. Chromatographic purification of the mother liquor (5.5 x 18 cm of DSH silica gel 40-63 mm, elution with 1 L of petroleum ether/ethyl acetate 4:1 followed by 1.5 L of 3:1 petroleum ether-ethyl acetate) gives 2.5 g of **4** as a pale yellow solid. All the material is combined and recrystallized from diethyl ether/petrol as above to yield 5.2 g (47%) of **4** in two crops.

18.   The enantiomeric purity of compound **4** was higher than 99.5% as determined by HPLC (2-cm Supelco Si precolumn, 250 x 4.6 mm Chiralpak AS, 10% isopropanol in heptane, 1mL/min, t_r = 14.5 min). The D enantiomer (t_r = 11.5 min) was not detected. Alternatively, the Boc protecting group of compound **4** (0.100 g) was removed by treatment with a solution prepared by addition of 100 μL of AcCl to 1.5 mL of methanol. Mosher's amides were formed using i-Pr$_2$NEt (1.1 equiv), EDCI (1.1 equiv), BtOH (1.1 equiv) and (R) or (S)-α-methoxy-α-(trifluoromethyl)phenylacetic acid (1.5 equiv), and ^{1}H NMR (500 MHz) analysis of both amides showed ee > 98%. Compound **4** exhibits the following spectroscopic and analytical properties: mp 82-83°C; $[\alpha]_D^{20}$ = -5.0 (acetone, c 1.0); IR (CH$_2$Cl$_2$) cm^{-1}: 3367, 2979, 2954, 1720, 1613, 1512, 1282, 1167; 1H NMR (500 MHz, CDCl$_3$) δ: 1.42 (s, 9 H), 3.10 (dd, J = 13.7, 6.4 Hz, 1 H), 3.20 (dd, J = 13.7, 5.8 Hz, 1 H), 3.91 (s, 3 H), 3.72 (s, 3 H), 4.61-4.63 (m, 1 H), 5.00 (d, J = 7.9 Hz, NH), 7.21 (app d, J = 7.9 Hz, 2 H), 7.98 (app d, J = 8.5 Hz, 2 H); ^{13}C NMR (125 MHz, CDCl$_3$) δ: 29.0 (3 C), 39.1, 52.8, 53.1, 54.9, 80.8, 129.6, 130.1 (2 C), 130.5 (2 C), 142.2, 155.7, 167.6, 172.7; MS (EI) m/z 337 (M$^+$, 0.3%), 306 [(M$^+$–OMe),

7%], 264 [(M$^+$–O^{t}Bu), 4%], 220 [(M$^+$–BocNH$_2$), 83%], 150 (79%), 57 (100%); HRMS (ESI) *m/z*: [M + Na]$^+$ calcd for C$_{17}$H$_{23}$NO$_6$ 360.1418; found, 360.1407; Anal. Calcd for C$_{17}$H$_{23}$NO$_6$: C, 60.51; H, 6.88; N, 4.15. Found: C, 60.3; H, 6.6; N, 4.0.

Waste Disposal Information

All toxic materials were disposed of in accordance with "Prudent Practices in the Laboratory"; National Academy Press; Washington, DC, 1995.

3. Discussion

The development of synthetic equivalents for the alanine β-anion is an important target for the synthesis of α-amino acids. A useful solution was provided by the discovery that organozinc reagents could be prepared from the iodoalanine derivative (**3**) by insertion of activated zinc into the carbon-iodine bond.[2] The resulting organometallic reagent could then be coupled with carbon electrophiles in the presence of a suitable palladium catalyst. Furthermore, treatment of the organozinc reagent with copper cyanide gave the organocuprate, which could then couple directly with reactive electrophiles.[3]

The iodoalanine derivative is available in both enantiomeric forms and this method offers an extremely simple route to large numbers of non-natural amino acids.[4] Therefore a reliable and practical method for the synthesis of fully protected iodoalanine **3** in a multigram scale is highly desirable.

Although the tosylation reaction of fully protected serine has already been described,[5] we have found that extended reaction times are required when the reaction is carried out on 100-200 mmol scale, taking from 3 to 5 days to go to completion. As a result of longer reaction times, the amount of by-products is increased making the purification step more difficult.

Recently Tanabe and co-workers have found that several alcohols were smoothly and efficiently tosylated using tosyl chloride/ triethylamine and a catalytic amount of trimethylamine hydrochloride as reagents.[6] Compared with the traditional method using pyridine as solvent, this procedure has the merit of much higher reaction rates, and it avoids the side reaction in which the desired tosylate is converted into the corresponding chloride.

TABLE

Entry	Electrophile	Catalyst	T(°C), t	Product	Yield[a]
1	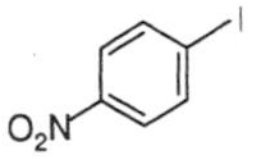 O_2N— (4-iodo)	$Pd_2(dba)_3$ 1% $P(oTol)_3$ 4%	rt, 2 h	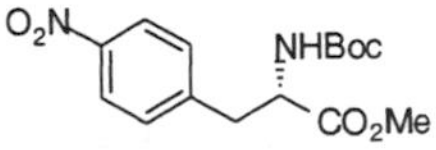 O_2N—...NHBoc, CO_2Me	72%[b]
2	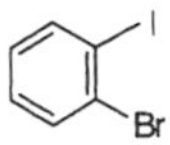 (2-bromo-iodobenzene) Br	$Pd_2(dba)_3$ 1.2% $P(oTol)_3$ 5%	40 °C, 5.5h	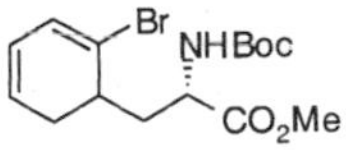 Br...NHBoc, CO_2Me	47%[b]
3	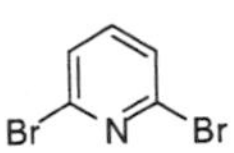 Br–N–Br	$Cl_2Pd(PPh_3)_2$ 5%	50 °C, 5h	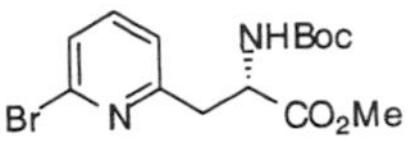 Br–N...NHBoc, CO_2Me	52%[c]
4	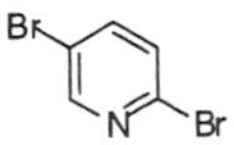 Br...N–Br	$Cl_2Pd(PPh_3)_2$ 2%	50 °C, 5.5h	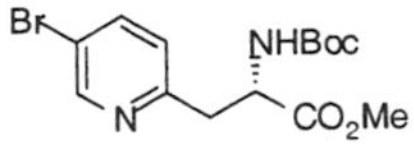 Br...N...NHBoc, CO_2Me	64%[d]
5	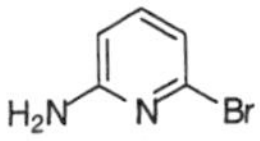 H_2N–N–Br	$Cl_2Pd(PPh_3)_2$ 5%	40 °C, 6h	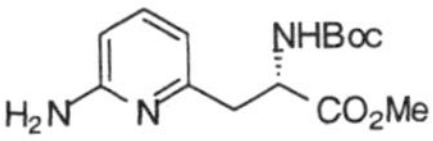 H_2N–N...NHBoc, CO_2Me	34%[e]

a) Isolated yield. b) 5 mmol scale. c) 10 mmol scale. d) 7 mmol scale. e) 2 mmol scale.

Stirring a mixture of tosylate and NaI in acetone at room temperature for four days produced the iodoalanine derivative 3 via a *pseudo*-Finkenstein reaction. Although this reaction has been reported to proceed in shorter reaction times when carried out at

85

reflux,[7] in our hands, an increase in the temperature of the reaction produced significant amounts of by-products and a poor mass balance. For instance, when the experimental procedure stated in step B was carried out at reflux of acetone only 4 g of crude material was recovered. Three water-soluble products account for the rest of the mass balance, and one of them has been identified as 4-methoxycarbonyloxazolidin-2-one.[8] Better yields can be obtained when the reaction was carried out at reflux on a smaller scale (2 g of compound **2**) and under much more dilute conditions as reported.[7] It is our contention that due to the instability of the iodoalanine derivative, high temperatures during the reaction or the work-up promote decomposition of the product *via* intramolecular nucleophilic attack by the carbamate moiety. Nevertheless, compound **3** could be kept in the refrigerator for one year without evidence of decomposition.

The procedure in Section C is representative of the synthesis of non-natural α-amino acids featuring the palladium cross coupling reaction of a β-alanine organozinc derivative with aromatic electrophiles. This methodology has been successfully extended with modifications to both the electrophile and the catalyst as shown in the Table.

1. Department of Chemistry, The University of Sheffield, Sheffield, S3 7HF, United Kingdom.
2. Jackson, R. F. W.; Wishart, N.; Wood, A.; James, K.; Wythes, M. J. *J. Org. Chem.* **1992**, *57*, 3397.
3. Dunn, M. J.; Jackson, R. F. W.; Pietruszka, J.; Turner, D. *J. Org. Chem.* **1995**, *60*, 2210.
4. For a review see: Gair, S.; Jackson, R. F. W. *Current Org. Chem.* **1998**, *2*, 527.

5. Hermkens, P. H. H.; Maarseveen, J. H. v.; Ottenheijm, H. C. J.; Kruse, C. G.; Scheeren, H. W. *J. Org. Chem.* **1990**, *55*, 3998.

6. Yoshida, Y.; Sakakura, Y.; Aso, N.; Okada, S.; Tanabe, Y. *Tetrahedron*, **1999**, *55*, 2183.

7. a) Stocking, E. M.; Schwarz, J. N.; Senn H.; Salzmann, M.; Silks, L. A. *J. Chem. Soc., Perkin Trans. 1* **1997**, 2443. b) Bajgrowicz, J. A.; A. El Hallaoui; Jacquier, R.; Pigiere, Ch.; Viallefont, P. *Tetrahedron* **1985**, *41*, 1833.

8. Sibi, M. P.; Rutherford, D.; Sharma, R. *J. Chem. Soc., Perkin Trans.1* **1994**, 1675.

Appendix

Chemical Abstracts Nomenclature (Registry Number)

N-(tert-Butoxycarbonyl)-L-serine methyl ester: L-Serine, N-[(1,1-dimethylethoxy)carbonyl]-, methyl ester; (2766-43-0)

N-(tert-Butoxycarbonyl)-β-iodoalanine methyl ester: L-Alanine, N-[(1,1-dimethylethoxy)carbonyl]-3-iodo-, methyl ester; (93267-04-0)

N-(tert-Butoxycarbonyl)-β-4-(methoxycarbonyl)phenyl]alanine methyl ester: L-Phenylalanine, N-[(1,1-dimethylethoxy)carbonyl]-4-(methoxycarbonyl)-, methyl ester; (160168-19-4)

4-Dimethylaminopyridine: 4-Pyridinamine, N,N-dimethyl-; (1122-58-3)

Trimethylamine hydrochloride: Methanamine, N,N-dimethyl-, hydrochloride; (593-81-7)

p-Toluenesulfonyl chloride: Benzenesulfonyl chloride, 4-methyl-; (98-59-9)

Triethylamine: Ethanamine, N,N-diethyl-; (1221-44-8)

Sodium iodide: Sodium iodide (NaI); (7681-82-5)

Sodium thiosulfate: Thiosulfuric acid ($H_2S_2O_3$), disodium salt; (7772-98-7)

Methyl 4-iodobenzoate: Benzoic acid, 4-iodo-, methyl ester; (619-44-3)

Tris(dibenzylideneacetone)dipalladium: tris[μ-[(1,2-η:4,5-η)-(1E,4E)-1,5-diphenyl-1,4-pentadien-3-one]]di-; (51364-51-3)

Tri-o-tolylphosphine: Phosphine, tris(2-methylphenyl)-; (6163-58-2)

Zinc; (7440-66-6)

Iodine; (7553-56-2)

Dimethyl biphenyl-4,4'-dicarboxylate: [1,1'-Biphenyl]-4,4'- dicarboxylic acid, dimethyl ester; (792-74-5)

SYNTHESIS OF 3-PYRIDYLBORONIC ACID AND ITS PINACOL ESTER. APPLICATION OF 3-PYRIDYLBORONIC ACID IN SUZUKI COUPLING TO PREPARE 3-PYRIDIN-3-YLQUINOLINE

[Quinoline, 3-(3-pyridinyl)-]

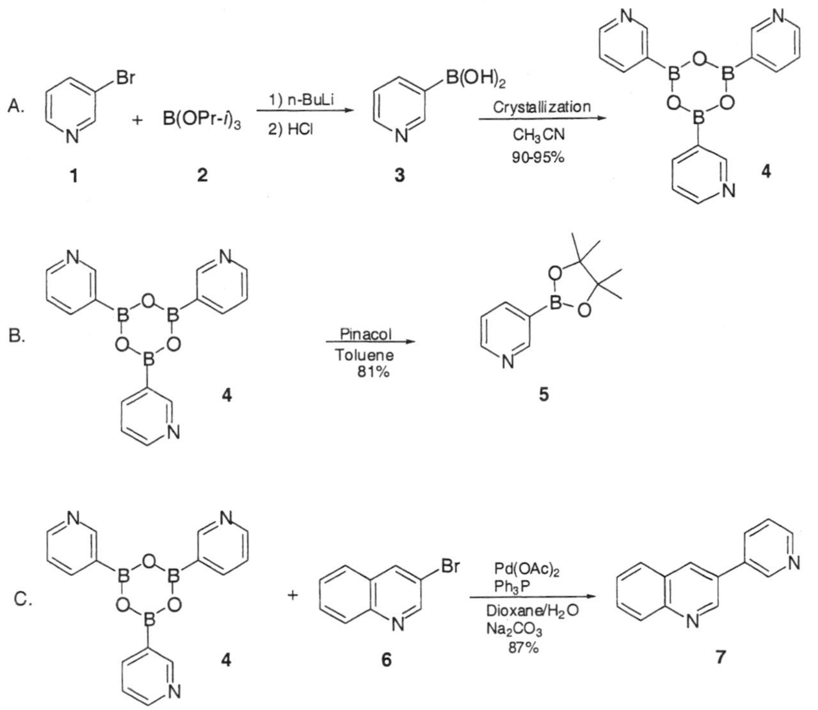

Submitted by Wenjie Li, Dorian P. Nelson, Mark S. Jensen, R. Scott Hoerrner, Dongwei Cai and Robert D. Larsen.[1]

Checked by Scott E. Denmark, Geoff T. Halvorsen, and Jeffrey M. Kallemeyn.

1. Procedure

A.   *3-Pyridylboronic acid [tris(3-pyridyl)boroxin].* A 1-L, 3-necked flask equipped with a temperature probe, overhead stirrer, and a nitrogen inlet adaptor capped with a rubber septum is charged with 320 mL of toluene, 80 mL of THF, triisopropyl borate (55.4 mL, 240 mmol), and 3-bromopyridine (19.3 mL, 200 mmol) (Notes 1, 2). The mixture is cooled to –40°C using a dry ice/acetone bath and 96 mL of *n*-butyllithium solution (2.5M in hexanes, 240 mmol) (Note 3) is added dropwise with a syringe pump over 1 hr. The reaction mixture is stirred for an additional 30 min maintaining the temperature at –40°C. The acetone/dry ice bath is then removed, and the reaction mixture is allowed to warm to –20°C whereupon a solution of 200 mL of 2N HCl solution is added. When the mixture reaches room temperature, it is transferred to a 1-L separatory funnel and the aqueous layer (pH ≈ 1) is drained into a 500-mL Erlenmeyer flask equipped with a magnetic stir bar. The pH of the aqueous layer is adjusted to 7.6-7.7 using 5N aqueous NaOH (ca. 30 mL) (Note 4). A white solid precipitates out as the pH approaches 7. The aqueous mixture is then saturated with solid NaCl, transferred to a 1-L separatory funnel, and extracted with three 250-mL portions of THF. The combined organic phases are concentrated on a rotary evaporator to leave a solid residue which is suspended in 80 mL of acetonitrile for crystallization. The mixture is heated to 70°C in an oil bath, stirred for 30 min, and then allowed to cool slowly to room temperature and then to 0°C in an ice bath. After being stirred at 0°C for 30 min, the mixture is filtered through a fritted-glass funnel.

The solid is washed with 15mL of cold (5°C) acetonitrile, and then dried under vacuum to afford 18.9 g of tris(3-pyridyl)boroxin·1.0 H_2O as a white solid (Note 5).

B. *3-(4,4,5,5-Tetramethyl-[1,3,2]dioxaborolan-2-yl)pyridine.* A 250-mL, one-necked, round-bottomed flask equipped with a magnetic stirbar and a Dean-Stark trap fitted with a condenser capped with a nitrogen inlet adaptor is charged with tris(3-pyridyl)boroxin·0.85 H_2O (3.0 g , 9.1 mmol), pinacol (4.07 g, 34.4 mmol) (Note 6), and 120 mL of toluene. The solution is heated at reflux for 2.5 hr in a 120°C oil bath. The reaction is complete when the mixture changes from cloudy-white to clear. The solution is then concentrated under reduced pressure on a rotary evaporator to afford a solid residue. This solid is suspended in 15 mL of cyclohexane (Note 7) and the slurry is heated to 85°C, stirred at this temperature for 30 min, and then allowed to cool slowly to room temperature. The slurry is filtered, rinsed twice using the mother liquors, washed with 3 mL of cyclohexane, and dried under vacuum to afford 4.59 g (82%) of 3-pyridylboronic acid pinacol ester as a white solid (Note 8).

C. *3-Pyridin-3-yl-quinoline.* A 100-mL, round-bottomed Schlenk flask equipped with a magnetic stir bar is charged with tris(3-pyridyl)boroxin·0.85 H_2O (3.80 g, 11.5 mmol), 3-bromoquinoline (6.24 g, 30.0 mmol) (Note 9), 30 mL of a 2M aqueous solution of Na_2CO_3, and 30 mL of 1,4-dioxane (Note 10). Palladium(II) acetate (0.336 g, 1.5 mmol) and triphenylphosphine (1.57 g, 6.0 mmol) are added (Note 10), the mixture is degassed using five vacuum/nitrogen back-fill cycles, and then is heated to 95°C for 2.5 hr with vigorous stirring (Note 11). The mixture is allowed to cool to room temperature, transferred to a 500-mL separatory funnel, and diluted with 100 mL of water and 150 mL of ethyl acetate. The aqueous layer is separated and back-extracted with 50 mL of ethyl acetate, and the combined organic layers are extracted with three 50-mL portions of 1M HCl solution. The combined acidic aqueous layers are treated with 40 mL of 5M aqueous NaOH resulting in a pH of ca. 9. The cloudy aqueous layer is then extracted with three 50-mL portions of ethyl acetate. The combined organic layers are dried over $MgSO_4$, filtered into a 500 mL, round-bottomed flask, and concentrated on a rotary evaporator under reduced pressure to afford a brown solid. Isopropyl acetate (10 mL) is added and the slurry is heated to reflux at which point 40 mL of heptane is added. The mixture is allowed to cool to room temperature over 3 hr, and then is stirred at room temperature for 2 hr. The solid is isolated by filtration, washed with 10 mL of 4:1 heptane/isopropyl acetate, and dried under vacuum at room temperature to give 5.36-5.40 g (87%) of 3-pyridin-3-ylquinoline as a white solid (Note 12).

2. Notes

1.   Toluene and THF were purchased from Fisher Scientific and dried over 4Å molecular sieves overnight prior to use. The water content of the solvents was <50 µg/mL by Karl Fischer titration.

2.   Triisopropyl borate was purchased from Aldrich Chemical Company, Inc. 3-Bromopyridine was purchased from Lancaster Synthesis. Both compounds were used without further purification.

3.   n-Butyllithium (2.5M in hexanes) was purchased from Aldrich Chemical Company, Inc. and was titrated prior to use.

4.   Measurement of pH was performed using a Metrohm model 691 pH meter equipped with a Metrohm combined LL micro pH glass electrode calibrated prior to use with pH = 2 and 9 buffers. The checkers found that adjustment to a lower pH led to product with higher amounts of inorganic impurities. The checkers also found that the use of pH paper results in different pH values as compared to the pH meter.

5.   The checkers obtained trispyridylboroxin in various hydration levels (0.85 – 1.0 H_2O). A satisfactory melting point for this solid could not be obtained. ^{1}H NMR (400 MHz, CD_3OD): δ 7.66 (br s, 1H), 8.38 (d, J=6.6, 1H), 8.51 (dd, J=1.2, 4.4, 1H), 8.61 (br s, 1 H). ^{1}H NMR spectra were complicated in other solvents such as $CDCl_3$ and DMSO-d_6. Anal Calcd. for $C_{15}H_{12}B_3O_3N_3$ • 1.0H_2O: C, 54.15; H, 4.24; N, 12.63. Found: C, 53.95; H, 3.91; N, 12.35. Yield based on this formula is 85%. The submitters report obtaining the product in 91% yield.

6.   Pinacol was purchased from Aldrich Chemical Company, Inc. and was used without further purification.

7.   Cyclohexane was purchased from Aldrich Chemical Company, Inc. and was used without further purification.

8.   The checkers obtained the product in 74-82% yield in different runs. The product exhibits the following physical properties: mp 102–105°C; IR (KBr pellet) cm[-1] 2994, 2968, 2932, 1609, 1572, 1476, 1410, 1361, 1209, 1154, 1063, 1017, 953, 926, 859, 833, 800, 759, 705; ^{1}H NMR (500 MHz, $CDCl_3$): δ 1.33 (s, 12H), 7.25 (ddd, J=1.1, 4.9, 7.5, 1H), 8.03 (dt, J=1.8, 7.5, 1H), 8.64 (dd, J=1.9, 4.9, 1H), 8.93 (d, J=1.1, 1H); ^{13}C NMR (126 MHz, $CDCl_3$) δ 24.8, 84.1, 123.0, 142.2, 152.0, 155.5; MS (EI, 70 eV):

205 (M+, 46), 204 (15), 191 (12), 190 (100), 189 (25), 162 (10), 148 (44), 147 (14), 120 (18), 119 (11), 106 (100), 105 (35), 85 (15), 59 (17), 58 (19); HRMS (EI) m/z 205.1280 , calcd for $C_{11}H_{16}NO_2B$ 205.1274). Anal. Calcd for : $C_{11}H_{16}BO_2N$: C, 64.43; H, 7.86; N, 6.83. Found: C, 64.23; H, 7.99; N, 6.88.

9. 3-Bromoquinoline was purchased from Acros Organics and was used without further purification.

10. 1,4-Dioxane, palladium (II) acetate, and triphenylphosphine were purchased from Aldrich Chemical Company, Inc. and were used without further purification.

11. A heterogeneous mixture that is difficult to stir may be formed during the reaction. The checkers found that rapid stirring from the onset of the reaction prevented loss of stirring. Inefficient stirring was found to lower the yield of the reaction.

12. The following characterization data was obtained: mp: 122–125°C; IR (KBr pellet): cm^{-1} 3044, 1568, 1495, 1410, 1338, 1298, 1187, 1126, 1059, 1022, 952, 932, 816, 784, 758, 709 909, 808, 786, 752, 709, 61; ^{1}H NMR (500 MHz, CDCl$_3$): δ 7.45 (ddd, J=0.9, 4.9, 7.9, 1H), 7.60 (dt, J=0.9, 7.9, 1H), 7.75 (ddd, J=1.2, 6.8, 8.3, 1H), 7.90 (dd, J=0.9, 8.1, 1H), 8.00 (dt, J=2.2, 7.8, 1H), 8.15 (d, J=8.4, 1H), 8.32 (d, J=2.3, 1H), 8.68 (dd, J=1.5, 4.9, 1H), 8.97 (d, J=2.1, 1H), 9.15 (d, J=2.1, 1H); ^{13}C NMR (126 MHz, CDCl$_3$) δ 123.8, 127.3, 127.8, 128.0, 129.3, 129.9, 130.6, 133.57, 133.63, 134.6, 147.7, 148.5, 149.27, 149.29; MS (EI, 70 eV): 207 (16), 206 (M+, 100), 205 (40);

HRMS (EI) m/z 206.0847, calcd for $C_{14}H_{10}N_2$ 206.0944). Anal. Calcd for $C_{14}H_{10}N_2$: C, 81.53; H, 4.89; N, 13.58. Found: C, 81.43; H, 4.86; N, 13.40.

Waste Disposal Information

All toxic materials were disposed of in accordance with "Prudent Practices in the Laboratory"; National Academy Press; Washington, DC, 1995.

3. Discussion

3-Pyridylboronic acid is a useful compound for the introdution of a 3-pyridyl moiety into a molecule by the Suzuki reaction.[2-3] It is commercially available only in small quantities at high cost. Typical preparation of this class of compounds involves reaction between boronic esters and an organometallic reagent (Li or Mg)[4-7] usually prepared by magnesium insertion[4] or lithium-halogen exchange[5-7] of 3-bromopyridine. The literature protocols, however, require conditions that are not suitable for large scale, including using dibromoethane as the solvent, running the reaction at very low temperature (–78°C), and affording poor to modest yields. Recently, we reported a revised procedure for the preparation of 3-pyridylboronic acid via lithium-halogen exchange and *in situ* quench with triisopropyl borate.[8] In this protocol, *n*-butyllithium is added to a solution of 3-bromopyridine and triisopropyl borate in THF/toluene based on the reasoning that lithium-halogen exchange is much faster than the reaction between *n*-butyllithium and triisopropyl borate. The 3-lithiopyridine intermediate thus generated, reacts rapidly with the borate in the reaction mixture, thereby minimizing the chance for 3-lithiopyridine to undergo side reactions.[9] The new procedure allows the reaction to be run at much higher temperatures, giving the best yields (90–95%) at

−40°C and a respectable 80% yield even at 0°C. The reaction can be carried out at kilogram scale. The product is isolated by crystallization from acetonitrile in the form of boroxin **4**. The characterization of **4** is difficult, however, due to the presence of varying amounts of water. Therefore, boroxin **4** is converted to its pinacol ester **5**, which is fully characterized by spectroscopic data and elemental analysis. The utility of **4** in palladium-catalyzed cross-coupling reaction is demonstrated in a Suzuki reaction with 3-bromoquinoline.[10]

1. Process Research Department, Merck Research Laboratories, P. O. Box 2000, Rahway, NJ 07065.

2. For reviews of Suzuki reaction, see: (a) Miyaura, N.; Suzuki A. *Chem. Rev.* **1995**, *95*, 2457-2483. (b) Suzuki, A. *J. Organomet. Chem.* **1999**, *576*, 147-168.

3. For examples of 3-pyridylboronic acid in cross-coupling reactions, see: (a) Enguehard, C.; Hervet, M.; Allouchi, H.; Debouzy, J.-C.; Leger, J.-M.; Gueiffier, A. *Synthesis*, **2001**, *4*, 595-600. (b) Li, J. J.; Yue, W. S. *Tetrahedron Lett.* **1999**, *40*, 4507-4510. (c) Bower, J. F.; Guillaneux, D.; Nguyen, T.; Wong, P. L.; Snieckus, V. *J. Org. Chem.* **1998**, *63*, 1514-1518.

4. Fischer, F. C.; Havinga, E. *Recl. Trav. Chim. Pays-Bas* **1965**, *84*, 439-440.

5. Terashima, M.; Kakimi, H.; Ishikura, M.; Kamata, K. *Chem. Pharm. Bull.* **1983**, *31* (12), 4573-4577.

6. Ishikura, M.; Kamada, M.; Oda, I.; Terashima, M. *Heterocycles* **1985**, *23* (1), 117-120.

7. Coudret, C. *Syn. Comm.* **1996**, *26* (19), 3543-3547.

8. (a) Li, W.; Nelson, D. P.; Jensen, M. S.; Hoerrner, R. S.; Cai, D.; Larsen, R. D.;
   Reider, P. J. *J. Org. Chem.* **2002**, *67*, 5394-5397. (b) Cai, D.; Larsen, R. D.; Reider,
   P. J. *Tetrahedron Lett.* **2002**, *43*, 4285-4287.

9. Gilman, H.; Spatz, S. M. *J. Org. Chem.* **1951**, *16*, 1485-1494.

10. (a) Ishikura, M.; Oda, I.; Terashima, M. *Heterocycles*, **1985**, *23 (9)*, 2375-2386. (b)
    Ishikura, M.; Kamada, M.; Terashima, M. *Synthesis*, **1984**, *11*, 936-938.

Appendix

Chemical Abstracts Nomenclature (Registry Number)

3-Bromopyridine: Pyridine, 3-bromo-; (626-55-1)

Triisopropyl borate: Boric acid (H_3BO_3), tris(1-methylethyl) ester; (5419-55-6)

n-Butyllithium: Lithium, butyl-; (109-72-8)

3-Pyridylboronic acid: Boronic acid, 3-pyridinyl-; (1692-25-7)

Tris(3-pyridyl)boroxin: Pyridine, 3,3',3''-(2,4,6-boroxintriyl)tris-; (160688-99-3)

Pinacol: 2,3-Butanediol, 2,3-dimethyl-; (76-09-5)

3-(4,4,5,5-Tetramethyl-[1,3,2]dioxaborolan-2-yl)-pyridine: Pyridine,
    3-(4,4,5,5-tetramethyl-1,3,2-dioxaborolan-2-yl)-; (329214-79-1)

3-Bromoquinoline: Quinoline, 3-bromo-; (5332-24-1)

Palladium (II) acetate: Acetic acid, palladium(2+) salt; (3375-31-3)

Triphenylphosphine: Phosphine, triphenyl-; (603-35-0)

3-Pyridin-3-ylquinoline: Quinoline, 3-(3-pyridinyl)-; (96546-80-4)

SYNTHESIS OF 5-BROMOISOQUINOLINE AND 5-BROMO-8-NITROISOQUINOLINE

(Isoquinoline, 5-bromo- and Isoquinoline, 5-bromo-8-nitro-)

A.

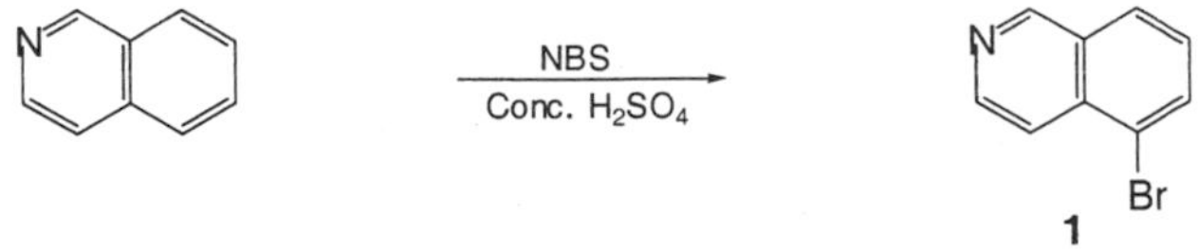

$$\xrightarrow[\text{Conc. } H_2SO_4]{\text{NBS}}$$

1

B.

$$\xrightarrow[\text{Conc. } H_2SO_4]{\begin{array}{l}\text{1) NBS}\\\text{2) } KNO_3\end{array}}$$

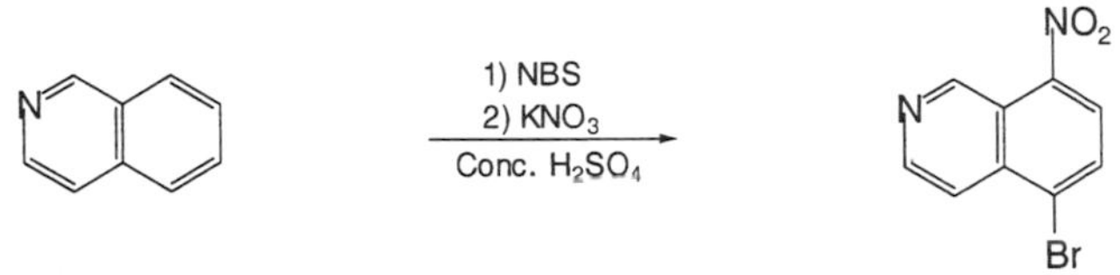

2

Submitted by William Dalby Brown[1a] and Alex Haahr Gouliaev[1b]

Checked by Steven Wolff and Walter Burger.

3. Procedure

A. 5-Bromoisoquinoline. A 1-L, three-necked, round-bottomed flask equipped with an internal thermometer, mechanical stirrer, and an addition funnel fitted with a nitrogen inlet is charged with concentrated sulfuric acid (96%, 340 mL, Note 1) and cooled to 0°C. Isoquinoline (40 mL, 44.0 g, 330 mmol) is slowly added to the well-stirred acid at a rate such that the internal temperature is maintained below 30°C. The solution is cooled to –25°C in a dry ice-acetone bath and N-bromosuccinimide (64.6 g, 363 mmol; Note 2) is added to the vigorously stirred solution in portions such that the internal temperature is maintained between –22 and –26°C (Note 3). The suspension is efficiently stirred for 2 hr at −22 ± 1°C and then for 3 hr at −18 ± 1°C. The resulting homogeneous reaction mixture is poured onto 1.0 kg of crushed ice in a 5-L flask placed in an ice water bath. The reaction flask is quickly washed with ice-cold water, which is added to the 5-L flask. The resulting mixture is stirred while the pH is adjusted to 9.0 using 25% aq NH_3 with the internal temperature maintained below 25°C. The resulting alkaline suspension is added to 800 mL of diethyl ether and the biphasic system is vigorously mixed (Note 4). The two clear phases are separated and the aqueous phase is extracted with two 200-mL portions of diethyl ether. The combined organic phases are washed with 200 mL of 1M NaOH (aq) and 200 mL of H_2O, dried over anhydrous $MgSO_4$, filtered, and concentrated to afford 47 g of a light brown solid. Fractional distillation under reduced pressure (bp 145-149°C at 14 mm, Note 5) furnishes 34-36 g (47-49%) of 5-bromoisoquinoline as a white solid (Notes 6 and 7).

B. *5-Bromo-8-nitroisoquinoline.* A 1-L, three-necked, round-bottomed flask fitted with an internal thermometer, mechanical stirrer, and an addition funnel fitted with a nitrogen inlet is charged with concentrated sulfuric acid (96%, 340 mL, Note 1) and cooled to 0°C. Isoquinoline (40 mL, 44.0 g, 330 mmol) is slowly added from the addition funnel to the well-stirred acid such that the internal temperature is maintained below 30°C. The solution is cooled to −25°C in a dry ice-acetone bath and N-bromosuccinimide (76.4 g, 429 mmol; Note 2) is added to the vigorously stirred solution in portions such that the internal temperature is maintained between −22 and −26°C (Note 3). The suspension is efficiently stirred for 2 hr at −22°C ± 1°C and then for 3 hr at −18°C ± 1°C. Potassium nitrate (35.0 g, 346 mmol) is then added at a rate such as to maintain the internal temperature below −10°C and the mixture is then stirred at −10°C for 1 hr. The cooling bath is removed and the solution is stirred overnight. The resulting homogeneous reaction mixture is poured onto 1.0 kg of crushed ice in a 5-L flask and the reaction flask is quickly washed with ice-cold water, which is added to the 5-L flask. The resulting mixture is stirred while the pH is adjusted to 8.0 using 25% aq NH_3 with the internal temperature maintained below 30°C. The resulting suspension is stirred in an ice water bath for 2 hr and the precipitated solids are isolated by filtration using a glass filter funnel. The solids are thoroughly washed three times with 1-L portions of ice-cold water and then air-dried to constant weight to afford 65 g of a slightly yellow solid. This material is suspended in 1000 mL of heptane and 250 mL of toluene (Note 8) in a 2-L, round-bottomed flask and heated at reflux for 1.5 hr with stirring. The hot solution is then filtered through Celite using vacuum suction (Note 9). The volume of the filtrate is reduced by distillation to 1000 mL and the resulting orange solution is allowed to slowly cool with stirring overnight. The solids are isolated by filtration, washed with 350 mL of ice-cold heptane, and air-dried to constant weight to afford 40-44 g (47-51%) of 5-bromo-8-nitroisoquinoline (Notes 10 and 11).

2. Notes

1. Sulfuric acid 96% (technical quality) and diethyl ether (technical quality) were purchased from Bie & Berntsen A/S, Sandbaekvej 7, DK-2610 Roedovre, Denmark and used without further purification. Isoquinoline (97%) and potassium nitrate (99%) were purchased from Aldrich Chemical Company, Inc. and used without further purification.

2. Isoquinoline must be completely dissolved prior to the addition of N-bromosuccinimide. N-Bromosuccinimide (99%) was purchased from Aldrich Chemical Company, Inc. and recrystallized[2] and air-dried prior to use. Recrystallization is essential in order to obtain high yield and pure product. The use of more NBS than stated (i.e., more than 1.1 equiv for the synthesis of 5-bromoisoquinoline and 1.3 equiv for the synthesis of 5-bromo-8-nitroisoquinoline) to obtain complete transformation of isoquinoline should be avoided as this leads to

formation of 5,8-dibromoisoquinoline, which cannot easily be separated from 5-bromoisoquinoline and which will also lead to a lower yield of 5-bromo-8-nitroisoquinoline.

3. The temperature is controlled throughout the reaction by intermittently adding additional pieces of dry ice to the dry ice-acetone bath. Strict temperature control throughout the bromination reaction is important to obtain high regioselectivity and purity of the product as the side products cannot be removed with ease.

4. Vigorous shaking for at least 5 min is required.

5. A 40-cm column insulated with cotton and aluminum foil is used for the fractional distillation. The condenser is heated to a constant temperature of 80°C by attaching a 80°C water bath and a circulatory pump. As distillation progresses, the temperature rises smoothly, settling at 126-128°C for some time, and then rising to 144°C (at 14 mm), at which temperature and pressure the product starts to solidify in the condenser. Increased heating (heating the water bath to 85-90°C) allows for a steady flow of product through the condenser. All fractions boiling below 144°C/14 mm are discarded. As an alternative to the preheated condenser, a heat gun may be used to melt the solidifying product.

3. The isolated product typically contains 0-3% isoquinoline. Physical data for 5-bromoisoquinoline: TLC: R_f = 0.30 (9:1 dichloromethane/diethyl ether); IR (CHCl$_3$) cm^{-1}: 3053, 1582, 1509, 1484, 1352, 1263, 1222, 1112, 1000; ^{1}H NMR (500 MHz, DMSO-d$_6$) δ: 7.62 (t, 1 H, J = 7.8), 7.91 (d, 1 H, J = 6.0), 8.15 (d, 1 H, J = 7.5), 8.19 (d, 1 H, J = 8.2), 8.66 (d, 1 H, J = 5.9), 9.37 (s, 1 H); ^{13}C NMR (DMSO-d$_6$) δ: 118.5, 120.3, 127.9, 128.4, 129.3, 133.9, 134.3, 144.6, 152.9.

7. Alternatively, the product can be isolated in >99% purity by column chromatography on 63-200 μm silica gel eluting with 9:1 → 5:1 dichloromethane/ethyl acetate and then 9:1 → 4:1 dichloromethane/diethyl ether. The product obtained in this manner

has mp 81-82°C (lit. 82-84°C,[3] 79.5-80.5°C,[4] 82-83°C[5]). Anal. Calcd for C_9H_6BrN: C, 51.96; H, 2.91; Br, 38.40; N, 6.73. Found: C, 52.06; H, 2.74; Br, 38.21; N, 6.65.

8. Heptane was purchased from Fisher Chemicals and toluene (technical quality) was purchased from SvedaKemi A/S, Rosenoerns Allé 9, DK-1970 Frederiksberg, Denmark.

9. The glass filter funnel containing Celite was preheated using a heat gun to avoid precipitation of product in the funnel. After filtration, an additional preheated mixture of 120 mL of heptane and 30 mL of toluene was passed through the Celite using suction. A precipitate forms in the filtrate, which is redissolved by heating to reflux.

3. The isolated product typically contains 0-1% 8-bromo-5-nitroisoquinoline, 0-1% 5,8-dibromoisoquinoline, and 0-1% 5-nitroisoquinoline. Physical data for 5-bromo-8-nitroisoquinoline: mp 137-139°C (lit. 139-141°C,[3] 138-140°C,[4] 139-141°C[5]); TLC R_f = 0.57 (9:1 dichloromethane/ethyl acetate); IR (CHCl$_3$) cm^{-1}: 3053, 1619, 1580, 1485, 1374, 1265, 1201; ^{1}H NMR (500 MHz, DMSO-d$_6$) δ: 8.12 (dd, 1 H, J_A = 0.8, J_B = 6.0), 8.33[†] (d, 1 H, J = 8.3), 8.35 (AB, 1 H, J = 8.2), 8.84 (d, 1 H, J = 5.9), 9.78 (s, 1 H); ^{13}C NMR δ: 118.9, 120.0, 125.8, 127.8, 133.4, 134.5, 145.3, 145.7, 148.1.

11. Alternatively, the product can be isolated in >99% purity by column chromatography on 63-200 μm silica gel, elution with 9:1 → 6:1 dichloromethane/diethyl ether followed by recrystallization to yield 5-bromo-8-nitroisoquinoline with mp 139-141°C. Anal. Calcd for C_9H_6BrN: C, 42.72; H, 1.99; Br, 31.58; N, 11.07. Found: C, 43.03; H, 1.82; Br, 31.19; N, 10.94.

Waste Disposal Information

All toxic materials were disposed of in accordance with "Prudent Practices in the Laboratory"; National Academy Press; Washington, DC, 1995.

3. Discussion

Previously published methods for electrophilic bromination of isoquinoline[3-5] lead to mixtures of isomers only separable with difficulty, use expensive additives or large excesses of reactants, or involve multistep procedures.

The present procedure describes conditions, which allow for the formation of 5-bromoisoquinoline in good yield and high purity using easily available and inexpensive starting materials. In order to obtain the desired product, it is important to ensure careful temperature control to suppress the formation of 8-bromoisoquinoline, which is difficult to remove. By choosing sulfuric acid as solvent for the bromination, a convenient one-pot procedure to prepare 5-bromo-8-nitroisoquinoline, without prior isolation of 5-bromoisoquinoline, has been developed. Finally, the method can easily be scaled up from grams to kilograms of the title compounds.

Many pharmacologically active compounds have been synthesized using 5-bromoisoquinoline or 5-bromo-8-nitroisoquinoline as building blocks.[6-11] The haloaromatics participate in transition-metal couplings [8,10,12] and Grignard reactions. The readily reduced nitro group of 5-bromo-8-nitroisoquinoline provides access to an aromatic amine, one of the most versatile functional groups. In addition to *N*-alkylation, *N*-acylation and diazotiation, the amine may be utilized to direct electrophiles into the *ortho*-position.

The heterocyclic ring may be reduced under very mild conditions after *N*-alkylation, giving access to bicyclic amines[7-10,13] or enamines[5] Use of 5-bromoisoquinoline in a metalation reaction yielded 6-aminoisoquinoline, a compound otherwise accessed with difficulty.[14]

1. (a) NeuroSearch A/S, DK-2750 Ballerup, Denmark; (b) Nuevolution A/S, DK-2100 Copenhagen.

2.   Perrin, D. D.; Armarego, W. L. F. *Purification of Laboratory Chemicals,* 3rd Edition, Pergamon Press: Oxford, England, **1988,** p 105.

3.   Osborn, A. R.; Schofield, K.; Short, L. N. *J. Chem. Soc.* **1956,** 4191.

4.   Gordon, M.; Pearson, D. E. *J. Org. Chem.* **1964,** *29,* 329.

5.   Rey, M.; Vergnani, T.; Dreiding, A. S. *Helv. Chim. Acta* **1985,** *68,* 1828.

6.   Ortwine, D. F.; Malone, T. C.; Bigge, C. F.; Drummond, J. T.; Humblet, C.; Johnson, G.; Pinter, G. W. *J. Med. Chem.* **1992,** *35,* 1345-1370.

7.   Bigge, C. F.; Humblet, C.; Johnson, G.; Malone, T.; Ortwine, D. F.; Pinter, G. W. *Trends Med. Chem.* ΄*90,* Proc. Int. Symp. Med. Chem., 11th **1992,** 153-159.

8.   Moldt, P.; Wätjen, F. (NeuroSearch A/S, DK) **1996** WO pat. Appl. WO96/08495.

9.   Bigge, C. F.; Malone, T.; Schelkun, R. M.; Yi, C. S. (Warner-Lambert Company, US) **1996** WO pat. Appl. WO96/28445.

10. Wätjen, F.; Drejer, J. (NeuroSearch A/S, DK) **1994** WO pat. Appl. WO94/26747.

11. Srivastava, S. K.; Chauhan, P. M. S.; Agarwal, S. K.; Bhaduri, A. P.; Singh, S. N.; Fatma, N.; Chatterjee, R. K.; Bose, C.; Srivastava, V. M. L. *Bio. Med. Chem. Lett.* **1996,** *6,* 2623.

12. Pridgen, L. N. *J. Heterocycl. Chem.* **1980,** *17,* 1289.

13. Mathison, I. W.; Morgan, P. H. *J. Org. Chem.* **1974,** *39,* 3210-3214.

14. Poradowska, H.; Huczkowska, E.; Czuba, W. *Synthesis* **1975,** *11,* 733*

Appendix

Chemical Abstracts Nomenclature (Registry Number)

Isoquinoline; (119-65-3)

N-Bromosuccinimide: 2,5-Pyrrolidinedione, 1-bromo-; (128-08-5)

5-Bromoisoquinoline: Isoquinoline, 5-bromo-; (34784-04-8)

5-Bromo-8-nitroisoquinoline: Isoquinoline, 5-bromo-8-nitro-; (63927-23-1)

Potassium nitrate; (7757-79-1)

PREPARATION OF 2,4-DISUBSTITUTED IMIDAZOLES:
4-(4-METHOXYPHENYL)-2-PHENYL-1H-IMIDAZOLE
[1H-Imidazole, 4-(4-methoxyphenyl)-2-phenyl-]

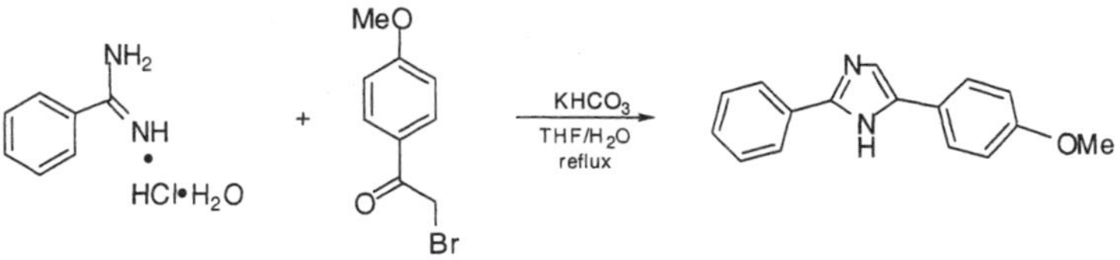

Submitted by Bryan Li, Charles K-F Chiu, Richard F. Hank, Jerry Murry, Joshua Roth, and Harry Tobiassen.

Checked by Renee Kontnik and Steven Wolff.

1. Procedure

4-(4-Methoxyphenyl)-2-phenyl-1H-imidazole. A 2-L, three-necked, round-bottomed flask equipped with an addition funnel, reflux condenser, and mechanical stirrer is charged with 500 mL of tetrahydrofuran (THF) and 125 mL of water. Benzamidine hydrochloride monohydrate (50 g, 0.29 mol) (Note 1) is added, followed by the slow, portionwise addition of potassium bicarbonate (54.4 g, 0.57 mol) (Note 2). The reaction mixture is vigorously heated to reflux. A solution of 4-methoxyphenacyl bromide (65.3 g, 0.29 mol) in 325 mL of THF is then added dropwise via the addition funnel over a period of 30 min while the reaction is maintained at reflux. After completion of the addition, the mixture is heated at reflux for 18-20 hr (Note 3), then cooled in an ice bath (Note 4), and THF is removed under reduced pressure using a rotary evaporator. An additional 100 mL of water is added, and the resulting suspension is stirred at 50-60°C for 30 min. The mixture is cooled in an ice bath and the solids are collected by filtration. The filter cake is rinsed with two 100-mL portions of water and air-dried in the filter funnel for 2 hr. The crude product is transferred to a 500-mL flask and 150 mL of diisopropyl ether and 150 mL of hexanes are added. The mixture is stirred for 2 hr at room temperature, and the solids are again collected by filtration. The filter cake is dried in a vacuum oven for 48 hr (68°C/ca. 100 mm) to give 68.6 g (96%) of the desired imidazole as an off-white solid (Notes 5, 6).

2. Notes

1. All reagents were purchased from Aldrich Chemical Company and used without further purification.

2. Caution: liberation of carbon dioxide.

3. TLC analysis (elution with 1:1 hexanes:ethyl acetate) indicated that the reaction was not complete after 5 hr; heating was continued overnight.

4. Inorganic salts are observed upon cooling and are removed by decantation.

5. The dried product contained 0.10-0.54% H_2O w/w; and has the following physical properties: mp 144.5-145.6°C [lit. mp 178-179°C[2]]; IR (KBr): cm^{-1} 3003, 2836, 1617, 1567, 1499, 1461, 1403, 1298, 1248, 1176; ^{1}H NMR (300 MHz, DMSO-d[6]): δ 3.76 (s, 3H), 6.94 (d, J = 8.4 Hz, 2H), 7.46-7.30 (m, 3H), 7.58 (s, 1H), 7.74 (d, J = 8.7 Hz, 2H), δ 7.97 (d, J = 7.8 Hz, 2H); ^{13}C NMR (75 MHz, DMSO-d[6]): δ 55.0, 114.0, 115.8, 124.9, 125.7, 126.4, 127.9, 128.6, 130.7, 139.2, 145.8, 158.1. MS m/z 250 ([M][+]); 235

([M-CH$_3$]); 77 ([C$_6$H$_6$]);   Anal. Calcd for C$_{16}$H$_{14}$N$_2$O: C, 76.78; H, 5.64; N, 11.19.   Found

C, 76.49; H, 5.62; N, 11.03.

6. HPLC analysis indicated 97.1 area % purity. HPLC conditions: Zorbax XDB-

C8 column (3.0 x 100mm) eluting with 5-100% MeCN + 0.1%TFA (0.5 mL/min); 220 nm

wavelength.

Waste Disposal Information

All toxic materials were disposed of in accordance with "Prudent Practice in the

Laboratory"; National Academy Press; Washington, DC, 1995.

3. Discussion

The imidazole nucleus is often found in biologically active molecules,[3] and a

large variety of methods have been employed for their synthesis.[4] We recently needed

to develop a more viable process for the preparation of kilogram quantities of 2,4-

disubstituted imidazoles. The condensation of amidines, which are readily accessible

from nitriles,[5] with α-halo ketones has become a widely used method for the synthesis

of 2,4-disubstituted imidazoles. A literature survey indicated that chloroform was the

most commonly used solvent for this reaction.[6] In addition to the use of a toxic solvent,

yields of the reaction varied from poor to moderate, and column chromatography was

often required for product isolation. Use of other solvents such as alcohols,[7] DMF,[8] and

acetonitrile[9] have also been utilized in this reaction, but yields are also frequently been

reported as poor.

Our initial attempts to optimize this reaction focused on utilizing anhydrous reaction conditions due to stability concerns of α-bromo ketones under basic aqueous conditions. Condensations using a variety of bases (potassium *t*-butoxide, potassium carbonate, cesium carbonate, etc.) in THF, DMF, CH_3CN or CH_2Cl_2 gave low yields. Reactions in alcohols (ethanol, 2-propanol and *t*-butyl alcohol) were equally unsatisfactory. We then investigated mixed organic/aqueous reaction media, since we reasoned that amidines are stronger nucleophiles than water, and therefore the condensation rate of α-bromo ketones with amidines should be faster than the decomposition rate of the bromo ketones in water. A series of reactions using THF, DMF and alcohols as the organic solvent were conducted, and from these experiments we made a number of observations.

(1) Aqueous THF is a suitable solvent system to solubilize the very polar amidines and non-polar α-bromo ketones, and it is superior to aqueous DMF or alcohol.

(2) Higher reaction temperatures in aqueous THF accelerate the condensation.

(3) Bicarbonate is the base of choice, since it only scavenges the acid produced during the condensation reaction.

(4) Since α-bromo ketones decompose under the reaction conditions, their concentration in the reaction should be minimized.

We found that the optimal reaction protocol was to add a solution of α-bromo ketone in THF to the amidine in aqueous THF in the presence of potassium bicarbonate under vigorous reflux. Using this procedure, 2,4-disubstituted imidazoles were isolated

108

in excellent yields with >95% purity without column chromatography. Aromatic and aliphatic α-halo ketones participate in this reaction with a variety of aromatic amidines, as indicated in Table 1. Particularly noteworthy is that reactions involving pyridylamidines or chloroacetone are substantially more robust using this process (entries 3 and 4). We have successfully used this protocol on a multi-kilogram scale.

In conclusion, a scaleable process for the preparation of 2,4-subsituted imidazole from amidines and α-halo ketones is described. This method avoids the use of chloroform as solvent and affords the desired products in consistently good to excellent yields.

Table 1. Amidine and α-Halo ketone Condensations[a]

Amidine	α-Halo ketone	Product	Isolated Yields	Lit. Yields
Benzamidine	2-Bromoacetophenone	2,5-Diphenyl-1 H-imidazole	86%	62% [b 10]
Nicotinamidine	Chloroacetone	3-(5-Methyl-1 H-imidazol-2-yl)-pyridine	83%	15% [11]
Thiophene-2-carboxamidine	2-Bromo-4'-methoxy-acetophenone	5-(4-Methoxyphenyl)-2-thiophen-2-yl-1 H-imidazole	87%	80% [c 12]
Benzamidine	Chloroacetone	5-Methyl-2-phenyl-1 H-imidazole	87%	59% [d 13]

[a] The 2,4-disubstituted imidazoles in Table 1 were previously characterized in the literature. The spectroscopic data of all products are consistent to that originally reported.

[b] Irreproducible results were reported.

[c] 3 eq. of 2-thiophenylamidine was used; yield was based on the α-bromo ketone.

[d] From the condensation of amidine and an α-halo oxime.

109

1. Chemical Research and Development, Pfizer Global Research and Development, Groton Laboratories, Groton, CT 06340, USA. Jerry Murry is currenty located at Process Research Department, Merck Research Laboratories, Merck and Co. Inc., Rahway, NJ, 07065.

2. Lombardino, J. G.; Wiseman, E. H. *J. Med. Chem.* **1974**, *17*, 1182.

3. (a) Kudzma, I. V.; Turnvull, S. P. Jr. *Synthesis,* **1991**, 1021; (b) Compagnone, R. S.; Rapoport, H. *J. Org. Chem.* **1986**, *51*, 1713 and references cited therein; (c) Shapiro, S.; Enz, A. *Drugs of the Future,* **1992**, *17*, 489.

4. For examples, see (a) Grimmett, M. R. in *Comprehensive Heterocycle Chemistry,* Katritzky, A. R.; Rees, C. Ed. Pergamon Press, vol. 5, 1984 and references therein; (b) Grimmet, M. R. in *Comperhensive Heterocyclic Chemistry II*, Katritzky, A. R.; Rees, C.; Scriven, E. F.V. Ed. Pergamon Press, vol. 3, 1996 and references cited therein; (c) Varma, R. S.; Kumar, D. *Tetrahedron Lett.* **1999**, *40*, 7665; ((e) Lengeler, D.; Weisz, K. *Nuclesides Nucleotides,* **1999**, *18*, 1657; (f) Batanero, B.; Escudero, J.; Barba, F. *Org. Lett.* **1999**, *1*, 1521; (g) Bergemann, M.; Neidlein, R. *Helv. Chim. Acta* **1999**, *82*, 909.

5. (a) Boeré, R. T.; Oakley, R. T.; Reed, R. W. *J. Organomet. Chem.* **1987**, *331*, 161; (b) Thurkauf, A.; Hutchison, A.; Peterson, J.; Cornfield, L; Meade, R. *J. Med. Chem.* **1995**, *38*, 2251.

6. (a) Kempter, G.; Spindler, J.; Fiebig, H. J.; Sarodnick, G. *J. Prakt. Chem.* **1971**, *313*, 977; (b) Baldwin, J. J.; Christy, M. E.; Denny, G. H.; Habecker, C. N.; Freedman, M. B. *J. Med. Chem.* **1986**, *29*, 1065; (c) Nagao, Y.; Takahashi, K.; Torisu, K.; Kondo, K.; Hamanaka, N. *Heterocycles,* **1996**, *42*, 517; (d) Baldwin, J. J.; Engelhardt, E. L.; Hirschmann, R.; Lundell, G. F.; Ponticello, G. S. *J. Med. Chem.* **1979**, *22*, 687.

7. Baldwin, J. E.; Fryer, A. M.; Pritchard, G. J. *J. Org. Chem.* **2001**, *66*, 2588.

8. Kikuchi, K.; Hibi, S.; Yoshimura, H.; Tokuhara, N.; Tai, K.; Hida, T.; Yamauchi, T.; Nagai, M.; *J. Med. Chem.* **2000**, *43*, 409.

9. Moody, C. J.; Roffey, J. R. A. *Chem. Abstr.* **2000**, 134:71748.

10. Burtles, R. J.; Pyman, F. L. *J. Chem. Soc.* **1923**, *123*, 362.

11. Baldwin, J. J.; Lumma, P. K.; Novello, F. C.; Ponticello, G. S.; Sprague, J. M.; Duggan, D. E., *J. Med. Chem.* **1977**, *20*, 1189.

12. Baldwin, J. J.; Engelhardt, E. L.; Hirschmann, R.; Lundell, G. S.; Ponticello, G. S. *J. Med. Chem.* **1979**, *22*, 687.

13. Nakanish, S.; Nantaku, J.; Otsuji, Y. *Chem. Lett.* **1983**, *3*, 341.

Appendix

Chemical Abstracts Nomenclature (Registry Number)

4-(4-Methoxyphenyl)-2-phenyl-1H-imidazole: 1H-Imidazole, 4-(methoxyphenyl)- 2-phenyl-; (53458-08-5)

Benzamidine hydrochloride: Benzenecarboximidamide, monohydrochloride; (1670-14-0)

4-Methoxyphenacyl bromide; Ethanone, 2-bromo-1-(4-methoxyphenyl)-; (2632-13-5)

PREPARATION OF N-ARYL-5R-HYDROXYMETHYL-2-OXAZOLIDINONES FROM N-ARYL CARBAMATES: N-PHENYL-(5R)-HYDROXYMETHYL-2-OXAZOLIDINONE

[2-Oxazolidinone, 5-(hydroxymethyl)-3-phenyl-, (5R)-]

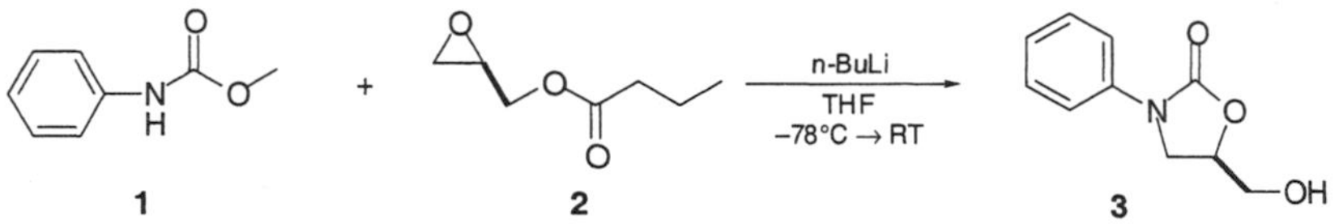

Submitted by Peter R. Manninen[1] and Steven J. Brickner.[2]

Checked by Ossama Darwish and Marvin J. Miller.

1. Procedure

N-Phenyl-(5R)-hydroxymethyl-2-oxazolidinone. A 2-L, three-necked, round-bottomed flask equipped with an addition funnel, rubber septum, nitrogen inlet, and a large magnetic stirbar is charged with 24.81 g (164.1 mmol) of N-phenylcarbamic acid methyl ester (1) (Notes 1, 2, 3). Freshly distilled tetrahydrofuran (THF, 750 mL)(Note 4) is added *via* syringe in 50-mL portions. The resulting solution is cooled to −78°C in a dry ice/acetone bath while 103 mL of butyllithium solution (1.6M in hexanes, 164.8 mmol) (Note 5) is added dropwise via the addition funnel over 60 min. The addition funnel is rinsed with 10 mL of distilled THF and the rinse is added to the reaction mixture. The reaction mixture is stirred for 38 min while the flask is cooled in the dry ice/acetone bath and then 23.4 mL (164.8 mmol) of R-(–)-glycidyl butyrate **2** (Note 6) is added dropwise via syringe over 6 min. After 15 min, the dry ice/acetone bath is removed and the reaction mixture is allowed to warm to room temperature and stirred for 22 hr. To the resulting thick slurry is then added 750 mL of saturated aqueous ammonium chloride solution and 20 mL of water (Note 7). The aqueous layer is separated and extracted with three 350-mL portions of ethyl acetate, and the combined organic layers are dried over magnesium sulfate, filtered, and concentrated under reduced pressure. The residual solid is dried in a vacuum oven at 80°C for 72 hr to provide 29.87 g (95%) of **3** as an off-white crystalline solid (Notes 8, 9).

2. Notes

1. All glassware was either dried in an oven or flame-dried and cooled under nitrogen.

2. A large football-shaped stir bar is recommended because the mixture becomes a thick slurry.

3. The N-phenylcarbamic acid methyl ester was purchased from TCI America.

4. Tetrahydrofuran was distilled from sodium/benzophenone under nitrogen.

5. Butyllithium was purchased from Aldrich Chemical Company, Inc.

6. R-(–)-Glycidyl butyrate was purchased from Lonza.

7. A little water is added to dissolve the precipitate that results in the aqueous layer.

8. The submitters report obtaining the product in 99% yield. The enantiomeric excess of the Mosher ester of **3** was measured to be 98% using a Chiralcel OD column (40% 2-propanol/hexane). This optical purity measurement substantiated the optical purity assessment made by ^{1}H NMR studies of **3** and racemic **3** prepared using a different method[3]. Addition of the chiral shift reagent tris[3-(heptafluoropropylhydroxymethylene)-(+)-camphorato]europium (III) resulted in clear resolution of the respective aromatic proton signals for the two enantiomers, which was demonstrated with the racemate. Under similar conditions, NMR analysis of **3** showed that within the detectable limits of the experiment (ca. <3%), there was none of the disfavored enantiomer.

9. The product has the following physical properties: mp 139-141°C; ^{1}H NMR (300 MHz, CDCl$_3$): δ 2.05 (s, broad, 1H), 3.77 (dd, J=12.6 Hz, J'=4.1 Hz, 1H), 4.02 (m, 3H), 4.76 (m, 1H), 7.15 (t, J=7.15 Hz, 1H), 7.38 (t, J=7.4 Hz, 2H), 7.54 (d, J=9.0 Hz, 2H); ^{13}C NMR (125 MHz, CDCl$_3$): δ 154.6, 138.1, 129.1, 124.2, 118.3, 72.7, 62.9, 46.3; IR (mineral oil mull): cm^{-1} 3391 (m), 1716 (s), 1424 (m), 1380 (m), 1307 (m), 1232 (m), 1146 (m); mass spectrum (EI): m/z (rel. abundance) 193 (100, M$^+$), 106 (42.9), 77 (38.0), (FAB): 194[100,(MTH)$^+$], 136(62.5), 106(50.1). Anal. Calcd for C$_{10}$H$_{11}$NO$_3$: C, 62.17; H, 5.74; N, 7.25. Found: C, 62.09; H, 5.80; N, 7.06. [α]$_D$= −61 (c = 0.969, CHCl$_3$). TLC: ethyl acetate:hexane (1:1); R$_f$ = 0.13.

Waste Disposal Information

All toxic materials were disposed of in accordance with "Prudent Practices in the Laboratory", National Academy Press; Washington, DC, 1995.

3. Discussion

N-Aryl-5-R-hydroxymethyl-2-oxazolidinones, represented by **3**, are important intermediates for the synthesis of oxazolidinone antibacterial agents. The procedure previously reported[3] and illustrated in Scheme 1 applied the method of Herweh–Kauffmann[4] in which an aryl isocyanate is reacted with kinetically resolved (R)-glycidyl butyrate in the presence of solubilized lithium bromide catalyst. While this method works well to provide the butyrate ester oxazolidinone **5** in high yield, an important limitation when utilized in a more general sense is the need to prepare non-commercially available isocyanates. This typically involves the reaction of anilines with phosgene, a reaction that generally terminates at 50% conversion, due to the formation of an equivalent of the aniline hydrochloride salt. In addition to the hazardous nature of phosgene, other limitations are the elevated temperatures required for the oxazolidinone ring formation, and the need for an additional step to cleave the butyrate ester to provide the requisite 5-(hydroxymethyl)oxazolidinone.

Scheme 1

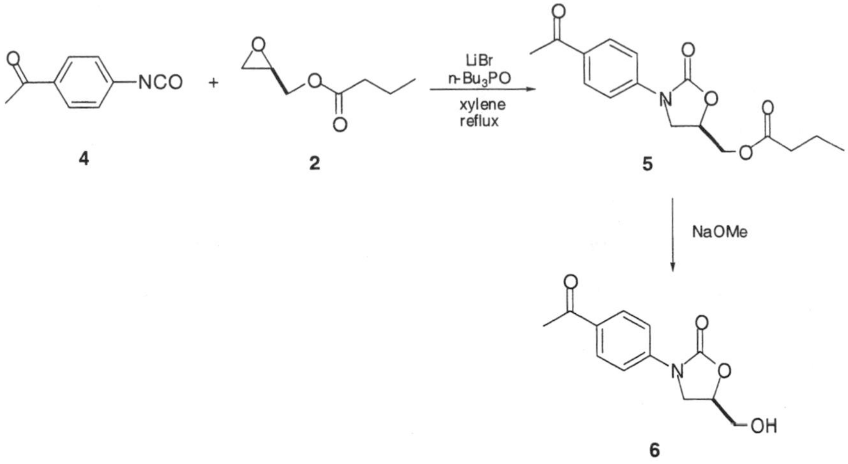

We have developed a novel and mild general approach to 5-(R)-hydroxymethyl-2-oxazolidinones that involves the alkylation of commercially available (R)-glycidyl butyrate with N-lithio-N-aryl carbamates generated by the deprotonation of aryl carbamates with *n*-butyllithium at –78°C. The *N*-aryl-5-hydroxymethyl oxazolidinone is directly obtained from this reaction, by virtue of the *in situ* transesterification of the oxazolidinone butyrate ester with the lithium alkoxide generated in the cyclization. This transesterification equilibrium between the two esters and alkoxides is driven toward the desired pathway as a consequence of the precipitation of the lithium salt **11** of the 5-(hydroxymethyl) oxazolidinone (Scheme 2). In the case of compound **3** derived from O-methyl N-phenyl carbamate, a very clean product (with purity assessed by an acceptable combustion analysis) is obtained simply by evaporative removal of the solvent following extractive aqueous workup, which conveniently removes the by-product methyl butyrate. When the O-benzyl N-phenyl carbamate is employed, **3** is obtained in high purity by taking the crude product derived from aqueous workup and triturating with ethyl acetate–hexane (1:1), which removes the benzyl butyrate. In other examples, it is usually necessary to chromatograph the product to attain analytical purity. Table 1 lists examples of N-aryl-(5*R*)-hydroxymethyl-2-oxazolidinones prepared in high yield and purity from the respective benzyl carbamate.

We have found in our studies that the use of the lithium counter ion in the base is essential for successful reaction with regiochemical control and allows cyclization to proceed under mild thermal conditions. In contrast, use of sodium (NaH, NaN(SiMe$_3$)$_2$) or potassium (KH, KN(SiMe$_3$)$_2$) bases require elevated temperatures, and results in poor yields of the desired product, and a mixture of several by-products, including the regioisomeric 4-hydroxymethy-2-oxazolidinone,[6] resulting from alternate processes. Thus, the lithium ion plays a very important role in the mechanism of this reaction.

Scheme 2

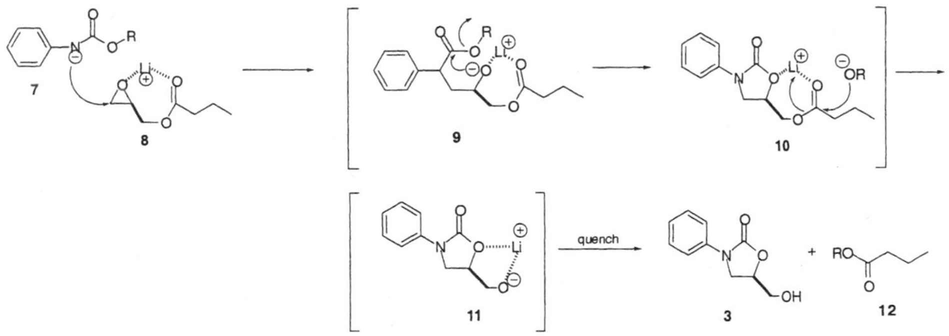

The preparation reported here represents a very general method[7] for the asymmetric synthesis of N-aryl-(5R)-hydroxymethyl-2-oxazolidinones with several distinct advantages over the previous method: (1) the oxazolidinone ring can be formed from readily available N-aryl carbamates, and avoids the hazards and limitations associated with the preparation and isolation of aryl isocyanates. This method significantly broadens the scope of aryl or heteroaryl substitution found in the oxazolidinone. The requisite carbamates are easily prepared from substituted anilines or heteroaryl amines and alkoxy carbonyl chlorides[5]. (2) The desired 5-(hydroxymethyl)oxazolidinone is directly obtained without need for a subsequent saponification step. (3) The reaction conditions are very mild, high-yielding, and provide the 5-(hydroxymethyl)oxazolidinone product in high enantiomeric excess, often with simple work up conditions.

Table 1.

Carbamate	Product	Yield	Ref.
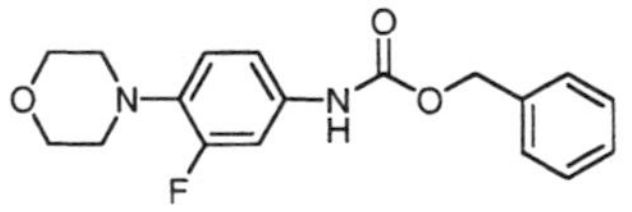	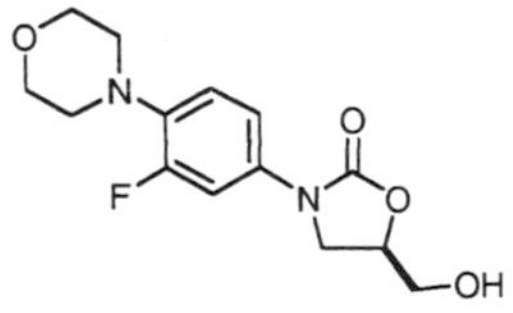	83%	5
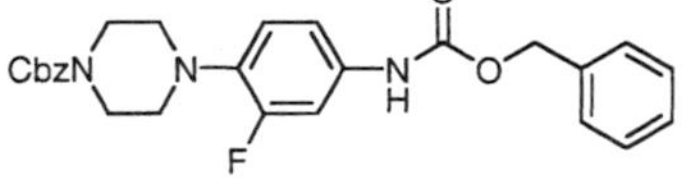	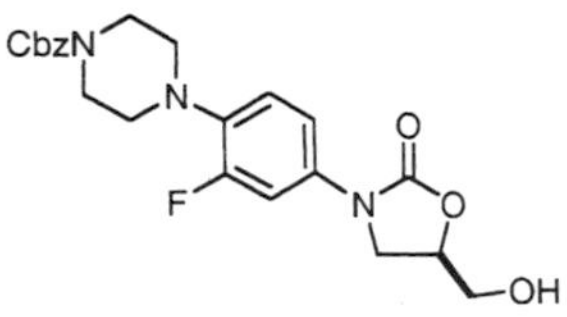	85%	5

118

1. This work was carried out in Medicinal Chemistry Research, Pharmacia, Kalamazoo MI 49001. Author's present address: Eli Lilly Corporation, Lilly Corporate Center, Drop Code 1523, Indianapolis, IN 46285

2. Co-author's present address: Pfizer Global Research & Development, Groton, CT 06340.

3. Gregory, W.A.; Brittelli, D.R.; Wang, C.L.J.; Wounola, M.A.; McRipley, R.J.; Eustice D.C.; Eberly, V.S.; Bartholomew, P.T.; Slee, A.M.; Forbes, M.; *J. Med. Chem.*, **1989**, *32*, 1673.

4. Herweh, J.E.; Kauffman, W.J.; *Tetrahedron Lett.* **1971**, *12*, 809.

5. Brickner, S.J.; Hutchinson, D.K.; Barbachyn, M.R.; Manninen P.R.; Ulanowicz, D.A.; Garmon, S.A.; Grega, K.C.; Hendges, S.K.; Toops, D.S.; Ford. C.W.; Zurenko, G.E.; *J. Med. Chem.* , **1996**, *39*, 673.

6. Manninen, P.R.; Little, H.A.; Brickner, S.J.; Division of Organic Chemistry, American Chemical Society, 212th ACS National Meeting, Orlando, FL, August, 1996.

7. Citations that have used this method: Barbachyn, M.R.; Hutchinson, D.K.; Brickner, S.J.; Cynamon, M.H.; Kilburn, J.O.; Klemens, S.P.; Glickman, E.S.; Grega, K.C.; Hendges, S.K.; Toops, D.S.; Ford. C.W.; Zurenko, G.E.; *J. Med. Chem.*, **1996**, *39*, 680.; Genin, M.J.; Allwine, D.A.; Anderson, D.J.; Barbachyn, M.R.; Emmert, D.E.; Garmon, S.A.; Graber, D.R.; Grega, K.C.; Hester, J.B.; Hutchinson, D.K.; Morris, J.; Reischer, R.J.; Ford. C.W.; Zurenko, G.E.; Hamel, J.C.; Schaadt, R.D.; Stapert, D.; Yagi, B.H.; *J. Med. Chem.*, **2000**, *43*, 953.; Tucker, J.A.; Allwine, D.A.; Grega, K.C.; Barbachyn, M.R.; Klock, J.L.; Adamski, J.L.; Brickner, S.J.; Hutchinson, D.K.; Ford.

C.W.; Zurenko, G.E.; Conradi, R.A.; Burton, P.S.; Jensen, R.M.; *J. Med. Chem.,* **1998**, *41*, 3727.

Appendix

Chemical Abstracts Nomenclature (Registry Number)

N-Phenyl-5*R*-hydroxymethyl-2-oxazolidinone: 2-Oxazolidinone;
    5-(hydroxymethyl)-3-phenyl-, (5R)-; (87508-42-7)

N-Phenylcarbamic acid methyl ester: Carbamic acid, phenylmethyl ester; (2603-10-3)

n-Butyllithium: Lithium, butyl-; (109-72-8)

(R)-(–)-Glycidyl butyrate: Butanoic acid, (2R)-oxiranylmethyl ester; (60456-26-0).

GENERATION AND CYCLIZATION OF 5-HEXENYLLITHIUM:

2-CYCLOPENTYLACETOPHENONE

(Ethanone, 2-cyclopentyl-1-phenyl-)

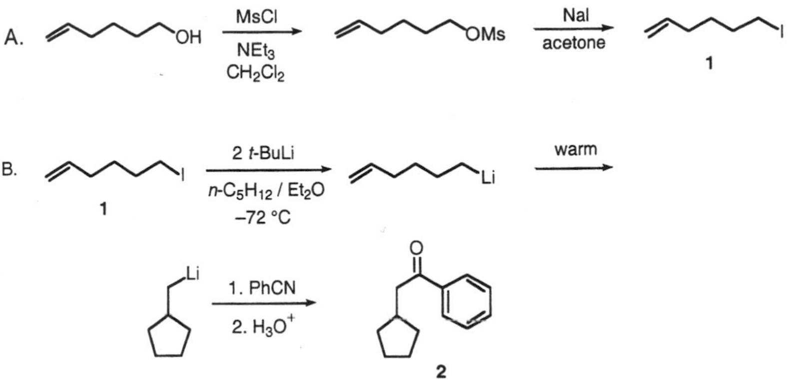

A. (5-hexen-1-ol) →[MsCl / NEt₃ / CH₂Cl₂]→ (OMs) →[NaI / acetone]→ (6-iodo-1-hexene) **1**

B. **1** →[2 t-BuLi / n-C₅H₁₂ / Et₂O / −72 °C]→ (Li) →[warm]→

(cyclopentylmethyllithium) →[1. PhCN / 2. H₃O⁺]→ (2-cyclopentylacetophenone) **2**

Submitted by William F. Bailey, Matthew R. Luderer, Michael J. Mealy, and Eric R. Punzalan.[1]

Checked by Scott E. Denmark and Stephen L. MacNeil.

1. Procedure

A. *6-Iodo-1-hexene* (**1**). A flame-dried, 1-L, one-necked, round-bottomed flask equipped with a magnetic stirbar and a pressure-equalizing addition funnel fitted with a nitrogen inlet adapter is charged with 10.0 g (0.100 mol) of 5-hexen-1-ol, 15.2 g (0.151

mol) of triethylamine, and 500 mL of dichloromethane (Note 1). The flask is cooled in an ice-salt bath (0 to –5°C) for 30 min and then 9.3 mL (0.12 mol) of methanesulfonyl chloride (Note 2) is added dropwise via the addition funnel. The reaction mixture is stirred at –5 to –10°C for an additional 1 hr before being transferred to a cold 1-L separatory funnel and washed successively with 150 mL of cold water, 150 mL of cold 10% (ca. 3.3N) aqueous hydrochloric acid, 150 mL of cold, saturated, aqueous sodium bicarbonate solution, and 150 mL of cold brine (Note 3). The organic layer is dried over $MgSO_4$ and then divided into three portions (Note 4). One portion is filtered into a 300-mL, round-bottomed flask and concentrated by rotary evaporation at 8-10 mm in a 20-25°C water bath, and this is repeated with the remaining two portions to give the desired mesylate as a clear, pale-yellow oil (Note 5).

Dry acetone (200 mL) and then 18.5 g (0.12 mol) of anhydrous sodium iodide (Note 6) are added to the above 300-mL flask, which is then equipped with a magnetic stirbar and a Friedrichs condenser fitted with a nitrogen inlet adapter. The pale-yellow solution is stirred in the dark under a positive pressure of nitrogen at gentle reflux for 4 hr (Note 7). The resulting mixture is allowed to cool to room temperature and the acetone is then removed by rotary evaporation at 8–10 mm in a 20-25°C water bath. The residue is partitioned between 50 mL of pentane and 50 mL of 10% aqueous sodium thiosulfate solution by swirling the flask until all of the precipitate dissolves. The aqueous phase is discarded and the organic layer is washed with 50 mL of brine, dried over $MgSO_4$, filtered, and concentrated by rotary evaporation at 8–10 mm in a 20-25°C water bath. The residue is then passed through a 60-mL, medium porosity, sintered-glass funnel containing ca. 20 g of alumina (Note 8) using ca. 100 mL of pentane as eluent. The pentane is removed by rotary evaporation at 8–10 mm in a 20-25°C water bath to give 17.44-17.97 g (83–86% overall from the alcohol) of **1** as a clear, colorless oil (Notes 9 and 10).

B. *2-Cyclopentylacetophenone* (**2**). A 500-mL, two-necked, round-bottomed flask (Note 11) equipped with an egg-shaped magnetic stir bar (4 cm x 1.5 cm), argon inlet adapter, and a rubber septum is flame-dried and allowed to cool to room temperature under a positive pressure of argon. A thermometer is inserted through the septum (Note 12) and the flask is charged with 114 mL of dry, alkene-free pentane and 76 mL of anhydrous diethyl ether (Notes 13, 14). The solution is cooled to approximately –72°C using a 2-propanol-dry ice bath and 52 mL of a 1.90M solution of *tert*-butyllithium (*t*-BuLi) in heptane (98.8 mmol) is added via teflon cannula at a rate of approximately 1.7 mL/min so as to maintain an internal temperature below ca. –68°C (Notes 15, 16). A two-necked, 100-mL, round-bottomed flask equipped with a rubber septum and argon inlet adapter is flame-dried under an atmosphere of argon, allowed to cool to room temperature, and then charged with 10.0 g (47.6 mmol) of oxygen-free 6-iodo-1-hexene (**1**) (Note 17) and 50 mL of dry, alkene-free pentane (Note 14). The iodide solution is cooled to –72°C using a 2-propanol-dry ice bath and then transferred via a teflon cannula under a positive pressure of argon to the stirred, –72°C solution of *t*-BuLi at a rate of approximately 1.4 mL/min so as to maintain an internal temperature below –65°C (Note 18). The last of the iodide solution is transferred through the cannula by addition of 5 mL of dry pentane to the flask. Upon completion of the addition, the mixture is stirred at −72°C for 15 min and the cooling bath is then removed. After 3.5 hr (Note 19), the opaque, pale-yellow solution is cooled to ca. 0°C in an ice-water bath, and 4.85 g (47.0 mmol) of benzonitrile (Note 20) is added dropwise via syringe (Note 21). The cooling bath is removed, and the bright-orange solution is allowed to stir for 1 hr. The reaction mixture is then re-cooled in an ice-water bath and 35.0 mL of 10% (ca. 3.3N) aqueous hydrochloric acid is added rapidly via syringe (Note 22). Once the exotherm abates, the cooling bath is removed, and stirring is continued for 2 hr. The contents of the flask are completely transferred to a 500-mL separatory funnel by repeated washings with pentane and ether, shaken vigorously, and the aqueous layer is discarded. The organic phase is washed with 50 mL of water and 50 mL of brine, dried over $MgSO_4$. filtered, and concentrated by rotary evaporation at 8–10 mm. Kugelrohr distillation of the residue (bath temperature ca. 180°C, 5 mm) affords 6.64-6.78 g (75-77 %) of pure **2** as a clear, colorless oil (Notes 10, 23, 24).

2. Notes

1.   5-Hexen-1-ol and triethylamine were purchased from Acros Organics and used without further purification. Alternatively, 5-hexen-1-ol may be prepared from 2-(chloromethyl)tetrahydropyran according to the literature procedure for the preparation of 4-penten-1-ol (Brooks, L. A.; Snyder, H. R. *Org. Synth. Coll. Vol. III* **1955**, 698).   Dichloromethane (certified ACS) was purchased from Fisher Scientific and was used as received.

2.   Methanesulfonyl chloride was purchased from Aldrich Chemical Co. and used as received. This reagent, when taken from a previously opened container, is distilled from phosphorus pentoxide prior to use.

3.   The separatory funnel, water, 10% hydrochloric acid, saturated sodium bicarbonate solution, and brine are cooled to ca. 0°C in a freezer prior to use.

4.   The dried organic layer is filtered and concentrated in ca. 150-mL portions.

5.   The mesylate is used in the next step without further purification and is essentially pure; a small quantity of residual dichloromethane has no effect on the yield of the Finkelstein reaction. The mesylate has the following spectroscopic properties: ^{1}H NMR (CDCl$_3$): δ 1.46 - 1.52 (m, 2H), 1.71 - 1.77 (m, 2H), 2.06 - 2.10 (q, J = 7.1 Hz, 2H), 2.98 (s, 3H), 4.21 (t, J = 6.6 Hz, 2H), 4.95 - 5.02 (m, 2H), 5.72 - 5.80 (m, 1H); ^{13}C NMR (CDCl$_3$): δ 24.5, 28.4, 32.9, 37.2, 69.9, 115.1, 137.8.

6.   Acetone (certified ACS) was purchased from Fisher Scientific and was dried over calcium sulfate and filtered prior to use. Anhydrous sodium iodide was purchased from Acros Organics and used as received.

7.   Aluminum foil is wrapped around the flask and lower part of the condenser during the reaction.

8.   Alumina, adsorption, ca. 150 mesh, was purchased from Aldrich Chemical Co. and used as received.

9.   The known iodide 1[2] is isolated in high purity (>99 % by GC analysis) and may be used without further purification. The material displays the following properties: n_D^{20} = 1.5121 (lit.[3] n_D^{20} = 1.5106); 1H NMR (CDCl$_3$: δ 1.46 - 1.55 (m, 2H), 1.80 - 1.87

(m, 2H), 2.05 - 2.11 (m, 2H), 3.19 (t, J = 7.0, 2H), 4.95 - 5.04 (m, 2H), 5.74–5.84 (m, 1H); 1H NMR (CDCl$_3$: δ 6.9, 29.6, 32.6, 32.9, 115.0, 138.1; IR (neat) 3076, 2931, 2856, 1641, 1441, 1215, 1176, 991, 912 cm^{-1}. The iodide may be stored indefinitely in a freezer if first saturated with dry argon.

10. The purity of **1** and **2** is assessed by analytical gas-liquid chromatography (GC) on a Hewlett-Packard 5890 gas chromatograph equipped with a flame-ionization detector and fitted with a 50 m x 0.2 mm HP-5 fused silica glass capillary column using linear temperature programming from an initial temperature of 150°C for 5 min to a final temperature of 200°C for 10 min at a rate of 5°C/min.

11. A one-necked Ace Glass 6935 flask with style-C side well (ordering code 6935-72) was used by the submitters, but the checkers found the use of a two-neck flask to be equally satisfactory.

12. An Omega model HH22 type J-K digital thermometer, connected to a type K thermocouple probe inserted directly into the flask, was used to measure the temperature.

13. Pentane (HPLC grade) was purchased from Fisher Scientific and used as received. Anhydrous diethyl ether was purchased from Mallinckrodt Inc. and distilled under nitrogen from a dark blue solution of sodium and benzophenone.

14. The appropriate volume of solvent is most conveniently transferred under argon from flame-dried graduated cylinders fitted with rubber septa using stainless steel double-tipped cannulas. Although the proportions of pentane and diethyl ether used as solvent are not crucial to the success of the lithium-iodine exchange, it is necessary to

run the reaction in a solvent system that contains enough diethyl ether to render the t-BuLi dimeric.[4] The submitters have found that a solvent mixture composed of 3:2 (by volume) of pentane and ether gives reproducible results.

15. The concentration of solutions of t-BuLi in heptane, purchased from FMC, Lithium Division, were determined immediately prior to use by titration with 2-butanol in xylene using 1,10-phenanthroline as the indicator following the procedure described by Watson and Eastham.[5] A typical procedure is as follows: a flame-dried, 25-mL, round-bottomed flask, equipped with a magnetic stirbar, is charged under a blanket of argon with 3 mL of a 0.10 % (w/v) solution of 1,10-phenanthroline in dry benzene and 0.40-0.50 mL of t-BuLi in heptane; the violet solution is titrated with a ca. 1.0M solution of 2-butanol in xylenes until the pale yellow end point is reached. It should be noted that the nominal concentration recorded on commercial samples of an organolithium is often in error, particularly once the original seal is breached, or if the material has been stored for an extended period, and it is essential that the actual concentration be determined immediately prior to use. At least two molar equivalents of t-BuLi must be used in the exchange reaction: a full equivalent of t-BuLi is consumed in reaction with the tert-butyl iodide, generated as a by-product of the exchange, to give lithium iodide, isobutylene, and isobutane.[4] A slight surplus of t-BuLi in excess of 2 molar equiv is not deleterious, since residual t-BuLi is rapidly consumed by proton abstraction from diethyl ether when the reaction mixture is warmed to room temperature,[4] but a large excess should be avoided.

16. A commercial solution of t-BuLi in pentane may be substituted for t-BuLi in heptane. However, the lower volatility and higher flash-point of the heptane solvent makes t-BuLi in heptane *much* easier and safer to handle than this alternative.

17. The 6-iodo-1-hexene is rendered free of oxygen by bubbling argon through the neat iodide for at least 5 min prior to use.

18. The exchange reaction is exothermic and too rapid an addition of the iodide will cause a temperature rise that may lead to the formation of unwanted side-products. A white precipitate of LiI is observed when approximately two-thirds of the iodide solution has been added.

19. The solution warms slowly to room temperature; after 1.5 hr the internal temperature is approximately 15-20°C.

20. Benzonitrile, purchased from Aldrich Chemical Co., was distilled from calcium hydride and saturated with dry argon for at least 5 min prior to use.

21. An approximately 5°C exotherm was noted during the addition.

22. The addition of the hydrochloric acid solution results in an exotherm of approximately 20°C, accompanied by the appearance of a white precipitate and a viscous, orange oil.

23. The known ketone **2**,[6] which is isolated in high purity (>99% by GC analysis), exhibits the following physical and spectroscopic properties: n_D^{20} = 1.5362; ^{1}H NMR (CDCl$_3$): δ 1.13 - 1.22 (m, 2H), 1.50 - 1.68 (m, 4H), 1.84 - 1.91 (m, 2H), 2.38 (7-line pattern, J = 7.7 Hz, 1H), 2.98 (d, J = 7.1 Hz, 2H), 7.43 - 7.46 (m, 2H), 7.52 - 7.56 (m, 1H), 7.94 -7.97 (m, 2H); ^{13}C NMR (CDCl$_3$): δ 24.9, 32.7, 36.0, 44.7, 128.0, 128.5,

128

132.8, 137.2 200.31; IR (neat) 3062, 2951, 2868, 1685, 1448, 1209, 752, 690 cm^{-1};

Anal. calcd for $C_{13}H_{16}O$: C, 82.94; H, 8.57, Found: C, 82.71; H, 8.50

24. Purification by short-path distillation (138-139°C/4.5 mm) resulted in a slightly diminished yield (6.25 g, 71%).

Waste Disposal Information

All toxic materials were disposed of in accordance with "Prudent Practices for Disposal of Chemicals from Laboratories"; National Academy Press; Washington, DC, 1995.

3. Discussion

The procedure described above may be used for the generation and 5-*exo* cyclization of a variety of substituted 5-hexenyllithiums and, with appropriate modification, provides a general route to a variety of other olefinic[7] and acetylenic[8] organolithiums that cyclize upon warming.[9]

Primary alkyllithiums may be prepared rapidly and in virtually quantitative yield by treatment of a primary alkyl iodide with slightly more than two molar equiv of *t*-BuLi in a hydrocarbon–ether solvent system at low temperature.[4] The exchange protocol outlined above for the preparation of 5-hexenyllithium (1) is a general one but its success often depends crucially on the appropriate choice of both halide and solvent. Primary alkyl iodides, rather than bromides, must be used to ensure that the reaction proceeds cleanly, most likely via a 10-I-2 ate-complex:[10] addition of *t*-BuLi to a primary alkyl bromide often initiates radical-mediated processes. The lithium-iodine exchange appears to involve a dimeric *t*-BuLi solvate and, for this reason, the best medium for the preparation of primary alkyllithiums by lithium-iodine exchange is a predominantly hydrocarbon solvent system that contains a quantity of a simple alkyl ether, such as Et$_2$O, MTBE, dibutyl ether, or the like, in which the *t*-BuLi is predominantly dimeric. Solvent systems containing THF, TMEDA, or other Lewis bases that render the *t*-BuLi monomeric should be avoided since elimination and coupling reactions often compete with exchange under these conditions.

Cyclization of an organolithium tethered to a suitably positioned carbon-carbon π-bond is a thermodynamically favorable process that proceeds in a totally regioselective *exo*-fashion with a high degree of stereocontrol via a transition state in which the lithium atom is intramolecularly coordinated with the remote π-bond.[9] The stereochemical outcome of the cyclization of a substituted 5-hexenyllithium follows from the preference of the substituent to occupy a pseudoequatorial position in the chair-like transition state depicted below.[7]

Despite the favorable thermodynamics associated with the cyclization of unsaturated organolithiums, the isomerization is often sluggish when the ring closure involves generation of a quaternary center or formation of a strained framework. In such cases it has been found that addition of lithiophilic Lewis bases such as THF or TMEDA facilitate the reaction.[7,9] The preparation of cuparene, a sterically congested sesquiterpene possessing two adjacent quaternary centers, illustrates the methodology.[11]

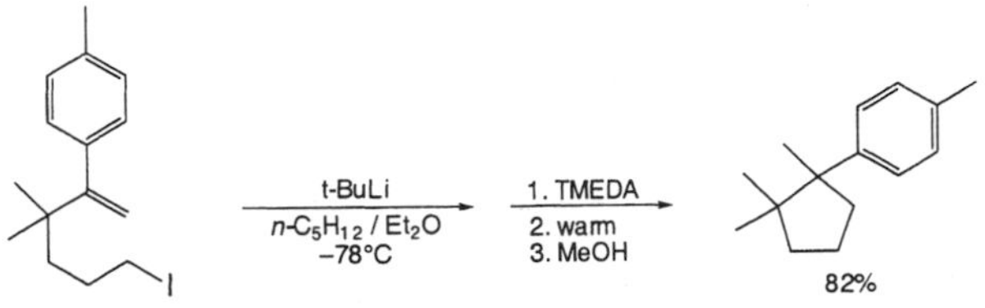

82%

The predictable stereochemistry of the ring closure of substituted 5-hexenyllithiums, coupled with the ease with which the product organolithium may be functionalized, permits rational design of synthetic routes to polycyclic systems by sequential anionic cyclizations of polyolefinic alkyllithiums.[12] The preparation of stereoisomerically pure endo-2-substituted bicyclo[2.2.1]heptanes is illustrative of this approach to polycyclic systems.[12]

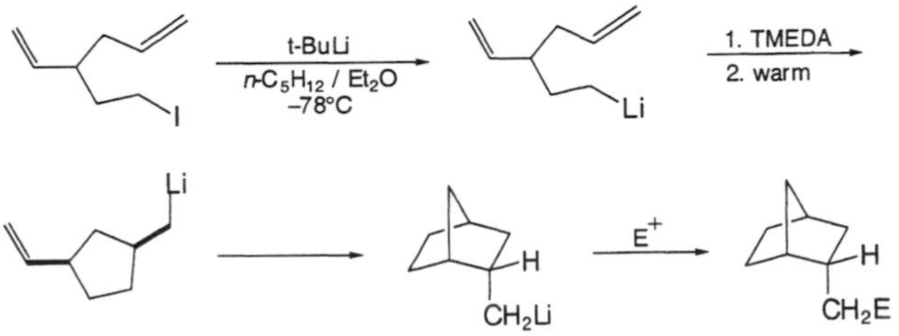

Heterocyclic systems, such as the substituted indoline illustrated below,[13] may also be constructed via cyclization of unsaturated organolithiums and a recent review of the preparation of nitrogen- and oxygen-containing heterocycles via this approach is available.[14]

131

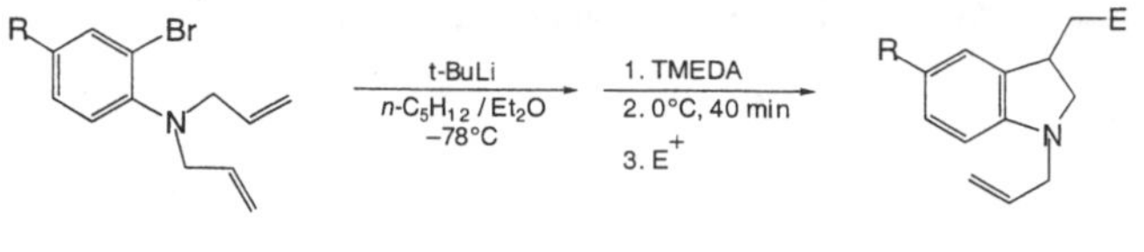

It should also be noted that the 5-exo-trig cyclization of achiral olefinic organolithiums has been found to proceed enantioselectively when conducted in the presence of a chiral ligand that serves to render the lithium atom stereogenic. Thus, for example, (R)-1-allyl-3-methylindoline has been prepared in 86 % ee by cyclization of an achiral aryllithium in the presence of an equivalent of (–)-sparteine.[15]

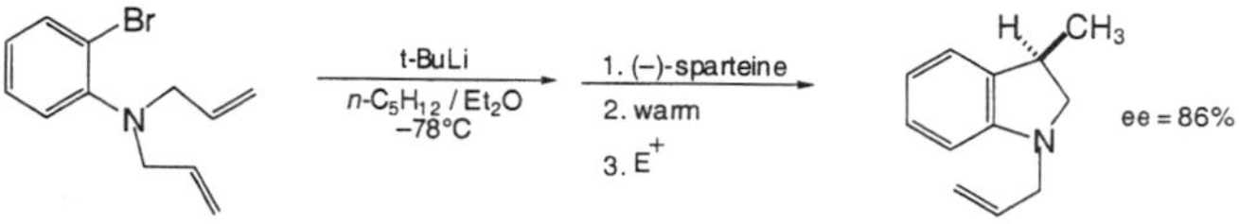

1.   Department of Chemistry, University of Connecticut, Storrs, CT, 06269.

2.   Bailey, W. F.; Gagnier, R. P.; Patricia, J. J. *J. Org. Chem.* **1984**, *49*, 2098.

3.   Ashby, E. C.; DePriest, R. N. Goel, A. B.; Wenderoth, B.; Pham, T. N. *J. Org. Chem.* **1984**, *49*, 3545.

4.   Bailey, W. F.; Punzalan, E. R. *J. Org. Chem.* **1990**, *55*, 5404.

5.   Watson, S. C.; Eastham, J. F. *J. Organomet. Chem.* **1967**, *9*, 165.

6.   Moureu, H.; Chovin, P; Bloch, G.; Rivoal, G. *Bull. Soc. Chim. Fr.* **1949**, *457*, 475.

7.   Bailey, W. F.; Khanolkar, A. D.; Gavaskar, K.; Ovaska, T. V.; Rossi, K.; Thiel, Y.; Wiberg, K. B. *J. Am. Chem. Soc.* **1991**, *113*, 5720.

8.   Bailey, W. F.; Ovaska, T. V. *J. Am. Chem. Soc.* **1993**, *115*, 3080.

9. Bailey, W. F.; Ovaska, T. V. In *Advances in Detailed Reaction Mechanisms*; Coxon, J. M., Ed.; JAI Press: Greenwich, CT, 1994; Vol. 3, *Mechanisms of Importance in Synthesis*; p. 251-273.

10. Wiberg, K. B.; Sklenak, S.; Bailey, W. F. *J. Org. Chem.* **2000**, *65*, 2014.

11. Bailey, W. F.; Khanolkar, A. D. *Tetrahedron* **1991**, *47*, 7727.

12. Bailey, W. F.; Khanolkar, A. D.; Gavaskar, K. V. *J. Am. Chem. Soc.* **1992**, *114*, 8053.

13. Bailey, W. F.; Jiang, X.-L. *J. Org. Chem.* **1996**, *61*, 2596.

14. Mealy, M. J.; Bailey, W. F. *J. Organomet. Chem.* **2002**, *646*, 59.

15. Bailey, W. F.; Mealy, M. J. *J. Am. Chem. Soc.* **2000**, *122*, 6787.

Appendix

Chemical Abstracts Nomenclature (Registry Number)

6-Iodo-1-hexene: 1-Hexene, 6-iodo-; (18922-04-8)

5-Hexen-1-ol; (821-41-0)

Triethylamine; Ethanamine, N,N-diethyl-; (121-44-8)

Methanesulfonyl chloride; (124-63-0)

5-Hexen-1-ol, methanesulfonate; (64818-36-6)

Sodium iodide; (7681-82-5)

2-Cyclopentylacetophenone: Ethanone, 2-cyclopentyl-1-phenyl-; (23033-65-0)

tert-Butyllithium; Lithium, (1,1-dimethylethyl)-; (594-19-4)

Benzonitrile; (100-47-0)

SYNTHESIS OF ORTHO SUBSTITUTED ARYLBORONIC ESTERS BY IN SITU TRAPPING OF UNSTABLE LITHIO INTERMEDIATES: 2-(5,5-DIMETHYL-1,3,2-DIOXABORINAN-2-YL)BENZOIC ACID ETHYL ESTER

[Benzoic acid, 2-(5,5-dimethyl-1,3,2-dioxaborinan-2-yl)-, ethyl ester]

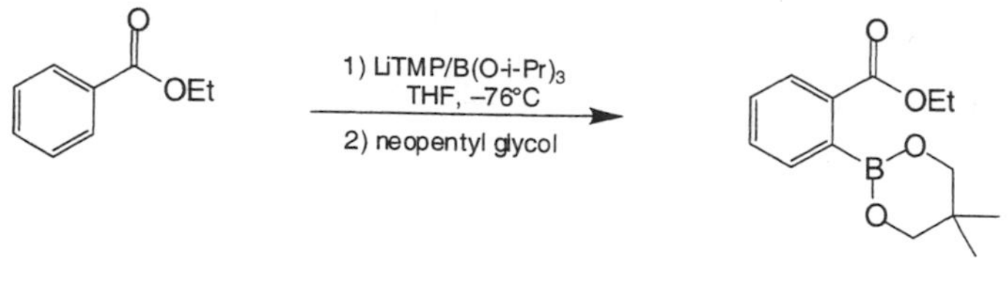

Submitted by Jesper Langgaard Kristensen, Morten Lysén, Per Vedsø and Mikael Begtrup.[1]

Checked by Günter Seidel and Alois Fürstner.

1. Procedure

2-(5,5-Dimethyl-1,2,3-dioxaborinan-2-yl)benzoic acid ethyl ester. An oven-dried, three-necked, 2-L, round-bottomed flask fitted with a thermometer, mechanical stirrer, and a 500-mL pressure-equalizing addition funnel (with volume graduation) capped with a rubber septum and nitrogen inlet is charged with 2,2,6,6-tetramethylpiperidine (51.2 g, 367 mmol) (Note 1) and 400 mL of anhydrous tetrahydrofuran (Note 2). The mixture is cooled to −30°C (internal temperature) in a dry ice-acetone bath and 243 mL of n-butyllithium solution (1.49M in hexanes, 362 mmol) (Note 3) is cannulated directly into the addition funnel from a 1-L sure-seal flask. The n-butyllithium solution is added dropwise to the stirred reaction mixture over 30 min while the temperature is maintained between −30 and −35°C resulting in an orange-yellow solution. The addition funnel is washed with two 10-mL portions of tetrahydrofuran and the reaction mixture is stirred an additional 10 min at −30°C before being cooled to −76°C (Note 4). Triisopropyl borate (112 mL, 483 mmol) (Note 5) is cannulated into the addition funnel directly from a sure-seal flask and then added dropwise (Note 6) to the creamy yellow suspension over 20 min while the internal temperature is maintained below $-73°C$. Ethyl benzoate (35.7 g, 238 mmol) (Note 7) is added via syringe to the addition funnel and added dropwise to the reaction mixture over 10 min, causing a slight reddening of the suspension. The addition funnel is washed with two 10-mL portions of tetrahydrofuran, and the reaction mixture is stirred for 3.5 hr at $-73°C$. The resulting deep red suspension is taken out of the cooling bath, and after approximately 20 min the internal temperature rises to $-30°C$. Glacial acetic acid (20.7 mL, 362 mmol) is then added dropwise over 5 min via the addition funnel, causing the internal temperature to rise to $-10°C$ while the color of the reaction mixture changes from red to yellow. The addition funnel is removed, neopentyl glycol (37.1 g, 356 mmol) (Note 8) is added in one portion, and stirring is continued for 2 hr. The mixture is decanted from a white precipitate into a 3-L separatory funnel, and 1 L of dichloromethane is added to the separatory funnel (Note 9). The resulting solution is washed with three 300-mL portions of a 1:1 mixture of saturated aqueous NH_4Cl and water, two 300-mL portions of saturated $NaHCO_3$ solution, and two 300-mL portions of water, dried over 100 g of Na_2SO_4, filtered into a 1-L round-bottomed flask, and concentrated by rotary evaporation at 40°C. When no more solvent can be distilled, a large teflon-coated magnetic stirbar is added to the flask, and the oily residue is stirred overnight at room temperature under reduced pressure (0.3 mm) to give 58.12 g (95%) of 2-(5,5-dimethyl-[1,3,2]dioxaborinan-2-yl)benzoic acid ethyl ester as a golden-brown oil which is pure enough for most purposes (Note 10). Analytically pure material is obtained as a colorless oil by distillation under high vacuum to give 50.3 g (82%), bp 105-110°C (1×10^{-5} Torr) (Notes 11, 12).

2. Notes

1.   2,2,6,6-Tetramethylpiperidine (99+%) was purchased from Aldrich Chemical Company, Inc. and used as received.

2. Tetrahydrofuran (THF) was distilled from sodium/benzophenone ketyl under a nitrogen atmosphere.

3. Butyllithium (1.6M solution in hexanes) was purchased from Aldrich Chemical Company, Inc. Three sequential titrations using *N*-pivaloyltoluidine (see: Suffert, J. *J. Org. Chem.* **1989**, *54*, 509) gave titers of 1.47M, 1.48M, and 1.51M, and on that basis the titer was assumed to be 1.49M.

4. At approximately –60°C the mixture becomes cloudy.

5. Triisopropyl borate (98+%) was purchased from Aldrich Chemical Company, Inc. and used as received.

6. Care should be taken that the triisopropyl borate drops directly into the solution and does not run down the side of the flask, as this will cause the triisopropyl borate to precipitate from the reaction mixture.

7. Ethyl benzoate (99+%) was purchased from Aldrich Chemical Company, Inc. and used as received.

8. Neopentyl glycol (2,2-dimethyl-1,3-propanldiol) (99%) was purchased from Aldrich Chemical Company, Inc. and used as received.

9. Ca. 200 mL of the dichloromethane is used to rinse the 2-L flask before being added to the separatory funnel.

10. GC-MS analysis showed the crude material to be 95% pure. The title compound is the only detectable species in the NMR-spectrum. The submitters obtained 61.8 g (99%)of the product which was 99.5% pure by GC-MS analysis.

11. High vacuum is necessary for the success of the distillation. Attempted distillation at 0.3 mm results in decomposition of the material at a bath temperature of approximately 150°C.

12. The product exhibitis the following properties: ^{1}H NMR (300 MHz, CDCl$_3$) δ: 1.11 (s, 6H), 1.39 (t, 3H, $J = 7.1$ Hz), 3.79 (s, 4H), 4.38 (q, 2H, $J = 7.1$ Hz), 7.42-7.35 (m, 1H), 7.52-7.48 (m, 2H), 7.93 (td, 1H, $J = 7.8$, 0.9 Hz); ^{13}C NMR (75 MHz, CDCl$_3$) δ: 14.3, 21.9, 31.6, 61.6, 72.4, 128.3, 128.6, 131.2, 131.9, 133.1, 168.6; ^{11}B NMR (96

MHz, CDCl$_3$) δ 28.1 ppm. Anal. Calcd for C$_{14}$H$_{19}$BO$_4$: C 64.15, H 7.31. Found: C 64.02, H 7.46.

2. Waste Disposal Information

All toxic materials were disposed of in accordance with "Prudent Practices in the Laboratory"; National Academy Press; Washington, DC, 1995.

3. Discussion

The procedure described herein is an improved version of our previously reported synthesis,[2] circumventing the need for an intermediate aqueous work-up, thereby providing the title compound in a one-pot procedure.

The transition metal catalyzed cross coupling of an organohalide with a boronic acid derivative, the Suzuki-Miyaura coupling, has become one of the most popular ways of preparing biaryls.[3] The reaction is very robust and can easily be scaled to provide multigrams of material.[4]

Arylboronic acids have traditionally been prepared via the addition of an organomagnesium or organolithium intermediate to a trialkyl borate. Subsequent acidic hydrolysis produces the free arylboronic acid. This limits the type of arylboronic acids one can access via this method, as many functional groups are not compatible with the conditions necessary to generate the required organometallic species, or these species may not be stable intermediates.

Complete characterization of arylboronic acids is often difficult because they are readily transformed into stable cyclic anhydrides called boroxines[5] and other polymeric species. Arylboronic acids are also known to be hygroscopic. Thus, arylboronic acids are often prepared and used directly as a mixture of different entities. Commercial arylboronic acids will very often contain varying amount of anhydrides.

The procedure described herein illustrates two important points:

(1) The in situ trapping of unstable lithio-intermediates as a way of circumventing the above mentioned limitations, and

(2) The direct isolation of the well defined and stable neopentyl glycol arylboronic esters, without the need for an intermediate aqueous work-up.

As first described by Krizan and Martin,[6] the in situ trapping protocol, i.e., having the base and electrophile present in solution simultaneously, makes it possible to lithiate substrates that are not applicable in classical *ortho*-lithiation reactions.[7] Later, Caron and Hawkins utilized the compatibility of lithium diisopropylamide and triisopropyl borate to synthesize arylboronic acid derivatives of bulky, electron deficient neopentyl benzoic acid esters.[8] As this preparation illustrates, the use of lithium tetramethylpiperidide instead of lithium diisopropylamide broadens the scope of the reaction, and makes it possible to functionalize a simple alkyl benzoate.[2]

The conversion of arylboronic acids to the corresponding neopentyl glycol arylboronic esters has several advantages: The esters are readily soluble in organic solvents, shelf stable, non-hygroscopic and easily characterized as a single entity.[9] Furthermore, boronic esters can be utilized in many of the transformations where arylboronic acids usually are employed, making them an attractive alternative from a practical point of view.

1. Department of Medicinal Chemistry, The Danish University of Pharmaceutical Sciences, Universitetsparken 2, DK-2100 Copenhagen, Denmark.

2. Kristensen, J.; Lysén, M.; Vedsø, P.; Begtrup, M. *Org. Lett.* **2001**, *3*, 1435.

3. (a) Miyaura, N.; Yanagi, T.; Suzuki, A. *Synth. Commun.* **1981**, *11*, 513; Reviews:
    (b) Miyaura, N.; Suzuki, A. *Chem. Rev.* **1995**, *95*, 2457;
    (c) Suzuki, A. *J. Organomet. Chem.* **1999**, *576*, 147.

4. (a) Huff, B.E.; Koenig, T.M.; Mitchell, D.; Staszak, M.A. *Org. Synth., Coll. Vol. X* **2004**, 102. (b) Goodson, F.E.; Wallow, T.I.; Novak, B.M. *Org. Synth., Coll. Vol. X* **2004**, 501. (c) Ruel, F.S.; Braun, M.P.; Johnson, C.R. *Org. Synth., Coll. Vol. X* **2004**, 467.

5. Beckmann, J.; Dakternieks, D.; Duthie, A.; Lim, A.E.K.; Tiekink, E.R.T. *J. Organomet. Chem.* **2001**, *633*, 149 and references cited therein.

6. Krizan, T.D.; Martin, J.C. *J. Am. Chem. Soc.* **1983**, *105*, 6155.

7. (a) Gschwend, H.W.; Rodriguez, H.R. *Org. React.* **1979**, *26*, 1;
    (b) Snieckus, V. *Chem. Rev.* **1990**, *90*, 879.

8. Caron, S.; Hawkins, J.M. *J. Org. Chem.* **1998**, *63*, 2054.

9. For a related approach to the isolation of arylboronic acids as the corresponding esters, see: Wong, K.-T.; Chien, Y.-Y.; Liao, Y.-L.; Lin, C.-C.; Chou, M.-Y.; Leung, M. *J. Org. Chem.* **2002**, *67*, 1041.

Appendix

Chemical Abstract Nomenclature (Registry Number)

2,2,6,6-Tetramethylpiperidine; Piperidine, 2,2,6,6-tetramethyl-; (768-66-1)

n-Butyllithium: Lithium, butyl-; (109-72-8)

Triisopropyl borate: Boric acid (H_3BO_3), tris(1-methylethyl) ester; (5419-55-6)

Ethyl benzoate: Benzoic acid ethyl ester; (93-89-0)

Neopentyl glycol: 1,3-Propanediol, 2,2-dimethyl-; (126-30-7)

2-(5,5-Dimethyl-1,3,2-dioxaborinan-2-yl)benzoic acid ethyl ester: Benzoic acid, 2-(5,5)-dimethyl-1,3,2-dioxaborinan-2-yl)-, ethyl ester; (346656-34-6)

SYNTHESIS OF 1,2:5,6-DIANHYDRO-3,4-O-ISOPROPYLIDENE-L-MANNITOL

A.

LiBH₄, MeOH, 0°C, then
2,2-dimethoxypropane, HCl
acetone

1   2

B.

60% AcOH, 45°C

2   3

C.

Ph₃P, DEAD, toluene
reflux

3   4

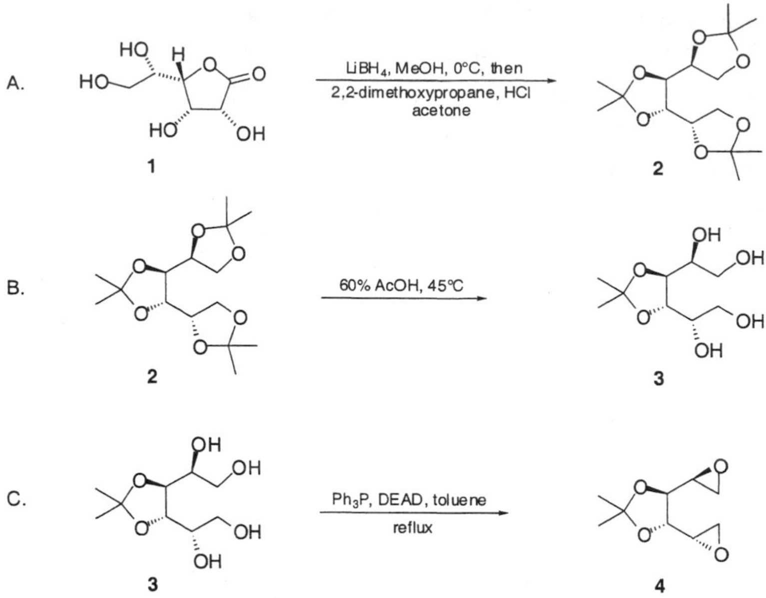

Submitted by David A. Nugiel[1], Kim Jacobs, A. Christine Tabaka, and Chris A. Teleha[2].

Checked by Peter A. Orahovats, Jason S. Newcom and William R. Roush.

1. Procedure

A. *1,2:3,4:5,6-Tri-O-isopropylidene-L-mannitol* (**2**). A 2-L, two-necked, round-bottomed flask equipped with a powder funnel and a magnetic stirbar is charged with L-mannonic acid γ-lactone (20 g, 0.11 mol) (Note 1) and 300 mL of methanol. The resulting suspension is cooled in an ice bath and lithium borohydride (4.12 g, 0.193 mol) is added in several portions via the powder funnel (Note 2) over 30 min. *Caution!* *Extensive foaming due to hydrogen gas evolution.* After the addition is complete, the ice bath is removed and the mixture is stirred for an additional 30 min. The reaction mixture is then recooled in an ice-bath and quenched by the addition of 4N HCl in dioxane (Note 3) until the mixture stops bubbling (1-2 mL). The mixture is then transferred to a 2-L, single-necked, round-bottomed flask and the solvent is concentrated at reduced pressure until a glassy solid is obtained (Note 4). The round-bottomed flask is fitted with a three-way stopcock to which a nitrogen-filled balloon is attached. The solid is then suspended in 100 mL of acetone and 2,2-dimethoxypropane (78 mL, 0.66 mol) is added in one portion. Next, 4N HCl in dioxane (82.5 mL, 0.33 mol) is added slowly with stirring. The reaction mixture is stirred at ambient temperature for 16 hr, during which time it becomes bright red in color. If TLC analysis indicates the reaction is not complete (Note 6), then an additional portion of 2,2-dimethoxypropane (13 mL, 0.11 mol) is added and stirring is continued for an additional 6 hr. The solvent volume is then reduced by 80% at reduced pressure and the reaction mixture is slowly poured into 400 mL of saturated sodium bicarbonate solution. The product precipitates from solution, and after 10-12 hr the solid is collected on a Büchner funnel and air-dried (Note 7). The waxy solid is dissolved in 350 mL of absolute ethanol and cooled to −78°C to give 23.1-24.3 g (69%) of the product as a white solid. A second crop of crystals is collected (5.7-5.9 g) and combined to give a final yield of 29.0-30.0 g (86-87%) (Notes 8, 9).

B. *3,4-O-Isopropylidene-L-mannitol* (**3**). A 2-L, single-necked, round-bottomed flask equipped with a magnetic stirbar is charged with 1,2:3,4:5,6-tri-O-isopropylidene-L-mannitol (**2**) (20 g, 0.066 mol) and 300 mL of 60% acetic acid. The flask is attached to a rotary evaporator (at atmospheric pressure) and the reaction mixture is stirred at 45°C for ca. 1.5 hr. At that point, monitoring by TLC indicates that the reaction has progressed approximately halfway to completion. The pressure in the rotary evaporator is then reduced to 1 mm (by attaching it to a vacuum pump in a hood), and the bath temperature is reduced to 40°C. Under these conditions, full removal of the solvent is

141

achieved in ca. 60 min to yield a highly viscous slurry. This residue is taken up in 200 mL of dichloromethane, stirred for 10 min, and any precipitate is removed by filtration through Celite (Note 11). The filtrate is concentrated at reduced pressure, and the residue is crystallized from 120 mL of diethyl ether and dried at 1 mm to give 9.8 g (67%) of the product as a white solid. The filtrate is concentrated to a volume of 50 mL and a second crop of 1.1 g (7%) is collected by filtration and combined to give a final yield of 10.9 g (74%) (Notes 12, 13).

C. *1,2:5,6-Dianhydro-3,4-O-isopropylidene-L-mannitol* (**4**). A 1-L, two-necked, round-bottomed flask equipped with a nitrogen inlet, addition funnel, and a magnetic stirbar is charged with 3,4-O-isopropylidene-L-mannitol (14.5 g, 0.065 mol), 160 mL of dry toluene (Note 14), and triphenylphosphine (42.9 g, 0.163 mol). The stirred suspension is cooled under a nitrogen atmosphere in an ice bath while diethyl azodicarboxylate (25.9 mL, 0.163 mol) (Note 15) is added dropwise over 20 min. During the addition the reaction mixture becomes homogeneous. After the addition is complete, the ice bath is replaced with an oil bath, the addition funnel is replaced with a reflux condenser, and the reaction mixture is heated at reflux with continued stirring for 1-2 hr (Note 16). The resulting pink reaction mixture is allowed to cool to room temperature, applied directly to a dry silica gel column, and eluted with 30-50% ether/hexane to give 9.4 g (78%) of the desired product as a volatile oil (Note 17).

2. Notes

1. L-Mannonic acid γ-lactone was purchased from Sigma Chemical Company, St. Louis , MO.

2. If the lithium borohydride sticks to the spatula or powder funnel, additional methanol can be used to wash the material into the reaction mixture without affecting the yield. On this reaction scale, an additional 25-50 mL of methanol could be used.

Using 2.0M LiBH$_4$ in THF instead of solid LiBH$_4$ complicates the following step resulting in lower yields.

3. 4N HCl in dioxane was purchased from Aldrich Chemical Company, Inc.

4. Caution should be used evaporating the solvent, as the residue tends to foam upon concentration. Concentration is carried out by rotary evaporation and then for 30 min at high vacuum (1 mm) while warming with a heat gun to remove all of the methanol.

5. Purchased from Aldrich Chemical Company, Inc. and used as received.

6. The reaction was monitored using TLC with 60% ether/hexane as the eluent. The desired product has Rf = 0.95 and the undesired diacetonide has Rf = 0.25. Visualization was done using p-anisaldehyde.

7. The checkers found that best results were obtained when the trisacetonide was allowed to precipitate from the sodium bicarbonate solution over 10-12 hr.

8. The checkers found that it was necessary to cool the crystallization solution to −78°C to induce crystallization.

9. The physical and spectral properties are as follows: mp 72-74°C; [α]$_D$ = −13.4 (CHCl$_3$, c 1.2) ; ^{1}H NMR (400 MHz, CDCl$_3$) δ: 1.4 (s, 6 H), 1.45 (s, 6 H), 1.5 (s, 6 H), 4.05 (m, 4 H), 4.15 (m, 2 H), 4.25 (m, 2 H); CIMS (NH$_3$) m/z: 303 (M+H$^+$,100%).

10. The reaction was monitored using TLC with 5% MeOH/CH$_2$C$_2$ as the eluent. The desired monoacetonide 3 has a Rf = 0.25. It was desirable to run the reaction to the point where most of the higher Rf components were consumed. This will produce some unwanted L-mannitol that can be filtered off as described above.

11. The L-mannitol recovered in this manner could be recycled if desired.

12. The checkers found that further hydrolysis of mannitol diacetonide can occur during the removal of solvent, and that removal of all acetic acid from the product was problematic. Residual acetic acid in the product complicates the next step. The procedure described reduces these problems.

143

13. The physical and spectral properties are as follows: mp 83-86°C; $[\alpha]_D = -26.4$ (c 3, H_2O); ^1H NMR (400 MHz, DMSO-d_6: δ 1.25 (s, 6 H), 3.35 (m, 2 H), 3.5 (m, 4 H), 3.85 (m, 2 H), 4.45 (t, J = 5.9 Hz, 2 H), 5.1 (d, J = 4.4 Hz, 2 H),; ^{13}C NMR (100 MHz,DMSO-d_6: δ 29.1, 64.8, 74.7, 80.9, 110.1,; CIMS (NH_3) m/z: 240 (M+NH_4^+,100%); Anal. calcd. for $C_9H_{14}O_4$: C, 48.64; H, 8.16. Found: C, 48.54; H, 8.16.

14. The toluene used was from a freshly opened bottle obtained from EM Science.

15. Neat DEAD was unavailable from commercial sources at the time the procedure was being checked. Therefore, the checkers used a commercially available 40% solution of DEAD in toluene with results comparable to that described in the original procedure.

16. The reaction was initially monitored using 10% MeOH/CH_2Cl_2 to follow the disappearance of tetraol **3**. The appearance of the diepoxide **4** is monitored using 50% ether/hexane.

17. The physical and spectral properties are as follows: $[\alpha]_D = + 0.38$ ($CHCl_3$ c 0.8); ^1H NMR ($CDCl_3$ δ: 1.40 (s, 6 H), 3.77 (dd, 2 H, J = 2.9, 1.4), 2.65 (dd, 2 H, J = 5.2, 3.0), 2.80 (t, 2 H, J = 9.5), 3.10 (m, 2 H); ^{13}C NMR ($CDCl_3$ δ: 16.6, 27.9, 46.3, 52.6, 79.5, 111.5; CIMS (NH_3 m/z: 204 (M+NH_4^+,100%); Anal calcd. for $C_9H_{14}O_4$: C, 58.05; H, 7.58. Found: C, 57.78; H, 7.52.

3. Discussion

Preparing the title compound from L-mannitol using methodology developed for the D-isomer[2] would be prohibitively expensive. Our approach uses a much less expensive, commercially available starting material. Current synthetic approaches to the D-isomer involve selective conversion of the two primary hydroxyl groups of 3,4-O-isopropylidene-D-mannitol into good leaving groups followed by base treatment to facilitate epoxide formation. We found prolonged exposure of the epoxide to base reduces the reaction's overall yield. To overcome this liability we used a one-pot conversion of 3,4-O-isopropylidene-L-mannitol to the title compound using Mitsunobu-based technology. This approach was found to be more reproducible and consistently gave yields in the 60-80% range.

The title compound is a key C_6 building block. Several labs have prepared novel α-amino acids, biological probes and other interesting compounds using the D–diepoxide as a key intermediate. An efficient route to the L-enantiomer provides a pathway to compounds with the opposite configuration, one not readily available from commercial sources, and a valuable probe of stereochemistry in biological systems and reaction mechanism.

1. Astrazeneca, Inc., B312, CNS Discovery, 1800 Concord Pike, Wilmington, DE, USA 19850-5437

2. Johnson and Johnson Pharmaceutical Research and Development, LLC, Welsh and McKean Roads, P.O. Box 776, spring House, PA 19477-0776.

3. Wiggins, L.F. *J. Chem. Soc.* **1946**, 384. Le Merrer, Y.; Dureault, A.; Greck, C.; Micas-Languin, D.; Gravier, C.; Depezay, J. *Heterocycles*, **1987**, *25*, 541. Ghosh, A.K.; McKee, S.P.; Thompson, W.J. *Tetrahedron Lett.* **1991**, *32(41)*, 5729.

Appendix

Chemical Abstracts Nomenclature (Registry Number)

1,2:3,4:5,6-Tri-O-isopropylidene-L-mannitol: L-Mannitol,1,2:3,4:5,6-tris-O-(1-methylethylidene)-; (153059-35-9).

L-Mannonic acid δ-lactone: L-Mannonic acid, δ-lactone; (22430-23-5).

Lithium borohydride: Borate(-1), tetrahydro-, lithium; (16949-15-8).

2,2-Dimethoxypropane: Propane, 2,2-dimethoxy-; (77-76-9).

3,4-O-Isopropylidene-L-mannitol: L-Mannitol, 3,4-O-(1-methylethylidene)-; (153059-36-0)

1,2:5,6-Dianhydro- 3,4-O-isopropylidene-L-mannitol: L-Mannitol,1,2:5,6-dianhydro-3,4-O-(1-methylethylidene)-; (153059-37-1).

Triphenylphosphine: Phosphine, triphenyl-; (603-35-0).

Diethyl azodicarboxylate: Azodicarboxylic acid diethyl ester; (1972-58-3)

PRACTICAL SYNTHESIS OF NOVEL CHIRAL ALLENAMIDES:

(R)-4-PHENYL-3-(1,2-PROPADIENYL)OXAZOLIDIN-2-ONE

(2-Oxazolidinone, 4-phenyl-3-(1,2-propadienyl)–, (4R)–)

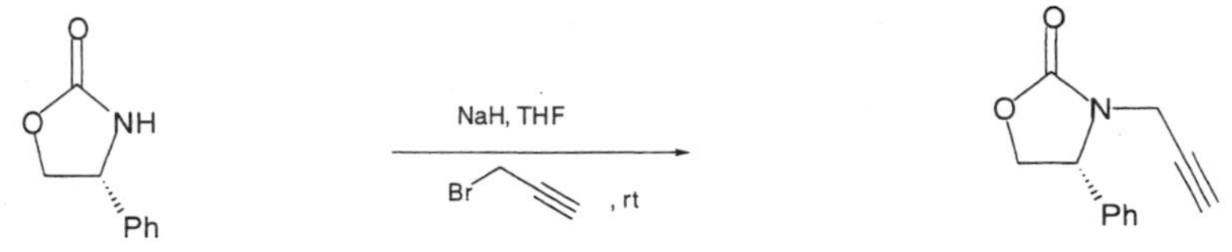

Submitted by H. Xiong, M. R. Tracey, T. Grebe, J. A. Mulder, and R. P. Hsung.[1]

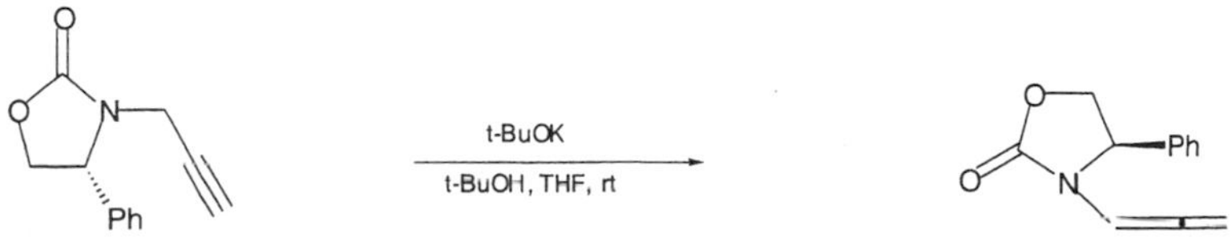

Checked by Peter Wipf and Jennifer Smotryski.

1. Procedure

A. *(R)-4-Phenyl-3-(2-propynyl)oxazolidin-2-one (2).* A flame-dried, 500-mL, round-bottomed flask equipped with a magnetic stirbar and a rubber septum fitted with a nitrogen inlet is charged with 9.90 g (61 mmol) of (R)-4-phenyl-2-oxazolidinone (Note 1) and 200 mL of anhydrous tetrahydrofuran (THF) under a nitrogen atmosphere (Note 2). Sodium hydride (NaH) (2.90 g, 60% w/w in mineral oil, 1.20 equiv, 73 mmol) is added in small portions (Note 3), and the resulting white slush is stirred for 1 hr at room temperature before carefully adding 8.00 mL of a solution of propargyl bromide in toluene (80% w/w in toluene, 72 mmol, 1.18 equiv) (Note 4) dropwise over ca. 10 min via syringe. Precipitation of sodium bromide is observed and does not affect the progress of the reaction. The reaction mixture is stirred at room temperature for 24 hr and then concentrated by rotary evaporation under reduced pressure. The residue is dissolved in 100 mL of anhydrous ether and filtered through a small bed of Celite washing with 1:1 ethyl acetate-hexane. The filtrate is concentrated by rotary evaporation, and the resulting residue is purified using silica gel column chromatography (gradient eluent with 0-33% ethyl acetate-hexane) to give 5.71-6.09 g (47-50%) of the desired propargyl amide **2** as a lightly yellow-colored oil (Note 5).

B. *(R)-4-Phenyl-3-(1,2–propadienyl)-2-oxazolidinone (3).* A flame-dried, 500-mL, round-bottomed flask equipped with a magnetic stirbar and a rubber septum fitted with a nitrogen inlet is charged with a solution of 6.67 g (33 mmol) of propargyl amide **2** in 300 mL of anhydrous THF (Note 2). Potassium *tert*-butoxide (1.26 g, 11 mmol, 0.33 equiv) (Notes 6, 7) is added in portions to the reaction mixture over ca. 10 min. The reaction mixture is stirred at room temperature for 24 hr and the progress of the reaction is monitored by TLC (elution with 50% ethyl acetate-hexane) and ^{1}H NMR analysis. Upon completion of the reaction, the solvent is removed by rotary evaporation under reduced pressure. The resulting crude residue is dissolved in 50 mL of ethyl acetate and vacuum filtered through a small bed of silica gel. The solids are washed with two 40-mL portions of 25-50% ethyl acetate-hexane, and the filtrate is then concentrated by rotary evaporation. The residue is purified using silica gel column chromatography

148

(gradient elution with 0% to 25% ethyl acetate-hexanes) to give 1.41-1.59 g (38-41%) of the desired allenamide **3** as a yellow brownish-red oil (Note 8).

2. Notes.

1.   (*R*)-4-Phenyl-2-oxazolidinone was purchased from Sigma-Aldrich or Urquima, S.A. Arnau de Vilanove, Barcelona, Spain and used as received.

2.   Anhydrous THF was freshly distilled from sodium/benzophenone under nitrogen.

3.   *Caution*: evolution of large amount of H_2 gas. NaH was obtained from Sigma-Aldrich and used as received. NaH free of oil can also be used with no difference in yield and was prepared via four cycles of washing and decanting (via pipette) with anhydrous pentane.

4.   Propargyl bromide was obtained as an 80% w/w solution in toluene from Sigma-Aldrich.

5.   The submitters report obtaining the product in 66% yield prior to purification at 0.050 mole-scale and suggest that the purification step is not necessary because in most cases simple filtration through a small bed of Celite or silica gel provided the desired propargyl amide with high purity as determined by GC analysis. Specifically, GC analysis of propargyl amide **2** shows its purify to be 98.2% prior to column chromatography, and 98.5% after chromatography. After column chromatography, the submitters obtained 6.40 g (52%) of the desired amide. The amide displays the following physical properties: $R_f = 0.49$ (50% EtOAc in hexane);

$[\alpha]_D^{20}$ 148.8 (c 10.2, EtOH); ^{1}H NMR (300 MHz, CDCl$_3$) δ: 2.25 (t, 1H, J = 2.4 Hz), 3.35

(dd, 1H, J = 2.4, 17.7), 4.12 (dd, 1H, J = 7.8, 8.7), 4.36 (dd, 1H, J = 2.4, 17.7), 4.65 (t,

1H, J = 8.7), 4.95 (dd, 1H, J = 7.8, 8.7, 7.32-7.44 (m, 5H); ^{13}C NMR (75 MHz, CDCl$_3$) δ:

32.0, 59.0, 70.0, 73.4, 76.6, 127.3, 129.3, 129.4, 136.6, 157.8; IR (neat) cm^{-1}: 3285,

2913, 2100, 1733, 1494, 1177, 860; mass spectrum (EI): m/e (%relative intensity) 201

(20) M$^+$, 156 (51), 143 (24), 124 (30), 116 (66), 104 (100), 91 (22), 77 (21); m/e calcd for

C$_{12}$H$_{12}$NO$_2$: 202.086; found 202.0871.

6.   Potassium tert-butoxide was obtained from Sigma-Aldrich and used as

received. Alternatively, equivalent results were obtained using t-BuOK that was

prepared from potassium metal and anhydrous tert-butyl alcohol (t-BuOH) followed by

removal of excess t-BuOH. In this case, the molecular weight of t-BuOK was calculated

based on a 1:1 ratio of t-BuOK to t-BuOH (i.e., 186.34 for C$_8$H$_{19}$O$_2$K).

7.   The submitters report that equivalent results were obtained using between

0.20-0.35 equiv of t-BuOK.

8.   The submitters report obtaining the allenamide in 63% yield on this scale

with its purity determined by GC analysis to be between 94%-96% in different runs. The

lower yield obtained by the checkers may be explained by partial decomposition of

product during the chromatographic purification on standard (40-60 nm) silica gel. The

physical properties of **3** are as follows: R$_f$ = 0.59 (50% ethyl acetate-hexane); $[\alpha]_D^{20}$

−156.4(c 0.225, CHCl$_3$); ^{1}H NMR (500 MHz, CDCl$_3$) δ: 4.14 (dd, 1H, J = 5.9, 9.8), 4.69

(t, 1H, J = 9.0), 4.87 (dd, 1H, J = 6.4, 9.5), 4.88 (dd, 1H, J = 6.4, 9.5), 5.16 (dd, 1H, J =

5.9, 9.8), 6.79 (t, 1H, J = 6.4), 7.23 – 7.37 (m, 5H); ^{13}C NMR (75 MHz, CDCl$_3$) δ: 59.0,

70.6, 87.7, 95.6, 126.5, 128.7. 129.0, 138.4, 155.5, 201.9; IR (neat) cm^{-1} 3063, 3035, 2979, 1963, 1767, 1494, 1462, 1216, 966, 911, 881; mass spectrum (EI): m/e (%relative intensity) 201 (17) M$^+$, 156 (100), 129 (17), 115 (20), 104 (54), 91 (17), 77 (17); m/e calcd for $C_{12}H_{11}NO_2$ 201.0790, found 201.0784.

Waste Disposal Information.

All toxic materials were disposed of in accordance with "Prudent Praactices in the Laboratory"; National Academy Press; Washington, DC, 1995.

3. Discussion.

A.   Preparation.

The level of purity during preparations of allenamides was unambiguously established using GC analysis for both the propargyl amide intermediates and the allenamides. The level of optical purity was established based on optical purity of the chiral auxiliaries from commercial sources. The $[\alpha]_D^{20}$ values for the commercial (R)-2-phenyloxazolidinone and (S)-2-phenyloxazolidinone that were used for this preparation are – 52.33 (c 2.0, $CHCl_3$) and +52.95 (2.0, $CHCl_3$), respectively. Based on the known $[\alpha]_D^{20}$ values for (R)-2-phenyloxazolidinone and (S)-2-phenyloxazolidinone from Aldrich (-48.0 (c 2.0, $CHCl_3$) and + 48.0(c 2.0, $CHCl_3$), respectively), the ee or optical integrity of these two enantiomeric auxiliaries should be very high. Since it is not likely to severely erode the ee of the auxiliary under the rather mild reaction conditions described in these two procedures, the level of *ee* for propargyl amide (*R*)-**2** and allenamide (*R*)-**3** should be comparable to that of the starting (*R*)-2-phenyloxazolidinone.

Furthermore, using (*S*)-2-phenyloxazolidinone led to the propargyl amide (*S*)-**2a** that has an opposite optical rotation of + 157.97 (c 10.0, EtOH). After based-induced isomerization, the allenamide (*S*)-**3a** also possesses an opposite optical rotation of + 157.13 (c 10.0, CH_2Cl_2).

B.   Applications.

1. *Stereoselective inverse-demand hetero (4 + 2) cycloadditions.* A Chiral Template for C-Aryl Glycoside Synthesis. Chiral allenamides[2,3,4] had been used in highly stereoselective inverse-demand hetero (4 + 2) cycloaddition reactions with heterodienes.[5] These reactions lead to stereoselective synthesis of highly functionalized pyranyl heterocycles. Further elaboration of these cycloadducts provides a unique entry to C-aryl-glycosides and pyranyl structures that are common in other natural products (**Scheme 1**).

Scheme 1

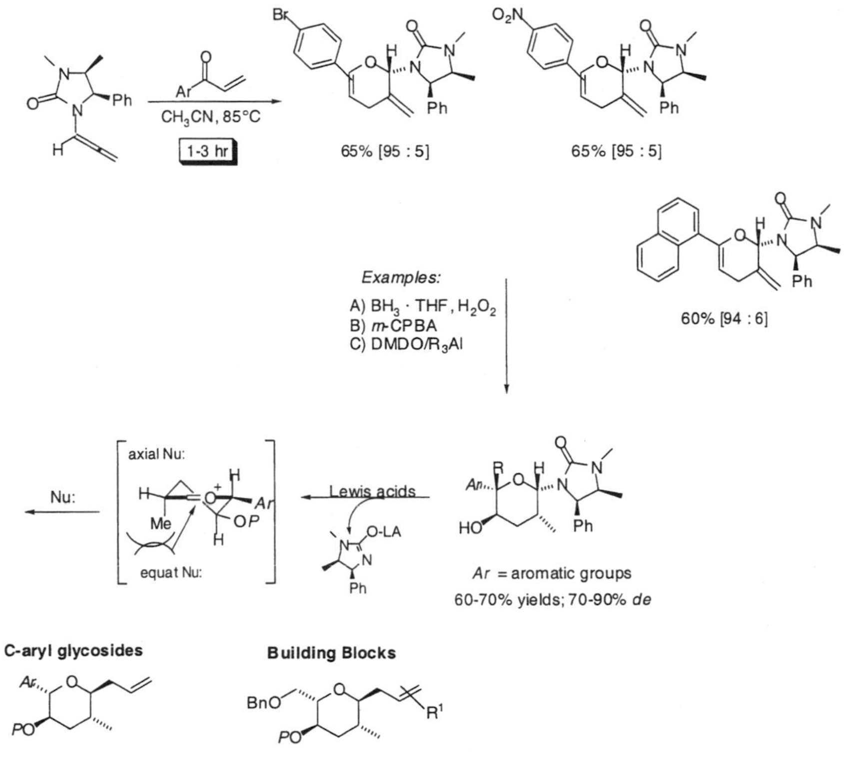

Examples:

A) $BH_3 \cdot THF$, H_2O_2
B) m-CPBA
C) DMDO/R_3Al

65% [95 : 5]   65% [95 : 5]

60% [94 : 6]

axial Nu:

equat Nu:

Nu:

Lewis acids

Ar = aromatic groups
60-70% yields; 70-90% de

C-aryl glycosides   **Building Blocks**

P = protecting groups
Lewis Acid: 1.5 equiv $SnBr_4$; **Nu:** = allylsilanes
56-75% yields; up to 55% to 80% de

2. Highly stereoselective [4 + 3] oxyallyl cycloadditions.

An endo-Selective Sequential Epoxidation-Oxyallyl Cycloaddition and the First Nitrogen-Stabilized Oxyallyl Cations.

Epoxidations of chiral allenamides lead to chiral nitrogen-stabilized oxyallyl catioins that undergo highly stereoselective (4 + 3) cycloaddition reactions with electron-rich dienes.[6] These are the first examples of epoxidations of allenes, and the first examples of chiral nitrogen-stabilized oxyallyl cations. Further elaboration of the cycloadducts leads to interesting chiral amino alcohols that can be useful as ligands in asymmetric catalysis (**Scheme 2**).

Scheme 2

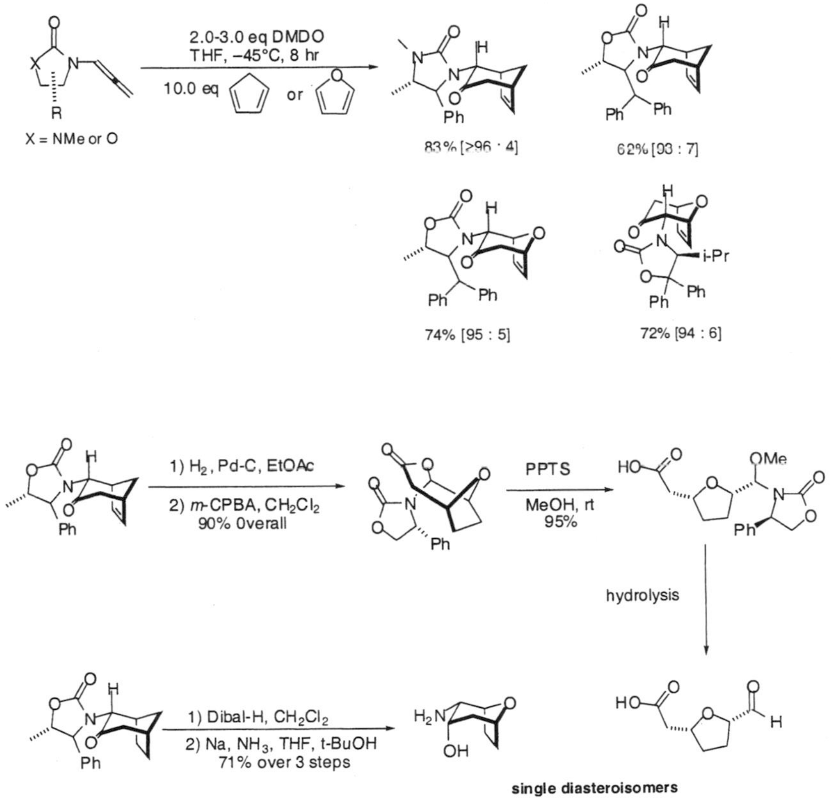

X = NMe or O
2.0-3.0 eq DMDO
THF, −45°C, 8 hr
10.0 eq
or
83% [>96 : 4]
62% [93 : 7]
74% [95 : 5]
72% [94 : 6]
1) H_2, Pd-C, EtOAc
2) m-CPBA, CH_2Cl_2
90% Overall
PPTS
MeOH, rt
95%
hydrolysis
1) Dibal-H, CH_2Cl_2
2) Na, NH_3, THF, t-BuOH
71% over 3 steps
single diasteroisomers

1. Department of Chemistry, University of Minnesota, Minneapolis, MN 55414, USA.

2. (a) Saalfrank, R. W.; Lurz, C. J., in *Methoden Der Organischen Chemie (Houben-Weyl)*, Kropf, H. and Schaumann, E., Eds. Georg thieme Verlag: Stuttgart, **1993**, 3093. (b) Schuster, H. E.; Coppola, G. M. *Allenes in Organic Synthesis*, John Wiley and Sons: New York, **1984**.

3. First preparations of allenamides or electron deficient allenamines: (a) Dickinson, W. B.; Lang, P. C. *Tetrahedron Lett.* **1967**, *8*, 3035. (b) Overman, L. E.; Marlowe, C. K.; Clizbe, L. A. *Tetrahedron Lett.* **1979**, *20*, 599.

4. (a) Gardiner, M.; Grigg, R.; Sridharan, V.; Vicker, N. *Tetrahedron Lett.* **1998**, *39*, 435, and references cited therein. (b) Noguchi, M.; Okada, H.; Watanabe, M.; Okuda, K.; Nakamura, O. *Tetrahedron* **1996**, *52*, 6581. (c) Farina, V.; Kant, J. *Tetrahedron Lett.* **1992**, *33*, 3559 and 3563. (d) Kimura, M.; Horino, Y.; Wakamiya, Y.; Okajima, T.; Tamaru, Y. *J. Am. Chem. Soc.* **1997**, *119*, 10869.

5. (a) Wei, L.-L.; Xiong, H.; Douglas, C. J.; Hsung, R. P. *Tetrahedron Lett.* **1999**, *40*, 6903. (b) Wei, L.-L.; Hsung, R. P.; Xiong, H.; Mulder, J. A.; Nkansah, N. T. *Organic Lett.* **1999**, *1*, 2145. (c) Xiong, H.; Hsung, R. P.; Wei, L.-L.; Berry, C. R.; Mulder, J. A.; Stockwell, B. *Organic Lett.* **2000**, *2*, 2869.

6.   (a) Xiong, H.; Hsung, R. P.; Berry, C. R.; Rameshkumar, C. *J. Am. Chem. Soc.* **2001**, *123*, 7174. (b) Rameshkumar, C.; Xiong, H.; Tracey, M. R.; Berry, C. R.; Yao, L. J.; Hsung, R. P. *J. Org. Chem.* **2002**, *67*, 1339.

Appendix
Chemical Abstracts Nomenclature (Registry Number)

(R)-4-Phenyl-2-oxazolidinone: 2-Oxazolidinone, 4-phenyl-, (4R)-; (90319-52-1).

Sodium hydride: Sodium hydride (NaH); (7646-69-7).

Propargyl bromide: 1-Propyne, 3-bromo-; (106-96-7).

R-4-Phenyl-3-(2-propynyl)-2-oxazolidinone: 2-Oxazolidinone, 4-phenyl-3-(2-propynyl); (4R)-; (256382-74-8).

R-4-Phenyl-3-(1,2-propadieny)-2-oxazolidinone: 2-Oxazolidinone, 4-phenyl-3-(1,2-propadienyl)-, (4R)-; (256382-50-0).

Potassium tert-butoxide: 2-Propanol, 2-methyl-, potassium salt; (865-47-4).

GENERATION OF NONRACEMIC 2-(*t*-BUTYLDIMETHYLSILYLOXY)-3-BUTYNYLLITHIUM FROM (*S*)-ETHYL LACTATE: (*S*)-4-(*t*-BUTYLDIMETHYLSILYLOXY)-2-PENTYN-1-OL

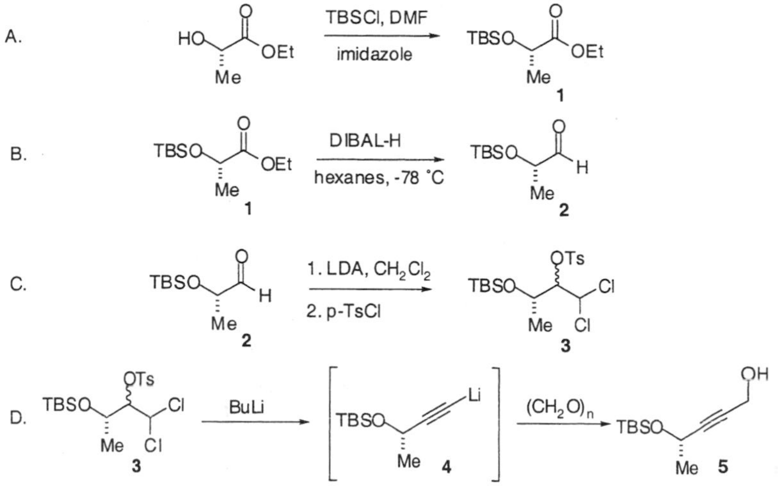

Submitted by James A. Marshall, Mathew M. Yanik, Nicholas D. Adams, Keith C. Ellis, and Harry R. Chobanian.[1]

Checked by Venugopal Gudipati, Tiffany Turner, and Dennis P. Curran.

1. Procedure

A.   *(S)-Ethyl 2-(t-Butyldimethylsilyloxy)propanoate (1).* A 2-L, two-necked, round- bottomed flask equipped with a mechanical stirrer and inert gas inlet (Note 1) is charged with (*S*)-ethyl lactate (118 g, 1.0 mol), 500 mL of dimethylformamide (DMF), and imidazole (102 g, 1.5 mol) (Note 2). The solution is cooled in a ice bath and *tert*-butyldimethylsilyl chloride   (TBDMSCl) (150 g, 1.0 mol) is added in three 50-g portions, at intervals of 30 min between each addition. After the addition of the third portion, a white precipitate forms. The ice bath is allowed to melt gradually overnight. After 18 hr, the reaction mixture is diluted with 300 mL of water and 500 mL of hexanes. The aqueous phase is separated and extracted with 300 mL of hexanes, and the combined hexane extracts are washed with three 50-mL portions of saturated brine, dried over MgSO$_4$, filtered, and concentrated by rotary evaporation to afford 240 g (103%) of the TBDMS ether as a colorless liquid. The product is distilled under vacuum (bp 70-78°C, 0.5 mm; bath temperature 95-105°C) (Note 3) to afford 222 g (96%) of ester **1** as a colorless liquid (Notes 4, 5).

B.   *(S)-2-(t-Butyldimethylsilyloxy)propanal* (**2**). A 2-L, single-necked, round-bottomed flask equipped with a large magnetic stirbar, rubber septum, and inert gas inlet (Note 1) is charged with (*S*)-ethyl 2-(*t*-butyldimethylsilyloxy)propanoate (**1**) (69.9 g, 300 mmol) and 600 mL of hexanes, and cooled in a dry ice-acetone bath at −78°C. A 500-mL, round-bottomed flask equipped with a rubber septum and inert gas inlet is charged with 310 mL of DIBAL-H (1.0M in hexanes, 310 mmol) (Note 6) and cooled to −78°C. The DIBAL-H solution is transferred by cannula into the well-stirred solution of ester over 20 to 25 min. After completion of the addition, the reaction mixture is stirred for 1 hr at −78°C and then quenched by addition of 30 mL of MeOH and stirred for 15 min at −78°C. The cold solution is transferred to a 2-L flask equipped with a mechanical stirrer containing 600 mL of saturated Rochelle salt and the resulting mixture is vigorously stirred for 3.5 hr (Note 7). The aqueous phase is separated and extracted with 200 mL of hexanes, and the combined organic extracts are washed with 100 mL of saturated brine, dried over MgSO$_4$, filtered, and concentrated by rotary evaporation (Note 8). After removal of the solvent, the residue is distilled (bp 45-52°C, 0.5 mm, bath temperature 65-80°C) to afford 50.3-50.9 g (91-92%) of aldehyde **2** as a colorless oil (Notes 9, 10).

158

C. *1,1-Dichloro-(3S)-(t-butyldimethylsilyloxy)-2-butyl p-toluenesulfonate* (**3**). A 2-L, single-necked, round-bottomed flask equipped with a large magnetic stirbar and a 500-mL pressure equalizing addition funnel (Note 1) is charged with aldehyde **2** (44.3 g, 235 mmol), dichloromethane (45.0 mL, 705 mmol), and 500 mL of tetrahydrofuran (Note 11) and the mixture is cooled in a dry ice-acetone bath at −78°C. A 500-mL, three-necked, round-bottomed flask equipped with a magnetic stirbar, a graduated 150-mL pressure-equalizing addition funnel, and two rubber septa is charged with 250 mL of THF and diisopropylamine (49.3 mL, 376 mmol) and then cooled with an ice bath. The addition funnel on the 500-mL flask is charged with 141 mL of BuLi solution (2.5M in hexanes, 352 mmol) which is added dropwise to the solution of amine over 30 min. The resulting solution of LDA is transferred by cannula to the addition funnel on the 2-L flask, and this solution is added dropwise over 1 hr to the reaction mixture, resulting in a color change to light yellow. After 30 min, the mixture is warmed to 0°C for 30 min with a change in color to dark brown. The addition funnel is replaced with a rubber septum, and *p*-toluenesulfonyl chloride (44.8 g, 235 mmol) is added in one portion through a funnel. After 10 min, the ice bath is removed and the mixture is stirred for 1.5 hr. The reaction is quenched by addition of 10 mL of water, stirred for 30 min, and then transferred to a separatory funnel and extracted with 250 mL of 10% HCl and 150 mL of 1N NaOH solution. Each of these aqueous extracts is separately back-extracted with 500 mL of Et_2O, and the combined organic phases are washed with 100 mL of brine, dried over $MgSO_4$, and filtered into a 1-L, round-bottomed flask for use in the next step. The solvent is removed under reduced pressure whereupon 80-85 g of the dichloro tosylate **3** is obtained as a dark brown oil which is used in the next step without purification (Note 12).

D. *(S)-4-(t-Butyldimethylsilyloxy)-2-pentynol* (**5**). THF (500 mL) is added to tosylate **3** (76 g, *ca* 177 mmol) prepared as described above, and the 1-L flask is equipped with a large magnetic stirbar and a pressure-equalizing addition funnel fitted

159

with a nitrogen inlet. The solution is cooled in a dry ice-acetone bath at −78°C for 20 min and the addition funnel is charged with 219 mL of BuLi solution (2.5M in hexanes, 548 mmol) which is then added dropwise over ca. 1 hr to the reaction mixture. When the addition is complete, the solution is stirred for 30 min at −78°C and warmed over 1 hr to 0°C. The reaction mixture is then recooled to −78°C. To the resulting black solution of lithium acetylide **4** is then added paraformaldehyde (10.6 g, 354 mmol) (Note 13) in one portion. After 15 min, the mixture is allowed to warm to room temperature over 4 hr during which time the suspension of paraformaldehyde gradually dissolves. The reaction is quenched by addition of 200 mL of aqueous ammonium chloride solution, and the aqueous layer is separated and extracted with three 200-mL portions of Et_2O. The combined organic extracts are washed with 100 mL of 1N NaOH and 50 mL of brine, dried over $MgSO_4$, filtered, and concentrated under reduced pressure to give the desired product contaminated with *tert*-butyldimethylsilanol. The product is purified by fractional distillation (Note 14) to give 21-24 g (44-53% overall from aldehyde **2**) of the alcohol **5** as a colorless oil (Notes 15, 16).

2. Notes

1.   All glassware and needles were dried in an oven at 120°C overnight and assembled under a nitrogen purge or flame-dried immediately prior to use. All reactions were performed under nitrogen (submitters) or argon (checkers).

2.   (*S*)-Ethyl lactate (98%) and imidazole (99+%) were purchased from Aldrich Chemical Co. TBDMSCl was purchased from FMC Corporation (submitters) or Aldrich (checkers). A newly opened bottle of dimethylformamide (ACS Reagent grade, 0.02% water) was used as received.

3.   The distillation was conducted in a 500-mL, round bottomed flask equipped with a magnetic stirring bar and a variable take-off distillation head.

4.   The ester **1** displayed the following properties: $[\alpha]_D$ −25.9 (c 1.56, $CHCl_3$); IR (thin film): cm^{-1} 1753; ^1H NMR (300 MHz, $CDCl_3$): δ 0.05 (s, 3H), 0.08 (s, 3H), 0.88 (s, 9H), 1.23 (t, J = 7.2 Hz, 3H), 1.36 (d, J = 6.6 Hz, 3H), 4.11 (m, 1H), 4.25 (q, J = 7.2 Hz, 2H); ^{13}C NMR (75 MHz,$CDCl_3$): δ −5.3, −4.9, 14.2, 18.3, 21.3, 25.7, 60.7, 68.4, 174.1.

5.   The submitters report that by the same procedures without modifications the (*R*)-enantiomer can be prepared from (*R*)-isobutyl lactate, available from Sigma Chemical Co.   The physical properties for (*R*)-isobutyl 2-(*t*-butyldimethylsilyloxy)-propanoate are: bp 85-88°C, 0.1 mm; $[\alpha]_D$ +28.7 (c 1.61, $CHCl_3$); IR (thin film) 1763 cm^{-1}; ^1H NMR (300 MHz, $CDCl_3$) δ 0.06 (s, 3H), 0.09 (s, 3H), 0.89 (s, 9H), 0.91 (d, J = 6.6 Hz, 6H), 1.38 (d, J = 6.6 Hz, 3H), 1.94 (septet, 1H), 3.88 (dq, J = 10.5, 6.9 Hz, 2H), 4.28 (q, J = 6.6 hz, 1H); ^{13}C NMR (125 MHz, $CDCl_3$) δ −5.4, −5.0, 18.3, 19.0, 21.4, 25.7, 27.7, 68.4, 70.8, 174.1

6.   DIBAL-H (1.0M in hexanes) was purchased from Aldrich Chemical Co. Hexanes from a freshly opened bottle (ACS Reagent Grade) was used as solvent.

7.   Potassium sodium tartrate (Rochelle salt) was purchased from Fluka (purum p.a. grade).   The quantity of salt solution specified was found to be optimal (2 mL/mmol of DIBAL-H).   Use of less salt resulted in incomplete complexation.   The submitters reported similar yields with stirring overnight.

8.   Concentration is carried out without heating.   Heating the rotary evaporator bath above 35°C results in lower yields due to the volatility of the aldehyde.

9.   The aldehyde contains small amounts of the starting ester and the overreduced alcohol along with other minor impurities.   It can be stored for short periods of time (1-2 days) in a freezer at −20°C without significant deterioration. However long term storage is not recommended.

10.   The enantiomeric excess of aldehyde **2** was estimated to be >96% by derivatization as the Schiff bases with (*S* and *R*)-α-methylbenzylamine as described below. Spectral characteristics for aldehyde **2**: $[\alpha]_D$ −12.1 (c 1.96, $CHCl_3$); IR (thin film): cm^{-1} 1741; ${}^1H$ NMR (300 MHz, $CDCl_3$): δ 0.07 (s, 3H), 0.08 (s, 3H), 0.9 (s, 9H), 1.26 (d, J = 6.9 Hz, 3H), 4.07 (dq, J = 6.9, 1.3 Hz, 1H), 9.59 (d, J = 1.3 Hz, 1H); ${}^{13}C$ NMR (125 MHz, $CDCl_3$): δ −4.8, 18.1, 18.5, 25.6, 73.8, 204.2.

(S) and (R)-α-Methylbenzylimines of (S)-4-(t-Butyldimethylsilyloxy)propanal.

To a solution of (*S*)-aldehyde **2** (0.150 g, 0.74 mmol), 4Å molecular sieves (0.100 g), in 1 mL of CH_2Cl_2 was added (*S*)-α-methylbenzylamine (100 μL, 0.77 mmol, Aldrich, 98% ee by GLC analysis). After 2 hr, the mixture was filtered through Celite, rinsed with CH_2Cl_2 and concentrated affording the (*S*,*S*)-imine as a colorless oil (0.215 g, 95%). The imine from (*R*)-α-methylbenzylamine (96% ee by GLC analysis) was prepared in identical fashion. The ratio of diastereoisomers determined through integration of the ${}^1H$ NMR spectra was 98:2 for the (*S*,*S*) derivative and 96.5:3.5 for the (*S*,*R*) derivative. **(*S*,*S*) Imine**: $[\alpha]_D$ −55.2 (c 1.20, $CHCl_3$); IR (thin film) 2968, 2951, 2864, 1675 cm^{-1}; ${}^1H$ NMR ($CDCl_3$, 300 MHz): δ 0.11 (s, 6H), 0.93 (s, 9H), 1.26, (d, J = 6.3 Hz, 3H), 1.50 (d, J = 6.6 Hz, 3H), 4.29 (q, J = 6.6 Hz, 1H), 4.38 (q, J = 6.3 Hz, 1H), 7.24-7.34 (m, 5H), 7.62 (d, J = 5.4 Hz, 1H); ${}^{13}C$ NMR (125 MHz, $CDCl_3$): δ −4.7, −4.6, 18.2, 21.8, 24.3, 25.8, 68.9, 70.7, 126.5, 126.8, 128.4, 144.6, 166.4. **(*S*,*R*) Imine**: $[\alpha]_D$ +46.2 (c 1.20 $CHCl_3$); ${}^1H$ NMR: δ −0.03 (s, 3H), 0.03 (s, 3H), 0.85 (s, 9H), 1.29 (d, J = 6.6 Hz, 3H), 1.48 (d, J = 6.6 Hz, 3H), 4.27 (q, J = 6.6 Hz, 1H), 4.32 (qd, J = 6.6 Hz, 5.4 Hz, 1H), 7.23-7.40 (m, 5H), 7.59 (d, J = 5.4 Hz, 1H); ${}^{13}C$ NMR (125 MHz, $CDCl_3$): δ −4.8, −4.7, 18.1, 21.7, 24.2, 25.8, 68.9, 70.7, 126.5, 126.8, 128.3, 144.4, 166.2.

11.   THF (99.9% anhydrous, inhibitor free) and CH_2Cl_2 (99.8% anhydrous) were obtained from Aldrich Chemical Co. and used as received. The submitters distilled diisopropylamine and stored it over KOH. The checkers used 99.5%

diisopropylamine as received from Aldrich. *p*-Toluenesulfonyl chloride (99%) was obtained from Acros Chemical Co. (submitters) or Avocado (checkers).

12. This material consisted of a 45:55 mixture of diastereomers based on ^{1}H NMR analysis. $R_f = 0.53$ (10% EtOAc/hexanes, phosphomolybdic acid stain). Spectral characteristics for tosylate **3**: ^{1}H NMR (300 MHz, CDCl$_3$): δ 0.08 (s, 6H), 0.09 (s, 6H), 0.88 (s, 18H), 1.24 (d, J = 6.3 Hz, 3H), 1.29 (d, J = 6.0 Hz, 3H), 2.44 (s, 6H), 4.04 (dq, J = 6.0, 7.5 Hz, 1H), 4.34 (dq, J = 3.3, 6.3 Hz, 1H), 4.70 (dd, J = 3.3, 6.0 Hz, 1H), 4.74 (dd, J = 2.1, 7.2 Hz, 1H), 5.81 (d, J = 6.3 Hz, 1H), 6.00 (d, J = 2.1 Hz, 1H), 7.31 (d, J = 8.4 Hz, 4H), 7.82 (dd, J = 5.4, 8.4 Hz, 4H).

13. Paraformaldehyde (95%) was obtained from Aldrich Chemical Co. (submitters) or Baker (checkers) and was dried azeotropically with benzene by concentrating a benzene solution (100 mL per gram of paraformaldehyde) at 45°C by rotary evaporation, repeating this process, and then drying the residue under high vacuum overnight. Dry paraformaldehyde was stored in a sealed container under argon. Butyllithium was obtained from Acros Chemical Co (submitters) or Aldrich (checkers).

14. The submitters reported that an immediate preliminary distillation is advisable to minimize contact time with the dark polymeric byproducts which results in decomposition and lowered yields. This is achieved by means of a Kugelrohr distillation apparatus preheated to 80°C. Care should be exercised to prevent bumping in the early stage of this distillation. However, the checkers had problems with bumping and preferred direct fractional distillation according to the following procedure. After concentration, the flask containing the crude product was equipped with a 30 x 1.5 cm distillation column and evacuated at 0.1 mm. After an initial period of distillation at room temperature to remove residual solvent, a first fraction of bp up to 45°C was collected; this is believed to be TBSOH. The main product fraction (21-24 g) was collected at bp 79-82°C.

15. The ee of alcohol **5** was determined to be >97% by derivatization with a chiral silylating reagent (Note 16). Physical characteristics for alcohol **5**: $R_f = 0.42$ (25% EtOAc/hexanes, phosphomolybdic acid); $[\alpha]_D$ –53.0 (c 1.42, $CHCl_3$); IR (thin film): cm^{-1} 3370; ^{1}H NMR (300 MHz, $CDCl_3$): δ 0.09 (s, 3H), 0.11 (s, 3H), 0.89 (s, 9H), 1.39 (d, J = 6.5 Hz, 3H), 2.44 (t, J = 6.2 Hz, 1H), 4.27 (dd, J = 6.2, 1.7 Hz, 2H), 4.54 (tq, J = 6.5, 1.7 Hz, 1H); ^{13}C NMR (75 MHz, $CDCl_3$): δ –5.0, –4.7, 18.2, 25.2, 25.7, 50.8, 59.0, 81.4, 88.0

16. The enantiomeric purity of alcohol **5** was determined by conversion to the silyl ether **8** via the following sequence:

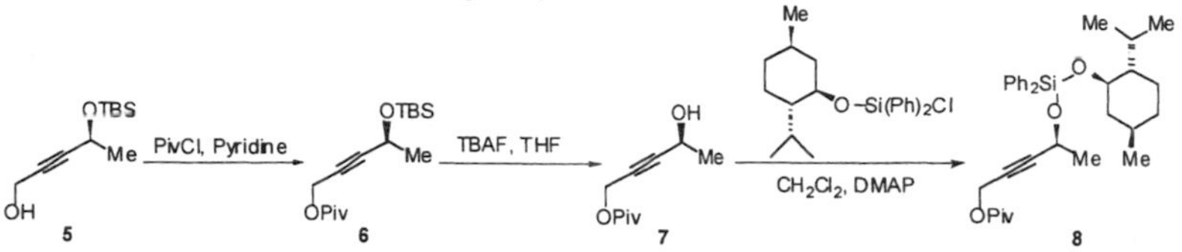

To a solution of (S)-pentynol **5** (0.107 g, 0.50 mmol) in pyridine (2.4 mL) is added trimethylacetyl chloride (PivCl, 0.1 mL, 0.8 mmol) (Note 17). After 3 hr, ice (ca 3 g) is added and the mixture is stirred vigorously. After 1 hr, 10% aqueous HCl (5 mL) is added and the resulting mixture is extracted with three 10-mL portions of ethyl acetate. The combined organic extracts are washed with 20 mL of brine, dried over $MgSO_4$, filtered, and concentrated under reduced pressure. The residue is purified by chromatography on silica gel (elution with 1% EtOAc/hexanes) to afford 0.129 g (87%) of ester **6** (Note 18).

To a solution of (S)-ester **6** (0.566 g, 1.90 mmol) in 20 mL of THF at 0°C is added tetrabutylammonium fluoride (3.0 mL, 3.0 mmol) (Note 19). After 15 min, the reaction mixture is warmed to room temperature. After 3 hr, 20 mL of aqueous saturated NH_4Cl is added and the mixture is extracted with three 10-mL portions of

Et$_2$O. The combined organic extracts are washed with 20 mL of brine, dried over MgSO$_4$, filtered, and concentrated under reduced pressure. The residue is purified by chromatography on silica gel (elution with 20% EtOAc/hexanes) to afford 0.271 g (78%) of alcohol **7** (Note 20).

To a solution of (S)-alcohol **7** (0.052 g, 0.28 mmol) in 2.8 mL of CH$_2$Cl$_2$ is added (–)-chloromenthyloxydiphenylsilane (0.105 g, 0.28 mmol) (Note 21) followed by DMAP (0.035 g, 0.29 mmol). After 2 hr, the mixture is concentrated and the residue is purified by chromatography on silica gel (elution with 1% EtOAc/hexanes) to afford 0.112 g (76%) of silylmenthol derivative **8**. The identical procedures are employed with the (R)-pentynol (Note 22).

17.   Pyridine was freshly distilled from calcium hydride and stored under nitrogen over potassium hydroxide. Trimethylacetyl chloride (99%) was obtained from Aldrich Chemical Company and used as received.

18.   Physical characteristics of **(S)-ester 6**: R$_f$ = 0.83 (25% EtOAc/hexanes, cerium molybdate); [α]$_D$ –36.0 (c 0.79, CHCl$_3$); IR (thin film): cm^{-1} 2968, 1745; ^{1}H NMR (500 MHz, CDCl$_3$): δ 0.11 (s, 3H), 0.13 (s, 3H), 0.90 (s, 9H), 1.21 (s, 9H), 1.40 (d, J = 6.5 Hz, 3H), 4.55 (qt, J = 6.0, 2.0 Hz, 1H), 4.68 (d, J = 2.0 Hz, 2H); ^{13}C NMR (125 MHz, CDCl$_3$): δ –5.0, –4.6, 18.2, 25.1, 25.8, 27.1, 38.7, 52.3, 58.9, 77.5, 88.9, 177.7.

19.   Tetrabutylammonium fluoride (1.0M in THF) was obtained from Aldrich Chemical Company and used as received.

20.   Physical characteristics of **(S)-alcohol 7**:   R$_f$ = 0.30 (25% EtOAc/hexanes, cerium molybdate); IR (thin film): cm^{-1} 3440, 2986, 1745; [α]$_D$ –18.1 (c 1.2, CHCl$_3$); ^{1}H NMR (500 MHz, CDCl$_3$): δ 1.22 (s, 9H), 1.46 (d, J = 7.0 Hz, 3H), 4.57 (dq, J = 1.5 Hz, 6.5 Hz, 1H), 4.69 (s, 2H); ^{13}C NMR (125 MHz, CDCl$_3$): δ 24.0, 27.0, 38.7, 52.2, 58.3, 78.3, 88.2, 177.9.

21.   The (−)-chloromenthyloxydiphenylsilane was prepared from (−)-menthol (Aldrich, >99% ee) according to the published procedure : Weibel, D. B.; Walker, T. R.; Schroeder, F. C.; Meinwald, J. *Org. Lett.* **2000**, *2*, 2381.

22.   Diagnostic ^{1}H NMR peaks (singlets) are located at δ 4.62 (*S*) and δ 4.57 (*R*) ppm, respectively, in the ^{1}H NMR spectra of the crude products prior to chromatography. Physical characteristics of the **(S)-silylmenthyl derivative 8**: R_f = 0.67 (25% EtOAc/hexanes, cerium molybdate); $[\alpha]_D$ −73.5 (*c* 0.63, CHCl$_3$); IR (thin film): cm^{-1} 2960, 1745; ^{1}H NMR (500 MHz, CDCl$_3$): δ 0.56 (d, J = 6.9 Hz, 3H), 0.87 (d, J = 6.3 Hz, 3H), 0.90 (d, J = 7.2 Hz, 3H), 0.80-0.94 (m, 2H), 1.22 (s, 9H), 1.12-1.33 (m, 3H), 1.42 (d, J = 6.6 Hz, 3H), 1.56-1.62 (m, 2H), 2.09 (m, 1H), 2.33 (m, 1H), 3.63 (dt, J = 10.2, 4.2 Hz, 1H), 4.62 (d, J = 1.5 Hz, 2H), 4.72 (dq, J = 6.6, 1.5 Hz, 1H), 7.33-7.46 (m, 6H), 7.65-7.70 (m, 4H); ^{13}C NMR (125 MHz CDCl$_3$): δ 15.6, 21.3, 22.3, 22.6, 24.9, 25.3, 27.1, 31.5, 34.4, 38.7, 45.2, 50.0, 52.3, 59.2, 73.5, 77.8, 88.4, 127.6, 130.1, 133.1, 133.5, 135.1, 177.8. **(R)-silylmenthyl derivative**: $[\alpha]_D$ +1.1 (*c* 0.94, CHCl$_3$); IR (thin film): cm^{-1} 2957, 1739; ^{1}H NMR (300 MHz, CDCl$_3$): δ 0.59 (d, J = 7.2 Hz, 3H), 0.85 (d, J = 6.3 Hz, 3H), 0.91 (d, J = 7.2 Hz, 3H), 0.79-0.92 (m, 2H), 1.12 (apparent q, J = 12.0 Hz, 1H) 1.21 (s, 9H), 1.24-1.34 (m, 2H), 1.46 (d, J = 6.3 Hz, 3H), 1.56-1.62 (m, 2H), 2.02 (m, 1H), 2.38 (m, 1H), 3.64 (dt, J = 10.2, 4.2 Hz, 1H), 4.57 (d, J = 1.5 Hz, 2H), 4.73 (qt, J = 6.6, 1.5 Hz, 1H), 7.33-7.45 (m, 6H), 7.64-7.72 (m, 4H); ^{13}C NMR (125 MHz CDCl$_3$): δ 15.6, 21.3, 22.2, 22.6, 24.9, 25.3, 27.1, 31.5, 34.4, 38.7, 45.2, 50.0, 52.3, 59.2, 73.6, 77.8, 88.3, 127.6, 130.1, 133.3, 135.1, 177.7.

Waste Disposal Information

All toxic materials were disposed of in accordance with "Prudent Practices in the Laboratory"; National Academy Press; Washington, DC, 1995.

166

3. Discussion

The present route to (S)-4-(t-butyldimethylsilyloxy)-2-pentyn-1-ol is based on a procedure for the preparation of terminal alkynes reported by a group from the Chemical Process Department at the DuPont Pharmaceutical Company.[2]   This alcohol serves as a convenient starting material for the preparation of 1-acyloxy 4-mesylates **10** (eq 1).

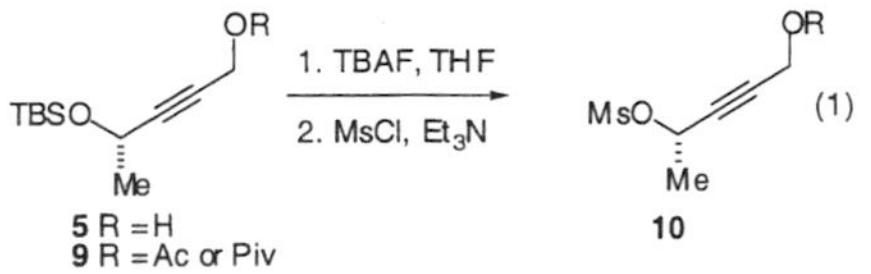

These mesylates, in turn, can be converted to enantioenriched allenyltin, zinc, and indium reagents which add to aldehydes with excellent diastereo-and enantioselectivity to afford either syn- or anti-homopropargylic alcohols or allenylcarbinols (eq 2, 3, and 4). [3,4] Adducts of this type serve as useful intermediates for the synthesis of polyketide and hydrofuran natural products.[5]

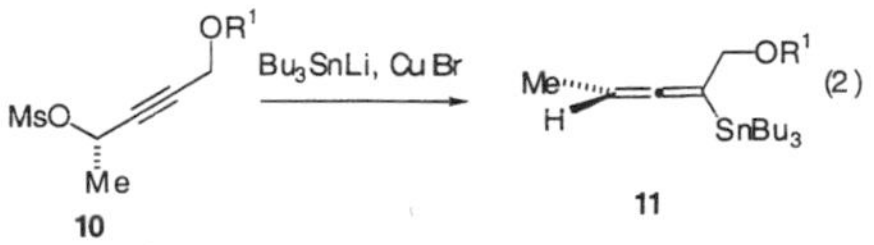

167

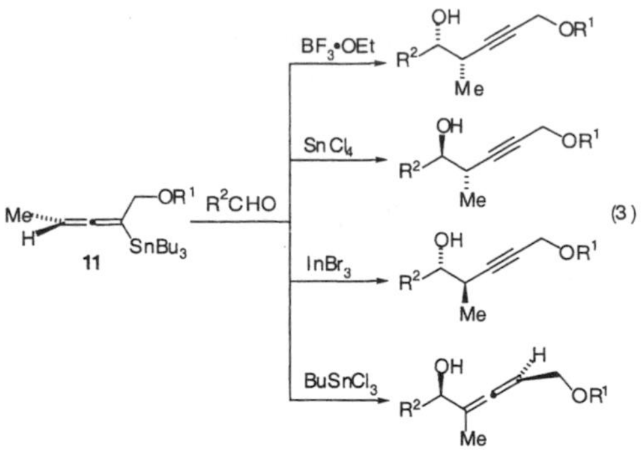

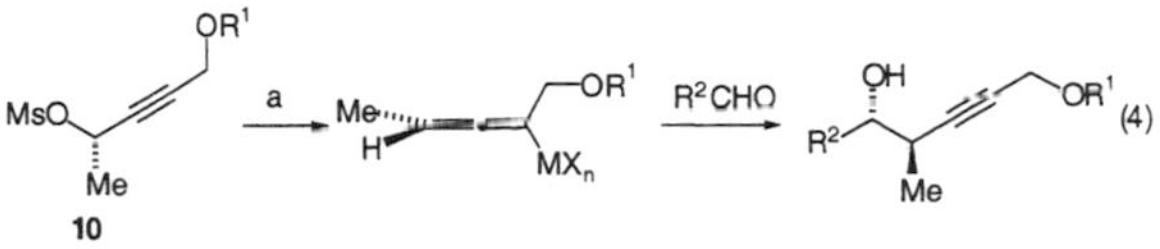

$$BF_3 \cdot OEt$$

$$SnCl_4$$

$$(3)$$

$$InBr_3$$

$$BuSnCl_3$$

11

a Pd(OAc)$_2$, Ph$_3$P, Et$_2$Zn (MX$_n$ = ZnOMs); Pd(dppf)$_2$, InI (MX$_n$ = In(OMs)I

10

$$(4)$$

$$R^2CHO$$

Previous syntheses of terminal alkynes from aldehydes employed Wittig methodology with phosphonium ylides and phosphonates.[6,7] The DuPont procedure circumvents the use of phosphorus compounds by using lithiated dichloromethane as the source of the terminal carbon. The intermediate lithioalkyne 4 can be quenched with water to provide the terminal alkyne or with various electrophiles, as in the present case, to yield propargylic alcohols, alkynylsilanes, or internal alkynes. Enantioenriched terminal alkynylcarbinols can also be prepared from allylic alcohols by Sharpless epoxidation and subsequent basic elimination of the derived chloro- or bromomethyl epoxide (eq 5). A related method entails Sharpless asymmetric dihydroxylation of an allylic chloride and base treatment of the acetonide derivative.[8] In these approaches the product and starting material contain the same number of carbons.

$$R \diagdown\diagdown \diagup^{OH} \xrightarrow{\text{Sharpless AE}} R \diagdown\overset{O}{\diagup}\diagdown X \quad \begin{matrix} X = OH \\ X = Cl \end{matrix}$$

(5)

1. Department of Chemistry, McCormick Road, P.O. Box 400319, University of Virginia, Charlottesville, VA 2290, E-mail: jam5x@virginia.edu.

2. Wang, Z.; Yin, J.; Campagna, S.; Pesti, J. A.; Fotunak, J. M. *J. Org. Chem.* **1999**, *64*, 6918.

3. a) Marshall, J. A. *Chem. Rev.* **1996**, *96*, 31. (b) Marshall, J. A. in "Lewis Acids in Organic Synthesis," H. Yamamoto Ed., Vol. 1, Wiley-VCH, Weinheim, **(2000)**. Chapter 10, pp. 453-520.

4. Marshall, J. A. *Chem. Rev.* **2000**, *100*, 3163.

5. a) Marshall, J. A.; Lu, S.-H.; Johns, B. A. *J. Org. Chem.* **1998**, *63*, 817. b) Marshall, J. A.; Palovich, M. R. *J. Org. Chem.* **1998**, *63*, 3701. c) Marshall, J. A.; Johns, B. A. *J. Org. Chem.* **1998**, *63*, 7885. d) Marshall, J. A.; Fitzgerald, R. A. *J. Org. Chem.* **1999**, *64*, 4477. e) Marshall, J. A.; Johns, B. A. *J. Org. Chem.* **2000**, *65*, 1501. f) Marshall, J. A.; Yanik, M. M. *J. Org. Chem.* **2001**, *66*, 1373.

6. a) Horner, L.; Hoffmann, H.; Wippel, H. G.; Klaher, G. *Chem. Ber.* **1959**, 2499. b) Wadsworth, W. S. Jr.; Emmons, W. D. *J. Am. Chem. Soc.* **1961**, *83*, 1733. c) Corey, E. J.; Achiwa,K.; Katzenellenbogen, J. A. *J. Am. Chem. Soc.* **1969**, *91*, 4318. d) Corey, E. J.; Fuchs, P. L. *Tetrahedron Lett.* **1972**, 3769. e) Gilbert, J. C.; Weerasooriya, U. *J. Org. Chem.* **1982**, *47*, 1837. f) Muller, S.; Liepold, B.; Roth, G. J.; Bestmann, H. J. *Syn Lett* **1996**, 521.

7. a) For a previous synthesis of the pivalic ester related to mesylate **10** from *(R)*-methyl lactate and $Ph_3P=CBr_2$, see Marshall, J. A.; Xie, S. *J. Org. Chem.*

1995, *60*, 7230. b) An analogous synthesis of *(S)*-3-butyn-2-ol derivatives has also been reported; Ku, Y.-Y.; Patel, R. R.; Elisseou, E. M.; Sawick, D. P. *Tetrahedron Lett.* **1995**, 36, *2738*.

8.   a) *Cf.* Yadav, J. S.; Chander, M. C.; Rao, C. S. *Tetrahedron Lett.* **1989**, *30*, 5455. b) Yadav, J. S.; Chander, M. C.; Joshi, R. V. *Tetrahedron Lett.* **1988**, *29*, 2737.

Appendix
Chemical Abstracts Nomenclature   (Registry Number)

(S)-Ethyl lactate: Propanoic acid, 2-hydroxy-, ethyl ester, (2S)-; (687-47-8)

Imidazole: 1H-Imidazole; (288-32-4)

tert-Butyldimethylsilyl chloride; Silane, chloro(1,1-dimethylethyl)dimethyl-; (18162-48-6)

(S)-Ethyl 2-(tert-butyldimethylsilyloxy)propanoate: Propanoic acid, 2-[[(1,1-dimethylethyl) dimethylsilyl]oxy], ethyl ester; (106513-42-2)

DIBAL-H: Diisobutylaluminum hydride; Aluminum, hydrobis(2-methylpropyl)-; (1191-15-7)

Rochelle salt: Butanedioic acid, 2,3-dihydroxy-(2R,3R)-, monopotassium monosodium salt ; (304-59-6)

(S)-2-(tert-Butyldimethylsilyloxy)propanal: Propanal, 2-[[(1,1-dimethylethyl)dimethyl silyl]oxy]-, (2S)-; (87727-28-4)

1,1-Dichloro-(3S)- (tert-butyldimethylsilyloxy)-2-butyl p-toluenesulfonate: 2-Butanol, 1,1-dichloro-3-[[(1,1-dimethylethyl)dimethylsilyl]oxy]-, 4-methylbenzene sulfonate, (3S)-; (329914-17-2)

Diisopropylamine: 2-Propanamine, N-(1-methylethyl)-; (109-72-8)

Butyllithium; Lithium, butyl-; (109-72-8)

Lithium diisopropylamide: 2-Propanamine, N-(1-methylethyl)-, lithium salt; (4111-54-0)

p-Toluenesulfonyl chloride: Benzenesulfonyl chloride, 4-methyl-; (98-59-9)

SYNTHESIS OF 9-SPIROEPOXY-*endo*-TRICYCLO[5.2.2.0²,⁶]UNDECA-4, 10-DIEN-8-ONE

(Spiro[4,7-ethano-1H-indene-8,2'-oxiran]-9-one, 3a,4,7a-tetrahydro-)

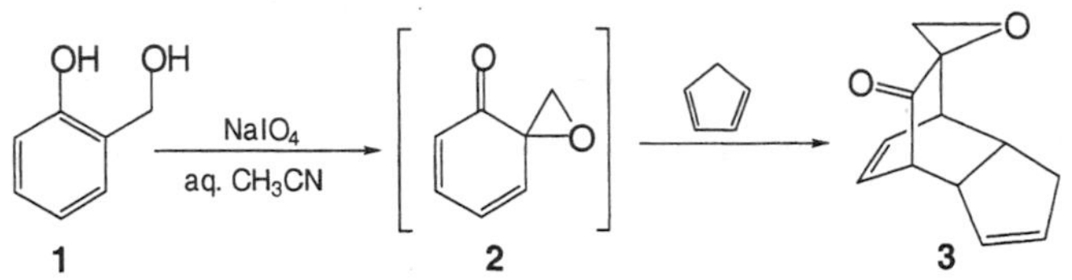

Submitted by Vishwakarma Singh*, Mini Porinchu, Punitha Vedantham and Pramod K. Sahu.[1]

Checked by Aaron Murray and Marvin J. Miller.

1. Procedure

9-Spiroepoxy-endo-tricyclo[5.2.2.0²,⁶]undeca-4,10-dien-8-one. A 2-L, single-necked, round-bottomed flask equipped with a magnetic stirbar is charged with salicyl alcohol 1 (20.0 g, 161 mmol) (Note 1) and 325 mL of acetonitrile (Note 2). The resulting solution is cooled at 0°C in an ice bath while freshly generated cyclopentadiene (20 mL, 15.6 g, 236 mmol) (Note 3) is added in one portion. A pressure-equalizing addition funnel is then attached to the flask and charged with a solution of sodium metaperiodate (60 g, 280 mmol) (Note 4) in 220 mL of water. This solution is added dropwise to the ice-cooled, vigorously stirred reaction mixture over a period of 2 hr (Note 5). The color of the reaction mixture initially turns yellow-orange. After the reaction mixture is stirred at 0°C for an additional 1 hr, additional cyclopentadiene (10 mL) is added and stirring is continued for 30 min at 0°C. The ice bath is then removed and after 30 min an additional 10-mL portion of cyclopentadiene (Note 6) is added and the reaction mixture is stirred overnight at room temperature (Note 7).

Sodium chloride (50 g) is added to the reaction mixture and the suspension is stirred for 1 hr. The reaction mixture is then filtered under vacuum through a 30 x 35 mm pad of Celite in a sintered glass funnel washing with two 50-mL portions of diethyl ether. Two phases form in the filtrate. The organic phase is separated and concentrated by rotary evaporation. The residue is combined with the aqueous phase and extracted with four 100-mL portions of ether, and the combined organic phases are washed with 50 mL of brine, dried over anhydrous sodium sulfate, filtered, and concentrated to afford 32.5 g of an orange-brown oil which is adsorbed onto 25 g of silica gel and charged onto a 430 mm x 45 mm column of 350 g of 60-120 mesh silica gel. The column is eluted with ca. 700 mL of petroleum ether (bp 60-80°C) to remove hydrocarbon impurities, and elution is then continued with 500 mL of 98:2 petroleum ether and then with 95:5 petroleum ether:ethyl acetate to give a small amount of salicylaldehyde (Note 8). Further elution with 3.0 L of 95:5 petroleum ether:ethyl acetate gives pure **3**. The column is further eluted with 90:10 petroleum ether-ethyl acetate to give 3.5 g of starting material (**1**). Recrystallization of **3** from a mixture of 25 mL of ether and 15 mL of petroleum ether gives 18.57-19.5 g (61-64%) of pure **3** as a white solid (Note 9).

2. Notes

1. Salicyl alcohol was purchased from Lancaster, UK or from Fluka and used as received.

2. The submitters purchased acetonitrile (AR grade) from Sisco Research Laboratories, India and used it as received. The checkers obtained acetonitrile from Aldrich Chemical Company.

3. Cyclopentadiene was obtained by pyrolysis of dicyclopentadiene (pract., 90%) which was purchased from Fluka. Pyrolysis was carried out following the previously described general procedure.[2] Dicyclopentadiene (80 mL) was placed in a round-bottomed flask equipped with a magnetic stirbar and a Vigreux column fitted with a distillation head through which cold water was circulated. The contents of the flask were slowly heated with stirring at 160°C in an oil bath, and ca. 60 mL of cyclopentadiene (bp 38-42°C) was collected in a receiver cooled in an ice-salt bath. The cyclopentadiene was used immediately and the residue left in the distillation flask was discarded.

4. Sodium metaperiodate was purchased by the submitters from S.D. Fine Chemicals, India and by the checkers from Aldrich Chemical Company. Occasionally a small amount of some material remains insoluble and is removed by filtration; the checkers observed that it begins to precipitate during the addition. An excess of the reagent was used so that oxidation is as complete as possible.

5. A powerful magnetic stirrer is required since a large amount of inorganic material precipitates after the addition of sodium metaperiodate. The submitters used a 35-mm stirbar to achieve vigorous stirring.

6. An excess of cyclopentadiene was used because of its volatility and tendency to dimerize. Cyclopentadiene was added intermittently in portions so that a sufficient amount is always present in the reaction mixture to efficiently intercept the spiroepoxycyclohexadienone (2) generated in situ.

7. Analysis by thin layer chromatography (TLC) indicates product formation and a small amount of starting material. TLC was performed on Merck Silica gel 60 F-$_{254}$ coated on aluminum sheets. Product 3 has R_f = 0.44 (elution with 70:30 petroleum ether-ethyl acetate; visualization with iodine vapor).

8. Some salicylaldehyde is also formed during the oxidation.

9. The physical properties of **3** were as follows: mp 40-41°C; IR (neat) cm^{-1}: 3058, 2931, 1734; ^{1}H NMR (300 MHz, CDCl$_3$): δ 2.01 (m, 1 H), 2.57 (m, 1 H), 2.63 (m, 1 H), 2.81 (m, 1 H), 3.03 (m, 1 H), 3.08 (m, 1 H), 3.34 (m, 2 H), 5.44 (m, 1 H), 5.69 (m, 1 H), 6.10 (t, superimposed dd, $J_1 = J_2$ =7.2, 1 H), 6.38 (t, superimposed dd, $J_1= J_2$=7.2, 1 H); ^{13}C NMR (75 MHz, CDCl$_3$) δ: 36.3, 38.5, 44.1, 50.4, 52.4, 52.8, 58.1, 126.4, 129.6, 132.2, 133.5, 205.3. These spectral features match data reported earlier.[4a] The orientation of the oxirane ring was suggested on the basis of general tendency of cyclohexa-2,4-dienones during their cycloaddition and comparison with other similar adducts.[3,4]

Waste Disposal Information

All toxic materials were disposed of in accordance with "Prudent Practices in the Laboratory"; National Academy Press; Washington, DC, 1995.

3. Discussion

This preparation is a modification of a general method developed in our laboratory for syntheses of a variety of tricycloundecadienones of type (**3**) which involves *in-situ* generation of spiroepoxycyclohexa-2,4-dienones and their interception with dienes.[3] This method is fairly versatile and a number of tricycloundecadienones have been prepared by oxidation of hydroxymethyl phenols and interception of the resulting cyclohexadienones with cyclopentadiene, spiro[4,2]cycloheptadiene and dimethylfulvene.[4] This synthesis also constitutes an example of generation of molecular complexity in a single step from simple starting materials.[5] Recently, interception of cyclohexa-2,4-dienones with electron-rich olefins has also been reported.[6] Oxidation of hydroxymethyl phenols with sodium metaperiodate has been extensively studied and is known to involve spiroepoxycyclohexa-2,4-dienones which are highly unstable and undergo dimerization instantaneously.[7] Recently, substituted spiroepoxycyclohexadienones were isolated and used as precursors in synthesis.[8,9] Oxidation of 5-bromosalicyl alcohol was also reported to give a stable 4-bromo-6-spiroepoxycyclohexa-2,4-dienone.[10]

Though the cycloaddition described here gives the adduct **3** in a regio- and stereoselective fashion, there exists a mechanistic dichotomy regarding the mode of the pericyclic reaction between cyclohexa-2,4-dienones with dienes.[11,12] The spiroepoxycyclohexa-2,4-dienone **2** may participate as a 4-π component (diene) and cyclopentadiene as a 2-π partner (dienophile) to give the adduct **3** directly. Alternatively, the cyclohexa-2,4-dienone may react as a 2-π partner (dienophile) and cyclopentadiene as a 4-π component (diene) to give the adduct of type **I** which may undergo a 3,3-shift[12] to give the product **3**. However, we were unable to isolate adducts of type **I** under our experimental conditions.

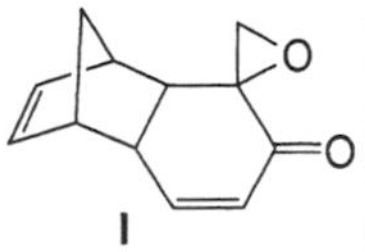

The epoxyketone **3** is a versatile precursor for a variety of tricycloundecane systems having a β,γ-unsaturated carbonyl chromophore, which are not so readily accessible. The contiguous epoxy ketone functionality and the double bond present in the five-membered ring provide opportunities for further manipulation. Adduct **3** may be transformed into a variety of molecular frameworks such as linearly fused *cis:anti:cis*

tricyclopentanoids, protoilludanes, and marasmanes in a stereoselective fashion after suitable chemical and photochemical manipulation.[13]

1.   Department of Chemistry, Indian Institute of Technology, Bombay, Mumbai, 400 076, India.

2.   Moffett, R. B. *Org. Synth., Coll. Vol. IV,* **1963**, 238-241. (b) Partridge, J. J.; Chadha, N. K.; Uskokovic, M. R. *Org. Synth., Coll. Vol. VII,* **1990**, 339-345.

3.   (a) Singh, V.; Porinchu, M. *J. C. S. Chem. Commun.* **1993**, 134-136. (b) Singh, V.; Thomas, B. *J. C. S. Chem. Commun.* **1992**,1211-1213.

4.   (a) Singh, V.; Porinchu, M. *Tetrahedron.* **1996**, *52,* 7087-7126. (b) Singh, V.; Thomas, B. *J. Org. Chem.* **1997**, *62,* 5310-5320.

5.   (a) Chanon, M.; Barone, R.; Baralotto, C.; Julliard, M.; Hendrickson, J. B. *Synthesis* **1998**, 1559-1583. (b) Corey, E. J.; Cheng, X. -M., *The Logic of Chemical Synthesis;* John Wiley & sons: New York, 1989.

6.   Bonnarme, V.; Bachmann, C.; Cousson, A.; Mondon, M.; Gesson, J. -P. *Tetrahedron,* **1999**, *55,* 433-448.

7.   Adler, E.; Brasen, S.; Miyake, H. *Acta. Chem. Scand.* **1971**, *25,* 2055-2069. (b) Adler, E.; Holmberg, K. *Acta. Chem. Scand.* **1974**, *28B,*465-472.

8.   Haseltine, J. N.; Cabal, M. P.; Mantlo, N. B.; Iwasawa, N.; Yamashita, D. S.; Coleman, R. S.; Danishefsky, S. J.; Schulte, G. K. *J. Am. Chem. Soc.* **1991**, *113,* 3850-3866.

9.   Corey, E. J.; Dittami, J. P. *J. Am. Chem. Soc.* **1985**, *107,* 256-257.

10. Mondon, M.; Boeker, N.; Gesson, J. -P. *J. Chem. Research(S)* **1999**, 484-485.

11. Gesson, J. -P.; Hervaud, L.; Mondon, M. *Tetrahedron Lett.* **1993**, *34,* 2941-2944.

12. Singh, V.; Sharma, U.; Prasanna, V.; Porinchu, M. *Tetrahedron,* **1995**, *51,* 6015-6032.

13. Singh, V. *Acc. Chem. Res.* **1999**, *32*, 324-333.

Appendix

Chemical Abtracts Nomenclature (Registry Number)

9-Spiroepoxy-endo-tricyclo[5.2.2.0^{2,6}]undeca-4,10-dien-8-one: Spiro[4,7-ethano-1H-indene-8,2'-oxiran]-9-one, 3a,4,7,7a-tetrahydro-;   (146924-02-9)

Salicyl alcohol: Benzenemethanol, 2-hydroxy-;   (90-01-7)

Cyclopentadiene: 1,3-Cyclopentadiene;   (542-92-7)

Sodium metaperiodate: Periodic acid (HIO$_4$), sodium salt; (7790-28-5)

Dicyclopentadiene: 4,7-Methano-1H-indene, 3a,4,7,7a-tetrahydro-; (77-73-6)

SYNTHESIS AND RU(II)-BINAP REDUCTION OF A KETOESTER DERIVED FROM HYDROXYPROLINE: 2(S)-(β-tert-BUTOXYCARBONYL-α-(S) and α-(R)-HYDROXYETHYL)-4(R)-HYDROXYPYRROLIDINE-1-CARBOXYLIC ACID, tert-BUTYL ESTER

(2-Pyrrolidinepropanoic acid, 1-[(1,1-dimethylethoxy)carbonyl]-β,4-dihydroxy-, 1,1-dimethylethyl ester, [2S-[2α(S*),4β]]-)

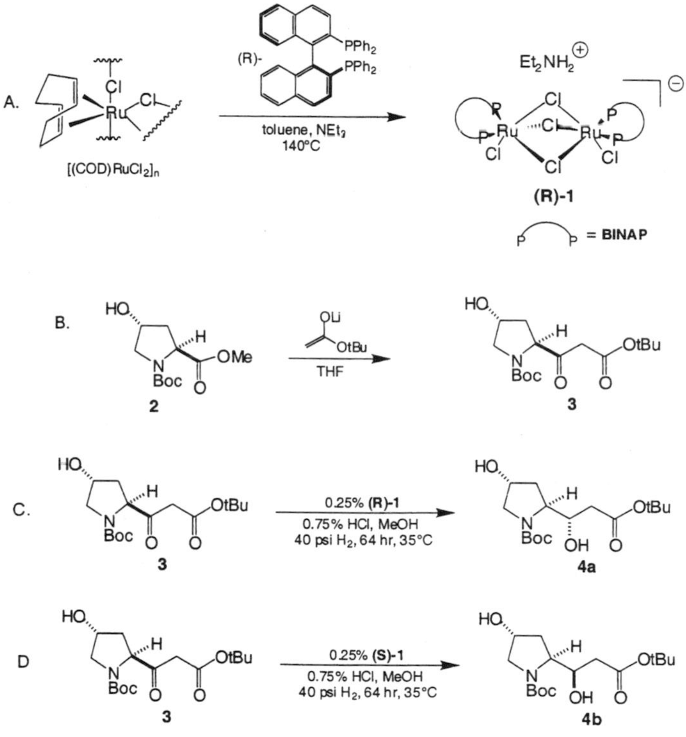

Submitted by Steven A. King, Joseph Armstrong, and Jennifer Keller.[1]

Checked by Amos B. Smith, III and Meinrad Brenner.

1. Procedure

A. *Preparation of catalyst.* Two 100-mL, round-bottomed flasks are connected to a double-ended filter which consists of two chambers, each equipped with a side arm, separated by a glass frit (Note 1). Vacuum grease is used to ensure an air-tight seal and rubber bands are used to secure the flasks to the filter assembly. A magnetic stirbar, (cyclooctadienyl)ruthenium dichloride polymer (0.426 g, 1.5 mmol), and (R)-BINAP (1.00 g, 1.60 mmol) (Note 2) are placed in one flask, and the entire apparatus is evacuated and filled with nitrogen. Toluene (34 mL) and triethylamine (3.4 mL) (Note 2), which have been deoxygenated by nitrogen sparging for several min, are added via the lower side arm. The vessel is sealed and the mixture heated to 140°C producing a deep brick red colored solution (Note 3). After 4 hr, the apparatus is allowed to cool to room temperature and the reaction mixture is vigorously stirred while the catalyst precipitates. The apparatus is vented to nitrogen and inverted to filter the product using vacuum on the lower side arm and nitrogen on the upper side arm (Note 4). If the frit cakes with product, slowing down filtration, momentarily reversing the flow of nitrogen will lift the cake away from the frit. The precipitate is washed with 34 mL of deoxygenated toluene and the flask containing the filtrate is replaced with an empty one. The entire apparatus is placed under vacuum and the product is dried overnight to give 0.850 g (67%) of a dark red solid (Notes 5, 6).

B. *(2S,4R)-2-tert-Butoxycarbonylacetyl-4-hydroxypyrrolidine-1-carboxylic acid, tert-butyl ester.* A 12-L, four-necked, round-bottomed flask equipped with a mechanical stirrer, addition funnel, thermocouple, and nitrogen inlet is charged with 3.3 L (3.82 mol) of 1.15M lithium hexamethyldisilylamide (LHS) solution in THF (Note 7). The solution is cooled to −40°C, and 514 mL (3.82 mol) of tert-butyl acetate is added via the addition funnel over 15 min. A solution of N-Boc hydroxyproline methyl ester **2** (187 g, 0.763 mol) in 940 mL of THF (Note 8) is cooled to 0°C and then added slowly via the addition funnel over 20 min while the temperature of the reaction mixture is maintained below –30°C. The resulting solution is stirred until the reaction is complete (30 min)(Note 9) and then poured over 20 min into an ice cold mixture of 3.8 L of 1M citric acid and 1.7 L of heptanes (Note 10). After stirring for an additional 30 min, the layers are separated (Note 11). The organic layer is concentrated by distillation under reduced pressure to a volume of 1.1 L (Note 12). Heptanes (3.6 L) are added and the mixture is concentrated by distillation under reduced pressure at 30°C to a volume of 1.1 L (Note 13). The resulting solution is diluted with 800 mL of heptanes and cooled at 0°C overnight. The resulting precipitate is filtered, washed with two 800-mL portions of heptanes, and dried overnight in a vacuum oven to provide 227 g (90%) of β-ketoester **3** as white crystals (Note 14).

C. *2(S)-(β-tert-Butoxycarbonyl-α-(S)-hydroxyethyl)-4-(R)-hydroxy-pyrrolidine-1-carboxylic acid, tert-butyl ester.* A 50-mL Parr shaker bottle capped with a rubber septum was charged with β-ketoester **3** (10.0 g, 30 mmol) and 15 mL of methanol and the solution was deoxygenated by sparging with nitrogen for 15 min. (*R*)-BINAP catalyst (*R*)-**1** prepared as described above (0.064 g, 0.035 mmol) was added followed by 8.7N HCl in methanol (26 μL, 0.23 mmol). The vessel was transferred to a standard Parr shaker apparatus and flushed by evacuating and refilling with nitrogen and then hydrogen several times (Note 15). The apparatus was heated at 35°C with shaking under 40 psi of hydrogen (Note 16). After 20 min, the reaction mixture became a homogeneous clear yellow solution. After 11 hr, the reaction mixture is transferred to a 250-mL, round-bottomed flask and concentrated to dryness. The residue is suspended in 50 mL of toluene and concentrated to dryness to ensure complete removal of methanol. The product is slurried in 25 mL of toluene/hexane/isopropyl acetate solution (90:5:5) and stirred overnight at room temperature. The precipitate is filtered, washed with 10 mL of toluene, and dried overnight in a vacuum oven (35°C, <0.5 mmHg) to provide 8.5 g (85%) of (*S*)-hydroxy ester **4a** (Notes 17, 18).

D. *2(S)-(β-tert-Butoxycarbonyl-α-(R)-hydroxyethyl)-4-(R)-hydroxy-pyrrolidine-* 1-*carboxylic acid, tert-butyl ester.* The identical procedure was followed, in this case using the *(S)*-BINAP catalyst **(S)-1**. Hydrogenation is conducted for 64 hr, and the reaction mixture is then transferred to a 250-mL, round-bottomed flask and concentrated to dryness. The residue is dissolved in 17 mL of methanol and cooled to 15°C. After the slow addition of 7 mL of DI water, the solution is aged for 15 min gradually forming a thin slurry. More DI water (75 mL) is added over 1 hr and the mixture is allowed to stand for an additional 1 hr at 15°C. The resulting crystals (Note 19) are filtered at 15°C, washed with 10 mL of 1:4-MeOH:water, and then dried overnight in a vacuum oven (35°C, 686 mm) to yield 7.0 g (70%) 0f (*R*)-hydroxy ester **4b** (Note 20).

2. Notes

1. A double-ended filler was purchased from Kontes (catalog #215500-6044) and is shown below.

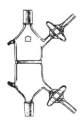

2. The checkers purchased [(COD)RuCl$_2$]$_n$ from Fluka (purum quality), BINAP from Aldrich (97 %), and toluene (HPLC grade) and triethylamine (reagent grade) from Fisher Scientific; the latter was distilled from CaH$_2$ under Ar prior to use. The submitters dried toluene and triethylamine over 4 Å molecular sieves. Karl-Fischer titration indicated <200 μg/mL water.

3. A dark green color of the solution indicates decomposition of the reaction mixture by oxygen.

4. ^{31}P NMR showed that the filtrate contained none of the desired product.

5. The checkers obtained 0.849-0.875 g of catalyst, but ^{1}H-NMR analysis revealed that the product was of low purity. Reactions run on a 0.75 mmol scale afforded the catalyst in 47-51% yield and in higher purity. Since the reaction is run in a sealed apparatus, the pressure built up in the apparatus (and therefore the boiling point and reaction temperature) may depend on the reaction scale and how far the flask is immersed into the oil bath.

6. ^{1}H-NMR (CD$_2$Cl$_2$, 500 MHz) δ: 1.47 (t, J = 7.4, 6 H), 3.20-3.30 (m, 4 H), 6.52-6.60 (m, 12 H), 6.65 (d, J = 8.6, 2 H), 6.69 (t, J = 7.4, 2 H), 6.74-6.81 (m, 4 H), 6.86 (t, J = 7.4, 2 H); 6.96 (t, J = 7.7, 2 H), 7.18-7.34 (m, 14 H), 7.38 (t, J = 7.2, 2 H), 7.47-7.61 (m, 12 H), 7.63-7.68 (m, 6 H), 7.84 (t, J = 8.3, 2 H), 8.08 (t, J = 8.4, 4 H), 8.57 (s br., 2 H); ^{31}P NMR (CD$_2$Cl$_2$, 101 MHz) δ: 52.3 (d, J = 38), 56.5 (d, J = 38).

7. The submitters obtained LHS in THF from Callery Chemical Co. and determined its concentration to be 1.15M by titration according to the method of Ireland (Ireland, R.; Meissner, R. *J. Org. Chem.* **1991**, *56*, 4566-4568). The checkers purchased LHS as a 1M solution in THF from Aldrich Chemical Co.

8. The submitters obtained N-Boc-4-hydroxyproline methyl ester from Synthetech, Inc. [Albany, OR, (503) 967-6575] or Bachem California [Torrance, CA, (310) 530-1571] and dried THF over 4Å molecular sieves for two days prior to use (Karl-Fischer titration gave 145 μg/mL water). The checkers purchased N-Boc hydroxyproline methyl ester (97%) and *tert*-butyl acetate (99+%) from Aldrich Chemical Co. and obtained THF (HPLC grade) from Fisher Scientific.

9. The progress of the reaction was followed by HPLC (Zorbax RX-C8 column, 1.5 mL/min, 50:50 CH$_3$CN:0.01M H$_3$PO$_4$ in water, room temperature, detection at 200

nm; retention times: methyl ester **2** 2.343 min, β-ketoester **3** 3.987 min. After 30 min, less than 0.5 area% methyl ester was found to be remaining.

10. There is an exotherm to 10°C during the quench.

11. GC assay of the organic layer showed no $HN(TMS)_2$ remaining after 15 min of stirring (GC conditions: Restek RTX-1 column (30 m x 0.53 mm, 1 m film thickness), 2.53 mL/min, initial temperature 50°C, final temperature 300°C, rate 20 deg/min, injection temperature 200°C, detector temperature 350°C, injection volume 1 μL, inject sample neat; retention times: *tert*-butyl alcohol 1.4 min, THF 1.7 min, heptane 2.1 min, $HN(TMS)_2$ 2.6 min, ethylbenzene (present in commercial LHS) 3.1 min, *tert*-butyl acetate 4.0 min). Volume percents were determined based on standard solution counts.

12. The checkers found the product to contain small amounts of citrates. Drying $(MgSO_4)$ or washing (H_2O) the organic phase prior to concentration prevents this contamination.

13. The distillation is carried out at ca. 60 mm and at a temperature below 40°C due to the instability of the β-ketoester in the crude solution. It is important to remove all THF prior to crystallization.

14. In smaller scale runs, the checkers obtained the product in 88-93% yield. The product exhibited the following properties: mp 98-99°C, $[\alpha]_D^{23°C}$: –73.2 (c = 0.95, MeOH); ^{13}C NMR (CD_3OD, 500 MHz, rotamers) δ: 28.4, 28.6, 28.7, 38.3, 39.1, 56.0, 56.3, 65.3, 65.6, 70.0, 70.8, 81.7, 82.2, 83.0, 155.9, 156.7, 168.1, 168.3, 204.7, 204.9. IR (film) cm^{-1}: 3439 (br. s), 2979 (s), 2934 (m), 1707 (s), 1395 (s), 1368 (s), 1326 (m), 1258 (m), 1162 (s), 1077 (m), 983 (m), 963 (m), 853 (m), 772 (m). Anal. Calcd for $C_{16}H_{27}NO_6$: C, 58.34; H, 8.26; N, 4.24. Found: C, 58.71; H, 8.52; N, 4.16.

15. Thorough removal of oxygen is imperative to prevent destruction of the catalyst.

16. The progress of the reaction is followed by HPLC (Zorbax RX-C8 column, 1.5 mL/min, gradient elution of 20:80 to 70:30 CH_3CN:0.01 M H_3PO_4 in water over 10 min, held for 20 min, room temperature, detection at 200 nm, retention times: β-keto *t*-butyl ester **2** 8.31 min, hydroxy *tert*-butyl ester **3** 7.10 min, hydroxy methyl ester 4.45 min, β-keto methyl ester 5.47 min).

17. The checkers obtained the product in 73-90% yield. The product exhibited the following properties: mp 99-100°C; $[\alpha]_D^{23°C}$: −50.5 (c = 0.93, MeOH); ^{1}H NMR (CDCl$_3$, 500 MHz) δ: 1.46 (s, 9 H), 1.47 (s, 9 H), 1.90-2.00 (m, 1 H), 2.00-2.10 (m, 1 H), 2.34 (dd, J = 15.7, 9.0, 1 H), 2.41 (dd, J = 15.7, 3.3, 1 H), 3.39 (dd, J = 12.1, 4.1, 1 H), 3.55-3.70 (m, 1 H), 4.11-4.17 (m, 1 H), 4.40-4.50 (m, 1 H). ^{13}C NMR (CD$_3$OD, 500 MHz, rotamers) δ: 28.4, 28.7, 36.7, 37.1, 39.3, 40.8, 56.6, 56.9, 61.0, 61.4, 70.1, 70.3, 70.7, 71.9, 81.3, 81.5, 81.8, 156.7, 157.2, 172.9. IR (film) cm^{-1}: 3418 (br. s), 2978 (s), 2933 (s), 1703 (s), 1403 (s), 1367 (s), 1257 (s), 1159 (s), 999 (m), 969 (m), 848 (m), 774 (m), 737 (m). Anal. Calcd for $C_{16}H_{29}NO_6$: C, 57.99; H, 8.82; N, 4.23. Found: C, 58.18; H, 9.09; N, 3.98.

18. The submitters determined the crystalline hydroxy esters to be >99.9% diastereomerically pure by supercritical fluid chromatography (EMdiol silica column and a Chiralcel (+) OD-(H) column (Chiral Technologies, Inc.) in tandem (100 bar CO_2; 1.0 mL/min; 8% MeOH, modifier 35°C; retention times: β-ketoester (keto form) **3** 15.83 min, β-ketoester (enol form) **3** 18.01 min, (*R*)-hydroxy ester **4b** 18.78 min, (*S*)-hydroxy ester **4a** 19.70 min).

19. The checkers found that the product was still slightly brown, indicating that the purification procedure is not effective in removing the catalyst.

20. The product exhibits the following properties: mp 135-137°C; $[\alpha]_D^{23°C}$: −70.2 (c = 0.34, MeOH); ^{1}H NMR (CDCl$_3$, 500 MHz) δ: 1.46 (s, 9 H), 1.47 (s, 9 H), 1.89-1.94 (m, 1 H), 2.00-2.20 (m br., 1 H), 2.25 (dd, J = 15.8, 9.5, 1 H), 2.32 (dd, J = 15.8, 3.5, 1 H), 3.40 (dd, J = 11.9, 4.0, 1 H), 3.45-3.80 (m br., 1 H), 3.80-4.15 (m br., 1 H), 4.30-4.45

(m, 1 H), 4.45-4.55 (m, 1 H); ^{13}C NMR (CD$_3$OD, 500 MHz, rotamers) δ: 28.4, 28.8, 34.4, 34.8, 41.1, 41.7, 56.1, 56.7, 61.8, 69.1, 69.2, 70.4, 70.7, 81.0, 81.3, 81.8, 156.7, 157.2, 172.3, 172.6. IR (film) cm^{-1}: 3419 (br. s), 2980 (s), 2919 (s), 1733 (s), 1673 (s), 1477 (m), 1418 (s), 1366 (s), 1244 (m), 1161 (s), 1080 (m), 1049 (m), 1002 (m), 952 (m), 876 (m), 852 (m), 775 (m), 749 (m); Anal. Calcd for C$_{16}$H$_{29}$NO$_6$: C, 57.99; H, 8.82; N, 4.23. Found: C, 58.33; H, 9.01; N, 3.81.

3. Discussion

The Ru(II)-BINAP complex,[2] [Et$_2$NH$_2$]$^+$[Ru$_2$Cl$_5$(BINAP)$_2$]$^-$,[3] is prepared as a toluene solvate in nearly pure form by this procedure. Typical crystallized product shows no other signals in the phosphorus NMR and gives a good combustion analysis. The material is quite stable and can be routinely handled in air. Storage under nitrogen will extend its shelf life, however.

The Claisen condensation of t-butyl acetate with a methyl ester is a general route for the preparation of complex β-ketoesters.[4] The reaction requires an excess of the enolate of t-butyl acetate to rapidly deprotonate the product and prevent tertiary alcohol formation. Some workers have also used excess LDA or t-butoxide for this purpose.

The reduction of β-ketoesters with Ru(II)-BINAP is the most efficient man-made catalytic asymmetric reaction.[5] Enantioselectivity is nearly always greater than 97% and 1,000 turnovers are typically accomplished in 2-8 hr. In the case of the optically active ketoester shown here, diastereoselectivity is >99:1 in the matched (C.) reaction and 94:6 in the mismatched (D.) case. Not surprisingly, the matched reaction is also considerably faster. Ketoester reductions are best run as a concentrated solution in methanol. (Methylene chloride/methanol mixtures have also been used; the presence of an alcoholic cosolvent is mandatory.) Reaction mixtures are extremely susceptible to poisoning by the simultaneous presence of hydrogen and oxygen. Thus, thorough degassing is necessary prior to reaction. The addition of a small amount of strong acid (e.g., HCl, H_2SO_4, CH_3SO_3H) is necessary to activate the catalyst; in its absence nearly no reduction occurs, The particular case of t-butyl esters demonstrates that the conditions are still quite mild, since acid-catalyzed dealkylation and transesterification are possible side reactions.

1. Merck Research Laboratories, Rahway, NJ 07065.

2. King, S. A.; Thompson, A. S.; King, A. O.; Verhoeven, T. R. *J. Org. Chem.* **1992,** *57,* 6689.

3. This structure is a revision of the originally assigned structure which was [Ru$_2$Cl$_4$(BINAP)$_2$]·Et$_3$N.   King, S. A.; DiMichele, L. In "Catalysis of Organic Reactions"; Scaraos, M. G.; Prunier, M. L., Ed.; **1995,** 157.

4. See, for example: Ohta, S.; Shimabayashi, A.; Hayakawa, S.; Sumino, S.; Okamoto, M. *Synthesis* **1985,** 45; Yamaguchi, M.; Nakamura, S.; Okuma, T.; Minami, T. *Tetrahedron Lett.* **1990,** *31,* 3913; Brower, P. L.; Butler, D. E.; Deering, C. F.; Le, T. V.; Millar, A.; Nanninga, T. N.; Roth, B. D. *Tetrahedron Lett.* **1992,** *33,* 2279.

5. Noyori, R.; Ohkuma, T.; Kitamura, M.; Takaya, H.; Sayo, N.; Kumobayashi, H.; Akutagawa, S. *J. Am. Chem. Soc.* **1987,** *109,* 5856; Noyori, R.; Takaya, H. *Acc. Chem. Res.* **1990,** 23, 345.

Appendix

Chemical Abstracts Nomenclature (Registry Number)

2(S)-(β-tert-Butoxycarbonyl-α-(S)-hydroxyethyl)-4-(R)-hydroxypyrrolidine-1-carboxylic acid, tert-butyl ester: 2-Pyrrolidinepropanoic acid, 1-[(1,1-dimethylethoxy)carbonyl]-β, 4-dihydroxy-, 1,1-dimethylethyl ester, [2S-[2α(S*),4β]]-; (167963-30-6).

(2S-4R)-2-tert-Butoxycarbonylacetyl-4-hydroxypyrrolidine-1-carboxylic acid, tert- butyl ester: 2-Pyrrolidinepropanoic acid, 1-[(1,1-dimethylethoxy)carbonyl]-4 hydroxy-β-oxo-,1,1-dimethylethyl ester, [2S-4R]-; (167963-29-3).

R-BINAP; Phosphine, (1R)-[1,1: binaphthalene]-2,2'-diylbis (diphenyl)-; (76189-55-4).

Triethylamine; Ethanamine, N,N-diethyl-; (121-44-8).

Lithium hexamethyldisilylamide: Silanamine, 1,1,1-trimethyl-N-(trimethylsilyl)-,   lithium salt; (4039-32-1)

tert-Butyl acetate; Acetic acid, 1,1-dimethylethyl ester; (540-88-5).

N-Boc hydroxyproline methyl ester: 1,2-Pyrrolidinedicarboxylic acid, 4-hydroxy, 1-(1, 1-dimethylethyl) 2-methyl ester, (2S,4R); (74844-91-0).

INDIUM/AMMONIUM CHLORIDE-MEDIATED SELECTIVE REDUCTION OF AROMATIC NITRO COMPOUNDS: ETHYL 4-AMINOBENZOATE

(Benzoic acid, 4-amino-, ethyl ester)

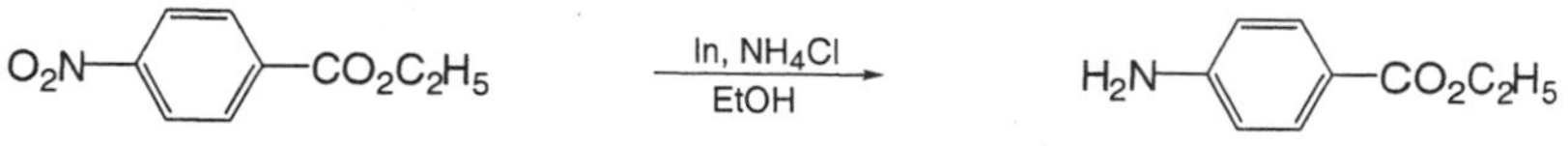

$$O_2N\text{—}\langle\bigcirc\rangle\text{—}CO_2C_2H_5 \xrightarrow[\text{EtOH}]{\text{In, NH}_4\text{Cl}} H_2N\text{—}\langle\bigcirc\rangle\text{—}CO_2C_2H_5$$

Submitted by Bimal K. Banik[1], Indrani Banik, and Frederick F. Becker.

Checked by Weiqiang Huang and Marvin J. Miller.

1. Procedure

Ethyl 4-aminobenzoate. A 1000-mL, round-bottomed flask equipped with a magnetic stirbar is charged with a suspension of 10 g (51 mmol) of ethyl 4-nitrobenzoate in 250 mL of ethanol, and a solution of 27.4 g (510 mmol) of ammonium chloride in 125 mL of water is then added (Note 1). Indium powder (23.5 g, 205 mmol) (Note 2) is added, and the resulting mixture is heated at reflux for 2.5 hr. The reaction mixture is allowed to cool to room temperature, diluted with 350-400 mL of water, and filtered under vacuum. The filtrate is extracted with 6-8 portions of 50-60 mL of dichloromethane, and the combined organic phases are washed with 100 mL of brine and dried over anhydrous sodium sulfate. The resulting solution is concentrated under reduced pressure and the crude product is dissolved in 100 mL of dichloromethane. The solution is concentrated by warming, and 50 mL of hexane is then added. The resulting solution is allowed to stand in a refrigerator overnight and then filtered under vacuum to give 7.63 g (90%) of ethyl 4-aminobenzoate (Note 3).

2. Notes

1. Ethyl 4-nitrobenzoate, ammonium chloride, ethanol were purchased from Aldrich Chemical Company and used as received.

2. Indium powder (99.99%) was obtained from Aldrich Chemical Company.

3. The spectral properties of ethyl 4-aminobenzoate are as follows: IR (film) cm^{-1}: 3424, 3345, 3224, 1685, 1636, 1598, 1515, 1367, 1312, 1281, 1173, 773; ^{1}H NMR (300 MHz, CDCl$_3$) δ: 1.36 (3 H, t, J = 6), 4.04 (2 H, brs), 4.31 (2 H, q, J = 7), 6.63 (2 H, d, J = 9), 7.85 (2 H, d, J = 9); ^{13}C NMR (75 MHz, CDCl$_3$) δ: 14.6 (CH$_3$), 60.4 (CH$_2$), 114.0 (Ar-CH), 120.5 (Ar-q), 131.7 (Ar-CH), 151.0 (Ar-q), 166.9 (Ar-q = quaternary carbon atom); m/e: 166 (M+H)$^+$; exact mass: m/e (M$^+$): 165.0764 (measured), 165.0790 (theoretical).

Waste Disposal Information

All toxic materials were disposed of in accordance with the policy of UTMDACC to handle and dispose of hazardous waste, which is in accordance with the regulations of the Environmental Protection Agency, Occupational Safety and Health Administration, Federal Department of Transportation, Texas Department of Health, and the Texas Water Commission.

3. Discussion

The synthesis of aromatic amines is an active and important area of research.[2] Many methods are available in the literature for the synthesis of these compounds. Though some of these are widely used, still they have limitations based on safety or handling considerations. For example, catalytic hydrogenation[3] of nitro or azido compounds in the presence of metals such as palladium on carbon or Raney nickel require stringent precautions because of their flammable nature in the presence of air. In addition, these methods require compressed hydrogen gas and a vacuum pump to create high pressure within the reaction flask. To overcome these difficulties, several new methods have been reported in the recent literature[4] involving such reducing agents as decaborane,[5] electrochemically generated Raney nickel,[6] dimethyl hydrazine/ferric chloride,[7] hydrazine hydrate/ferric oxide-magnesium oxide,[8] diethyl chlorophosphite,[9] and sodium borohydride-sodium-methoxide in methanol[10] In general, the main drawbacks of these methods are long reaction time and non-chemoselectivity. The submitters[11] have also described new methods for the reduction of aromatic nitro compounds and imines to aromatic amines by novel samarium-induced iodine catalyzed and ammonium chloride mediated reduction.[12] While our samarium-induced reduction of aromatic nitro compounds works well in the polycarbocyclic series, similar reaction with several heteroaromatic nitro compounds results in a mixture of products under identical conditions. Moreover, samarium-induced reduction of the nitro compounds requires anhydrous reaction conditions.

The submitters have been actively involved in the use of polyaromatic amines for the development of anticancer agents.[13] Therefore, the submitters began a research program aimed at developing methods to synthesize several aromatic amines rapidly and in high selectivity, by using ecologically friendly reagents. The present study describes a method for the selective reduction of aromatic nitro compounds to the corresponding amines by indium in the presence of ammonium chloride in aqueous ethanol.[14]

TABLE 1
SELECTIVE REDUCTION OF AROMATIC NITRO
COMPOUNDS BY INDIUM METAL IN THE
PRESENCE OF $NH_4Cl/H_2O/EtOH$

Entry	Nitro compound	Product	Reduction time (h)	%Yield	mp °C
1	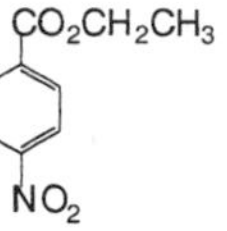 $CO_2CH_2CH_3$ — NO_2 (para)	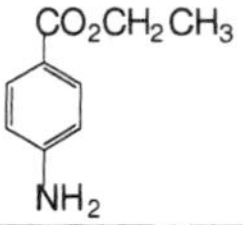 $CO_2CH_2CH_3$ — NH_2 (para)	2.5	94	88 (lit. 88-90)
2	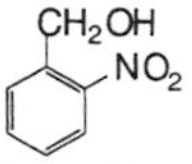 CH_2OH, NO_2 (ortho)	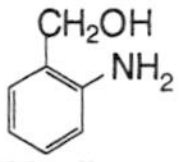 CH_2OH, NH_2 (ortho)	2h, 5 min	68.5	84 (lit. 83-85)
3	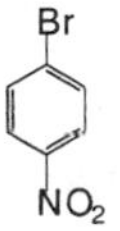 Br — NO_2 (para)	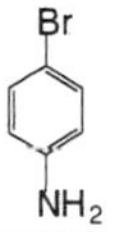 Br — NH_2 (para)	1.5	80	59 (lit. 60-64)
4	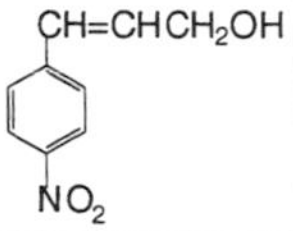 $CH=CHCH_2OH$ — NO_2 (para)	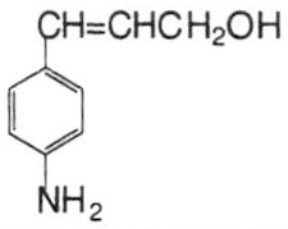 $CH=CHCH_2OH$ — NH_2 (para)	1	oil	oil
5	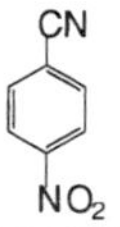 CN — NO_2 (para)	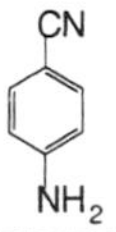 CN — NH_2 (para)	2	75	85 (lit 83-85)
6	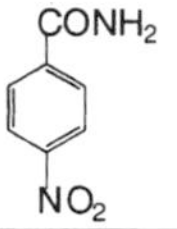 $CONH_2$ — NO_2 (para)	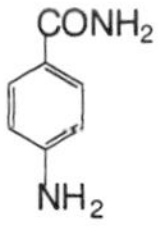 $CONH_2$ — NH_2 (para)	1.75	71	182 (lit. 181-183)
7	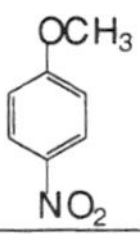 OCH_3 — NO_2 (para)	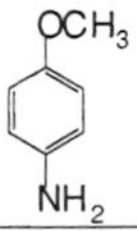 OCH_3 — NH_2 (para)	5	90	58-60 (lit. 57-60)
8	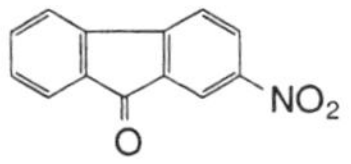 fluorenone-NO_2	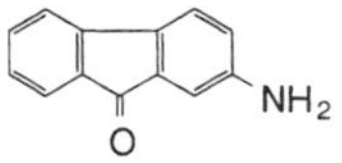 fluorenone-NH_2	8	80	157-160 (lit. 160)

The chemistry of indium metal is the subject of current investigation, especially since the reactions induced by it can be performed in aqueous solution.[15] The selective reductions of ethyl 4-nitrobenzoate (entry 1), 2-nitrobenzyl alcohol (entry 2), 1-bromo-4-nitrobenzene (entry 3), 4-nitrocinnamyl alcohol (entry 4), 4-nitrobenzonitrile (entry 5), 4-nitrobenzamide (entry 6), 4-nitroanisole (entry 7), and 2-nitrofluorenone (entry 8) with indium metal in the presence of ammonium chloride using aqueous ethanol were performed and the corresponding amines were produced in good yield. These results indicate a useful selectivity in the reduction procedure. For example, ester, nitrile, bromo, amide, benzylic ketone, benzylic alcohol, aromatic ether, and unsaturated bonds remained unaffected during this transformation. Many of the previous methods produce a mixture of compounds. Other metals like zinc, tin, and iron usually require acid-catalysts for the activation process, with resultant problems of waste disposal.

Because of the non-flammable nature of the process and ready availability of indium and ammonium chloride, the submitters believe this method is practical for the preparation of several aromatic amines. Further, this method is performed in aqueous ethanol, is extremely safe from the environmental point of view, and should prove useful in organic chemistry.

1. The University of Texas, M. D. Anderson Cancer Center, Department of Molecular Pathology, Box 89, 1515 Holcombe Blvd., Houston, TX 77030.

2. For a recent reference, see: Main, B. G.; Tucker, H. In *Medicinal Chemistry*, 2nd Ed.; Genellin, C. R.; Roberts, S. M.; Academic Press: New York, **1993**, p187.

3. (a) Rylander, P. N. *Hydrogenation Methods;* Academic Press: New York, **1985**; Chapter 8 (b) Johnstone, R. A. W.; Wilby, A. H.; Entistle, I. D. *Chem. Rev.* **1985**, *85*, 129.

4. (a) Banik, B. K.; Barakat, K. J.; Wagle, D. R.; Manhas, M. S.; Bose, A. K. *J. Org. Chem.* **1999**, *64*, 5746. (b) Weiner, H.; Blum, J.; Sasson, Y. *J. Org. Chem.* **1991**, *56,* 4481.

5. Bae, J. W.; Cho, Y. J.; Lee, S. H.; Yoon, C. M. *Tetrahedron Lett.* **2000**, *41*, 175.

6. Yasuhara, A.; Kasano, A.; Sakamoto, T. J. *Org. Chem.* **1999,** *64*, 2301.

7. Boothroyd, S. R.; Kerr, M. A. *Tetrahedron Lett.* **1995,** *36*, 2411.

8. Kumbhar, P. S.; Sanchez-Valente, J.; Figueras, F. *Tetrahedron Lett.* **1998,** *39*, 2573.

9. Fischer, B.; Sheihet, L. J. *Org. Chem.* **1998,** *63*, 393.

10. Suwinski, J.; Wagner, P.; Holt, E. M. *Tetrahedron* **1996,** *52,* 9541.

11. (a) Banik, B. K.; Mukhopadhyay, C.; Venkatraman, M. S.; Becker, F. F. *Tetrahedron Lett.* **1998**, *39*, 7243. (b) Banik, B. K., Zegrocka, O., Banik, I.; Hackfeld, L.; Becker, F. F. *Tetrahedron Lett.* **1999**, *40*, 6731.

12. Basu, M. K.; Becker, F. F.; Banik, B. K. *Tetrahedron Lett.* **2000,** *41*, 5603.

13. (a) Becker, F. F.; Banik, B. K. *Bioorg. Med. Chem. Lett.* **1998,** *8*, 2877. (b) Becker, F. F., Mukhopadhyay, C., Banik, I., Hackfeld, L., Banik, B. K. *Bioorg. & Med. Chem.* **2000**, *8*, 2693. (c) Banik, B. K. Becker, F. F. *Bioorg. & Med. Chem.* **2001**, *9*, 593. (d) Banik, B. K. Becker, F. F. *Current Med. Chem.* **2001**, *8*, 1513.

14. Banik, B. K.; Suhendra, M.; Banik, I.; Becker, F. F. *Synth. Commun.* **2000**, *30*, 3745. Also see: (a) Moody, C. J.; Pitts, M. R. *Synlett* **1998**, 1028. (b) Moody, C. J.; Pitts, M. R. *Synlett* **1998**, 1029.

15. Li has carried out a significant work on indium-mediated reactions, for example, see: (a) Li, C. J.; Chan, T.-H. *Organic Reactions in Aqueous Media*; J. Wiley & Sons: New York, **1997**. (b) Li, C. J. *Tetrahedron* **1996**, *52*, 5643. (c) Li, C. J. *Chem. Rev.* **1993**, *93*, 2023.

Appendix

Chemical Abstracts Nomenclature (Registry Number)

Ethyl 4-aminobenzoate: Benzoic acid, 4-amino-, ethyl ester; (94-09-7)

Ethyl 4-nitrobenzoate: Benzoic acid, 4-nitro-, ethyl ester; 99-77-4)

Ammonium chloride (NH_4Cl); (12125-02-9)

Indium; (7440-74-6)

OXIDATION OF PRIMARY ALCOHOLS TO CARBOXYLIC ACIDS WITH SODIUM CHLORITE CATALYZED BY TEMPO AND BLEACH: 4-METHOXYPHENYLACETIC ACID

(Benzeneacetic acid, 4-methoxy-)

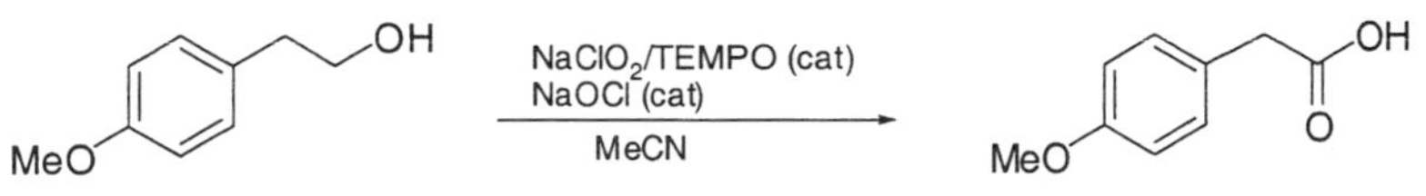

Submitted by Matthew M. Zhao,[1] Jing Li,[1] Eiichi Mano,[2] Zhiguo J. Song,[1] and David M. Tschaen.[1]

Checked by Arun Ghosh, Jamie Sieser, Weiling Cai, and Sarah E. Kelly.

1.   Procedure

A. 1-L, three-necked, round-bottomed flask equipped with a mechanical stirrer and two 100-mL addition funnels is charged with 4-methoxyphenethyl alcohol (**1f**) (6.09 g, 40 mmol), TEMPO (2,2,6,6-tetramethyl-1-piperidinyloxy free radical) (0.436 g, 2.8 mmol), 200 mL of acetonitrile, and 150 mL of 0.67M sodium phosphate buffer (pH 6.7) (Notes 1,2). A solution of sodium chlorite is prepared by dissolving 80% $NaClO_2$ (9.14 g, 80.0 mmol) in 40 mL of water and a solution of dilute sodium hypochlorite (NaOCl) is prepared by diluting household bleach (5.25% NaOCl, 1.06 mL, ca. 2.0 mol%) with 19 mL of water. The reaction mixture is heated to 35°C with stirring and approximately 20% of the $NaClO_2$ solution is added via one addition funnel followed by 20% of the dilute bleach solution via the other funnel. The remaining portions of both reagents are then added simultaneously over 2 hr (Note 3). The resulting mixture is stirred at 35°C until the reaction is complete (Note 4) (usually 6-10 hr). After cooling the reaction mixture to 25°C, 300 mL of water is added and the pH is adjusted to 8.0 by addition of

195

ca. 48 mL of 2.0N NaOH. The reaction mixture is then poured into ice-cold sodium sulfite solution (12.2 g in 200 mL of water) maintained below 20°C with an ice-water bath (Note 5). After stirring for 0.5 hr at rt, 200 mL of methyl t-butyl ether (MTBE) is added and the resulting mixture is stirred for 15 min. The organic layer is separated and discarded. More MTBE (100 mL) is added and the rapidly stirred mixture is acidified with 2.0N HCl to pH 3-4. The organic layer is separated and the aqueous layer is extracted with two 100-mL portions of MTBE. The combined organic phases are washed with two 100-mL portions of water, 100 mL of brine, and then concentrated to give 6.16-6.22 g (93-94%) of 4-methoxyphenylacetic acid (**2f**) as a colorless solid (Note 6).

2.   Notes

1.   All reactants and reagents were obtained from Sigma-Aldrich and used without purification.

2.   The sodium phosphate buffer consisted of a 1:1 mixture of 0.67M NaH_2PO_4 and 0.67M Na_2HPO_4 (pH = 6.5 at 22°C). For substrates not prone to chlorination, lower pH (3-4) can be employed to speed up the reaction.

3.   *Caution*: it is not advisable to mix sodium chlorite solution and bleach prior to the addition since the mixture appears to be unstable. On a small scale, it is acceptable to mix the substrate, sodium chlorite, TEMPO, acetonitrile, and buffer first and then add bleach in one portion.

4.   The progress of the reaction is monitored by HPLC: YMC ODS-AM column, 4.6x250mm, flow rate 1.00mL/min, 30°C, linear gradient 20-80% MeCN/15min, 0.1% H_3PO_4; retention time **1f**: 9.09 min, **2f**: 9.25 min.

5.   The pH of the aqueous layer should be 8.5 – 9.0.

6.   The product exhibits the following properties: mp 86-88.5°C (lit.[9] 86°C); ^{1}H NMR (400 MHz, CDCl$_3$) δ: 3.59 (s, 2H), 3.79 (s, 3H), 6.88 (d, J = 8.5, 2H), 7.19 (d, J = 8.5, 2H); ^{13}C NMR (100 MHz, CDCl$_3$) δ: 178.4, 159.1, 130.7, 125.5, 114.3, 55.5, 40.4; HPLC purity >99 area %.

Waste Disposal Information

All toxic material were disposed of in accordance with "Prudent Practices in the Laboratory"; National Academic Press; Washington, DC, 1996.

3.   Discussion

The procedure described here provides an efficient and environmentally benign method for oxidizing primary alcohols to carboxylic acids using stoichiometric NaClO$_2$, catalytic TEMPO, and NaOCl. Compared with the previously reported TEMPO/NaOCl/CH$_2$Cl$_2$ protocol,[5c] this new procedure gives significantly improved yields and purity of the desired product by reducing the chlorination of the aromatic groups. No racemization or epimerization is observed for substrates with labile chiral centers. Additionally, no chlorinated solvent is required. Similar to TEMPO/NaOCl, this procedure is not applicable to alkenic alcohols and substrates with exceedingly electron-rich aromatic groups.

The development of this procedure stems from our recent work involving the oxidation of primary alcohol **1** to the carboxylic acid **2**.[3]

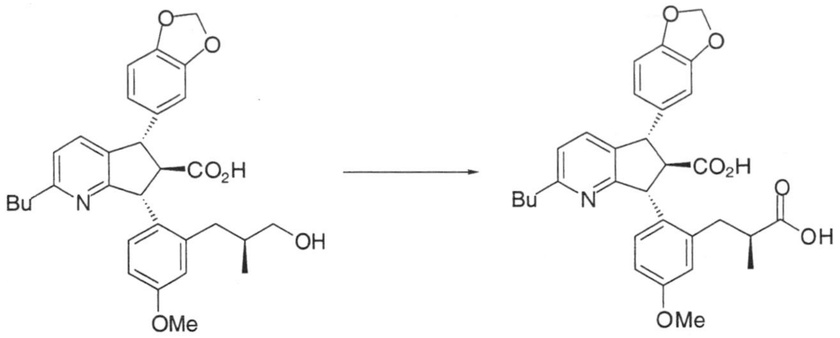

1 → **2**

The submitters found that the RuCl$_3$/H$_5$IO$_6$ protocol[4] gave low yields (<50%) due to the destruction of the electron-rich aromatic rings. TEMPO catalyzed oxidation[5] with NaOCl also gave unsatisfactory results because of the chlorination of the aromatic groups. Other oxidants, such as H$_2$O$_2$, MeCO$_3$H, t-BuO$_2$H, etc., were examined without much success. In the submitters' recent report[7] on a CrO$_3$ (1 mol%) catalyzed oxidation, although it is much improved over the Jones oxidation in terms of environmental impact, the heavy metal issue is not completely eliminated. Investigation of the TEMPO/NaOCl protocol using sodium chlorite (NaClO$_2$)[8] as the oxidant resulted in the development of this new procedure[3b]. The reaction appeared to be very slow (~2%/hr) initially. Subsequently, it was found that NaOCl dramatically accelerated the reaction. The conversion reached >50% in 1 hr and went to completion in 2-4 hr.

The remarkable acceleration can be explained by the proposed catalytic cycle shown in Scheme 1. TEMPO radical is first oxidized by NaOCl to the N-oxoammonium ion[5] **A**, which rapidly oxidizes the primary alcohol (**1**) to the aldehyde (**3**) and gives a molecule of the hydroxylamine **D**.[5] The aldehyde (**3**) is then oxidized by NaClO$_2$ to the carboxylic acid (**2**)[6] and regenerates a molecule of NaOCl. The hydroxylamine **D** can either be directly oxidized to **A** or undergo a syn proportionation with a molecule of **A** to

198

give two molecules of TEMPO radical which can be oxidized back to **A**[5a,b] Although the exact mechanism of TEMPO-catalyzed oxidation of alcohols is still unclear, previous work[5] has shown that N-oxoammonium ion **A** and hydroxylamine **D** are involved. It is also known that $NaClO_2$ can readily oxidize aldehydes to the carboxylic acids in the absence of TEMPO.[6] The long induction period of the reaction without a catalytic amount of bleach is likely due to the relatively slow oxidation of TEMPO radical or the hydroxylamine **D** by $NaClO_2$. Once the reaction is initiated, it becomes self-sustaining as NaOCl is continuously regenerated. The chlorination problem is greatly reduced because the concentration of NaOCl remained low throughout the reaction. Risks for epimerization of the neighboring chiral center are also reduced since the labile aldehyde intermediate is rapidly oxidized to the carboxylic acid by sodium chlorite.

Scheme 1. Catalytic Cycle for the TEMPO/NaOCl Catalyzed Oxidation

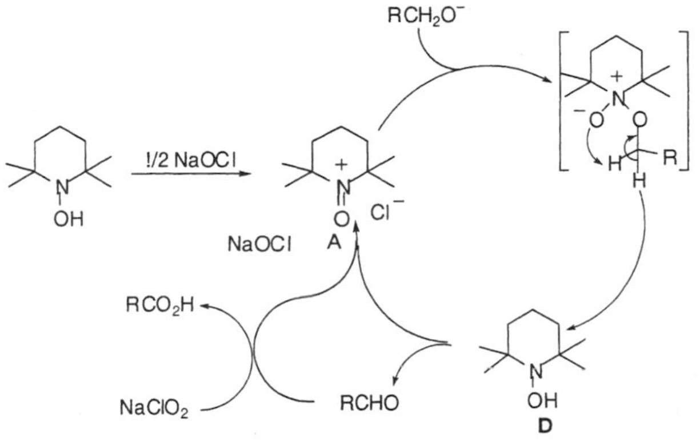

This method is mild and efficient and has been demonstrated on a variety of primary alcohols (Table 1). For comparison purposes, substrates with electron-rich aromatic rings were also subjected to the TEMPO/NaOCl oxidation[5c] (entries 3, 5-7, 10). In all of these cases, much improved yields were obtained with this new procedure. The most striking example is **1g**. The yield of the desired product **2g** was only 42% using stoichiometric NaOCl in contrast to the quantitative yield obtained using our procedure (entry 7). One of the major side products in the oxidation with stoichiometric NaOCl was isolated and identified as the chlorinated compound **4** (Figure 1) based on NMR studies (NOE). Similarly, 3-phenylpropargyl alcohol (**1h**) was oxidized to the acid **2h** in 90% yield using our procedure vs. < 5% with NaOCl. It appeared that carbon-carbon triple bonds can be tolerated, but substrates with ordinary carbon-carbon double bonds such as cinnamyl alcohol failed to react. This was likely due to quenching of the catalytic NaOCl, which shut down the catalytic cycle. Substrates with very electron-rich aromatic groups such as **1n-1o** (Figure 1) also failed for similar reasons. Surprisingly, oxidation of 4-methoxybenzyl alcohol (**1p**) was very sluggish. Substrate **1i**, which contained a cyclopropyl group, posed no problem (entry 9). Chiral alcohols, including protected amino alcohol **1l**, were oxidized to the corresponding carboxylic acids without racemization (or epimerization) of the labile chiral centers (entries 10-12).

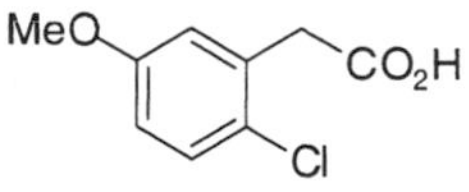

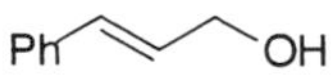

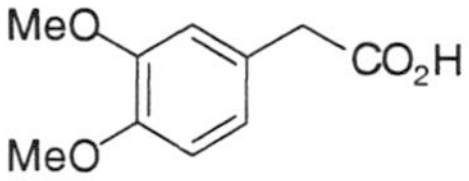

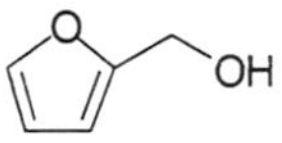

Figure 1

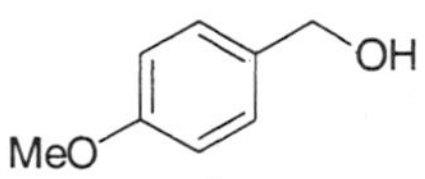

200

Table 1. **TEMPO** Catalyzed Oxidation of Primary Alcohols

Entry	Substrate	Product	Yield % (NaClO$_2$)	Yield % (NaOCl)
1	Ph⌒OH **1a**	Ph-CO$_2$H **2a**	98	–
2	O$_2$N–⟨⟩–CH$_2$OH **1b**	O$_2$N–⟨⟩–CO$_2$H **2b**	100	–
3	(Br, OMe aryl)–CH$_2$OH **1c**	(Br, OMe aryl)–CO$_2$H **2c**	96	80
4	Ph⌒⌒OH **1d**	Ph⌒CO$_2$H **2d**	100	–
5	(o-OMe aryl)⌒OH **1e**	(o-OMe aryl)⌒CO$_2$H **2e**	99	65
6	MeO–⟨⟩⌒OH **1f**	MeO–⟨⟩⌒CO$_2$H **2f**	100	86
7	(m-OMe aryl)⌒OH **1g**	(m-OMe aryl)⌒CO$_2$H **2g**	96	42
8	Ph—≡—⌒OH **1h**	Ph—≡—CO$_2$H **2h**	90	< 5
9	Ph(cyclopropyl)⌒OH **1i**	Ph(cyclopropyl)CO$_2$H **2i**	95	–
10	(Br, OMe aryl)⌒⌒OH **1j**	(Br, OMe aryl)⌒CO$_2$H **2j**	92	60
11	HO–(bicyclic, Ph) **1k**	HO$_2$C–(bicyclic, Ph) **2k**	95	–
12	Ph⌒OH NHCbz **1l**	Ph⌒CO$_2$H NHCbz **2l**	85	–

[a] All entries except **3, 9** and **10** were checked. The reactions took 6-10 h for completion in all cases. Additional charges of both NaOCl$_2$ and NaOCL, however, showed faster rate. Entry **11** was particularly sluggish towards oxidation. Yields and purities of products were generally comparable to those reported in the Table. For substrates not prone to chlorination, lower pH (3-4) can be employed to speed up the reaction.

1. Process Research, Merck Research Laboratories, Merck & Co. Inc., P.O. Box 2000, Rahway, NJ 07065

2. Pharm. R & D, Banyu Pharmaceutical Co. Ltd. Menuma, Saitama, Japan

3. (a) Song, Z. J.; Zhao, M.; Desmond, R.; Devine, P.; Tschaen, D. M.; Tillyer, R.; Frey, L.; Heid, R.; Xu, F.; Foster, B.; Li, J.; Reamer, R.; Volante, R.; Grabowski, E. J. J.; Dolling, U. H.; Reider, P. J.; Okada, S.; Kato, Y.; Mano, E. *J. Org. Chem.* **1999**, *64*, 9658-9667. (b) Zhao, M.; Li, J.; Mano, E.; Song, Z.; Tschaen, D. M.; Grabowski, E. J. J.; Reider, P. J. *J. Org. Chem.* **1999**, *64*, 2564-2566.

4. Carlsen, P. H. J.; Katsuki, T.; Martin V. S.; Sharpless, K. B. *J. Org. Chem.* **1981**, *46*, 3936–3938.

5. (a) Nooy, A. E. J. de; Besemer, A. C.; Bekkum, H. V. *Synthesis* **1996**, 1153–1174. (b) Bobbitt, J. M.; Flores, M. C. L. *Heterocycles* **1988**, *27*,509-533. (c) Anelli, P. L.; Biffi, C.; Montanari, F.; Quici, S. *J. Org. Chem.* **1987**, *52*, 2559–2562. (d) Miyazawa, T.; Endo, T.; Shiihashi, S.; Okawara, M. *J. Org. Chem.* **1985**, *50*, 1332–1334. (e) Rychnovsky, S. D.; Vaidyanathan, R. *J. Org. Chem.* **1999**, *64*, 310–312.

6. (a) Lindgren, B. O.; Nilsson, T. *Acta Chem. Scand.* **1973**, *27*, 888–890. (b) Dalcanale, E.; Montanari, F. *J. Org. Chem.* **1986**, *51*, 567–569.

7. Zhao, M.; Li, J.; Song, Z.; Desmond, R.; Tschaen, D. M.; Grabowski, E. J. J.; Reider, P. J. *Tetrahedron Lett.* **1998,** *39,* 5323-5326.

8. The use of $NaBrO_2$ has been reported; however, $NaClO_2$ was reported to be unsuccessful. Additionally, $NaBrO_2$ is not readily available commercially. Inokuchi, T; Matsumoto, S.; Nishiyama, T.; Torii, S. *J. Org. Chem.* **1990**, *55*, 462-466.

9.   CRC Handbook of Data on Organic Compounds, **1985**, HODOC
      No:04247.

Appendix

Chemical Abstracts Nomenclature (Registry Number)

4-Methoxyphenethyl alcohol: Benzeneethanol, 4-methoxy-; (702-23-8).

4-Methoxyphenylacetic acid: Benzeneacetic acid, 4-methoxy-; (104-01-8).

2,2,6,6-Tetramethyl-1-piperidinyloxy (TEMPO): 1-Piperidinyoxy, 2,2,6,6-
    tetramethyl-; (2564-83-2).

Sodium chlorite: Chlorous acid, sodium salt; (7758-19-2).

Sodium hypochlorite: Hypochlorous acid, sodium salt; (7681-52-9).

Methyl tert-butyl ether; Propane, 2-methoxy-2-methyl-; (1634-04-4).

METHYLTRIOXORHENIUM CATALYZED OXIDATION OF SECONDARY AMINES TO NITRONES: *N*-BENZYLIDENE-BENZYLAMINE *N*-OXIDE

[Benzenemethanamine, (N-phenylmethylene)-, N-oxide]

Submitted by Andrea Goti,[1] Francesca Cardona, and Gianluca Soldaini.
Checked by Xiaowu Jiang and Marvin J. Miller.

1. Procedure

A. 250-mL, two-necked, round-bottomed flask equipped with a magnetic stirbar, thermometer, and a reflux condenser fitted with a rubber septum and balloon of argon is charged with a solution of methyltrioxorhenium (MTO) (0.013 g, 0.05 mmol, 0.1% mol equiv) in 100 mL of methanol (Note 1). Urea hydrogen peroxide (UHP) (14.3 g, 152 mmol) is added (Notes 1, 2, 3, 4), the flask is cooled in an ice bath, and dibenzylamine (9.7 mL, 50.7 mmol) is then added dropwise via syringe over 10 min (Notes 1, 5). After completion of the addition, the ice bath is removed and the mixture is stirred at room temperature (Note 6). A white precipitate forms after approximately 5 min (Note 7) and the yellow color disappears within 20 min (Note 8). Another four portions of MTO (0.1% mol equiv, 0.013 g each) are added at 30-min intervals (2.5 hr total reaction time). After each addition, the reaction mixture develops a yellow color, which then disappears; only after the last addition does the mixture remain pale yellow (Note 9). The reaction flask is cooled in an ice bath and solid sodium thiosulfate pentahydrate (12.6 g, 50.7 mmol) is added in portions over 20 min in order to destroy excess hydrogen peroxide (Note 10). The cooled solution is stirred for 1 hr further, at which point a KI paper assay indicates that the excess oxidant has been completely consumed. The solution is decanted into a

500-mL flask to remove small amounts of undissolved thiosulfate. The solid is washed with 50 mL of MeOH and the methanol extract is added to the reaction solution which is then concentrated under reduced pressure by rotary evaporation. Dichloromethane (250 mL) is added to the residue and the urea is removed by filtration through cotton and celite. Concentration of the filtrate affords 10.3 g (97%) of the nitrone as a yellow solid (Note 11).

If desired, the nitrone can be recrystallized from diisopropyl ether (200 mL) (Notes 12, 13), affording 9.0 g (84%) of a white solid (Note 14). Concentration of the mother liquor and recrystallization of the residue from 40 mL of diisopropyl ether provides 0.600 g of additional nitrone. Total yield: 9.6 g (89%) of the nitrone as a white solid (Note 15).

2. Notes

1. Dibenzylamine, MTO, and UHP were purchased from Aldrich Chemical Company and used as received.

2. *Caution:* Potential hazards are associated with the use of the urea hydrogen peroxide complex (UHP). Due to the relatively high hydrogen peroxide content (ca. 36%) of this reagent, care should be exercised in its use. Additionally, concentrated solutions of H_2O_2 are potentially explosive, especially when contaminants such as metals are present. However, it has been reported that UHP is a safe source of hydrogen peroxide,[2] which makes it a good alternative to anhydrous or highly concentrated aqueous solutions of H_2O_2. UHP has been judged stable by negative impact and pressure-time tests for explosiveness on small samples[2] and the submitters have never experienced problems in its use.

3. Although the oxidation requires 2 mol equiv of oxidant with respect to amine, use of 3 mol equiv of UHP is advised in order to speed up the reaction and to reduce the amount of catalyst required. For example, use of a smaller excess of UHP (2.5 mol equiv) required an increase to 0.7% mol equiv of MTO and a 4 hr reaction time in order for the oxidation to proceed to completion.

4. Immediately upon addition of the urea-hydrogen peroxide adduct to the solution containing methyltrioxorhenium, a yellow color develops due to formation of the catalytically active rhenium peroxo complexes.[3]

5. The internal temperature of the solution rises from ca. 4°C to 12-13°C after all of the amine has been added.

6. The internal temperature varies from 24°C to 32°C during the whole reaction time, with the highest values being reached after each addition of the catalyst.

7. The white precipitate is presumably *N,N*-dibenzylhydroxylamine, an intermediate oxidation product, which disappears after the third catalyst addition.

8. The disappearance of the yellow color indicates a low concentration of the active catalytic species due to loss in efficiency of the cycle.

9. Monitoring of the reaction by TLC on silica gel showed complete consumption of the starting amine (elution with 20:1 CH_2Cl_2/CH_3OH, visualization by UV and by iodine; dibenzylamine: $R_f = 0.4$, nitrone: $R_f = 0.7$).

10. The submitters found that on small scale (1 mmol) this treatment is unnecessary. On a larger scale, it is recommended to destroy excess hydrogen peroxide for safety, although oxidations on larger scale without this treatment have been carried out without experiencing any problem. It was also observed that product of higher purity is obtained when hydrogen peroxide is destroyed.

11. The crude nitrone has the following physical data: mp 75-77°C; Anal. Calcd for $C_{14}H_{13}NO$: C, 79.59; H, 6.20; N, 6.63. Found: C, 79.39; H, 6.18; N, 6.78.

12. *Caution:* Use of diisopropyl ether from a freshly opened bottle is advised, due to possible formation of peroxides in aged ethereal solvents. Other solvents were less effective for this recrystallization.

13. The crude solid is readily dissolved in a smaller amount of solvent, but employing the indicated volumes is recommended in order to achieve better yields of recrystallization. With this volume of solvent, heating at 74°C is sufficient to have complete solubilization of the solid.

14. The nitrone exhibits the following spectral and physical data: mp (uncorr.) 79-81°C (lit.[4] 81-83°C); ^{1}H NMR (300 MHz, $CDCl_3$) δ: 5.09 (s, 2 H), 7.38-7.49 (m, 9 H),

8.17-8.20 (m, 2 H); ^{13}C NMR (75 MHz, CDCl$_3$) δ: 71.29, 128.41, 128.57, 128.93, 129.17, 130.35, 130.51, 133.35, 134.11; IR (KBr) cm^{-1}: 3061, 3030, 2995, 1581, 1497, 1459, 1153; MS (EI, 70 eV) m/z (%): 211 (M$^+$, 27), 195 (7), 181 (16), 91 (100), 83 (78), 66 (70), 65 (52); exact mass (FAB): Calcd (M+H)$^+$ 212.1075, found 212.1064. Anal. Calcd for C$_{14}$H$_{13}$NO: C, 79.59; H, 6.20; N, 6.63. Found: C, 79.57; H, 5.92; N, 6.51.

15. The submitters report a yield of 86% and the checkers obtained the product in 85-90% yield in several runs.

Waste Disposal Information

All toxic materials were disposed of in accordance with "Prudent Practices in the Laboratory"; National Academy Press; Washington, DC, 1995.

3. Discussion

Nitrones are very useful tools in the construction of structurally complex molecules and particularly nitrogen-containing biologically active compounds, usually allowing a high degree of diastereocontrol. In this context, both the [3+2] cycloaddition of nitrones to alkenes[5,6] and the alkylation of nitrones by organometallic reagents[7] have been extensively developed and have become extremely reliable synthetic procedures. In addition, nitrones are useful spin trap reagents, widely employed in biological systems.[8]

The most widely employed methods for the synthesis of nitrones are the condensation of carbonyl compounds with N-hydroxylamines[5] and the oxidation of N,N-disubstituted hydroxylamines.[5,9] Practical and reliable methods for the oxidation of more easily available secondary amines have become available only recently.[10-13] These include reactions with stoichiometric oxidants not readily available, such as dimethyldioxirane[10] or N-phenylsulfonyl-C-phenyloxaziridine,[11] and oxidations with hydrogen peroxide catalyzed by Na$_2$WO$_4$[4,12] or SeO$_2$.[13] All these methods suffer from limitations in scope and substrate tolerance. For example, oxidations with dimethyldioxirane seem to be limited to arylmethanamines and the above mentioned catalytic oxidations have been reported (and we have experienced as well) to give somewhat erratic and unreproducible results in some cases. Particularly, functionalized pyrrolidines and piperidines behaved as capricious substrates in those protocols. For example, a chiral optically active pyrrolidine underwent racemization to variable degrees upon oxidation to the corresponding nitrone with H_2O_2 catalyzed by Na_2WO_4[14] and oxidation with dimethyldioxirane of a chiral 2-morpholinone gave scarcely reproducible results.[15] In these cases, the described oxidation method using methyltrioxorhenium/urea H_2O_2 complex was demonstrated to be a better alternative, furnishing consistently good and reproducible yields (see Table, entries 9 and 11, respectively), without loss of enantiomeric purity, also on multi-gram scale. The examples reported in the Table are illustrative of the scope of the oxidation method described here. Most of the reactions reported in the Table have been carried out on a 0.5-1 mmol scale and are unoptimized, with a 2% mol equiv of MTO being employed. The procedure described above has been scaled-up and optimized, also in terms of minimal catalyst use. The amount of catalyst to be used in order to achieve complete conversion is not greatly influenced by the reaction scale. Indeed, essentially the same results as those reported above have been obtained on a 10 mmol scale with 0.4% mol equiv of MTO and in a further scale-up of the process on 0.1 mol of dibenzylamine with a minimal increase to 0.6% mol equiv of MTO. On a 10-12 mmol scale, 0.7% mol equiv of MTO were necessary for oxidizing completely dibutylamine and 1% mol equiv was sufficient to oxidize the more sterically congested N-(tert-butyl)-N-benzylhydroxylamine.

Due to its broad scope, as well as to its favorable features (commercial availability of the catalyst, use of a "green" oxidant, economy, extremely simple procedure and work-up), this method has been rapidly accepted, as demonstrated by its use by several different research groups, despite its recent disclosure.[16,17] This procedure employing CH_3ReO_3 and UHP appears to be the method of choice when optically active nitrones are prepared by oxidation of the corresponding amines.[14,15,18]

Table

Entry	Amine	Nitrone	Yield (%)
1[a]	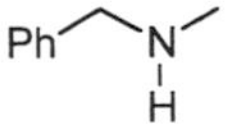 Ph⌒N(H)⌒ (N-methylbenzylamine)	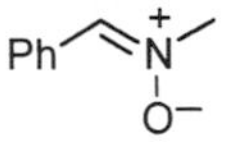 Ph⌒=N+(CH₃)–O⁻	70
2[a]	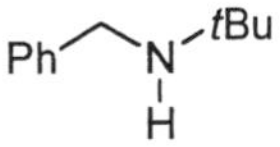 Ph⌒N(H)–tBu	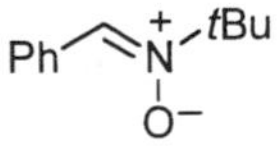 Ph⌒=N+–tBu, O⁻	86
3[a]	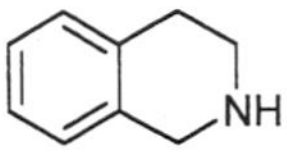 (tetrahydroisoquinoline) NH	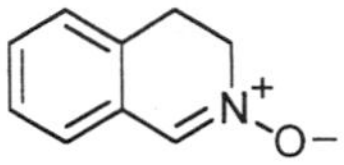 N+–O⁻ nitrone	82
4[a]	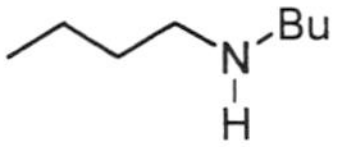 ⌒⌒N(H)–Bu	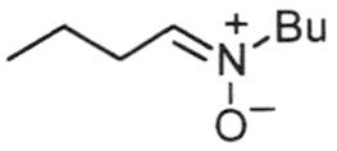 ⌒⌒=N+–Bu, O⁻	60
5[a]	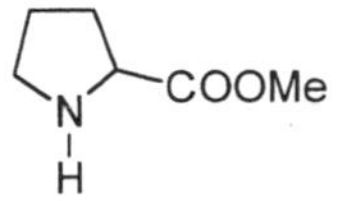 pyrrolidine N(H) –COOMe	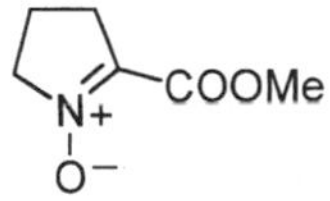 N+(O⁻) –COOMe	65
6[a]	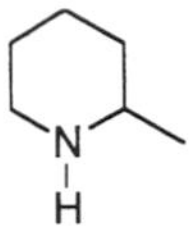 2-methylpiperidine N–H	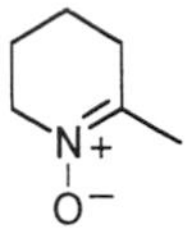 N+–O⁻	60[b]
7[a,c]	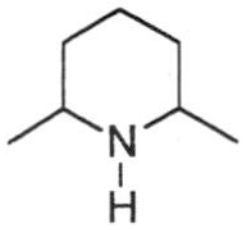 2,6-dimethylpiperidine N–H	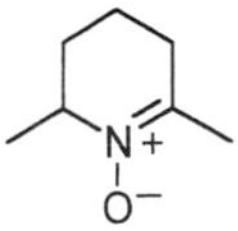 N+–O⁻	91
8[d]	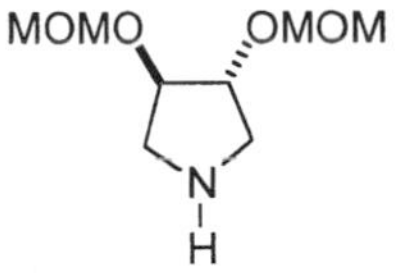 MOMO⟍ ⟋OMOM pyrrolidine N–H	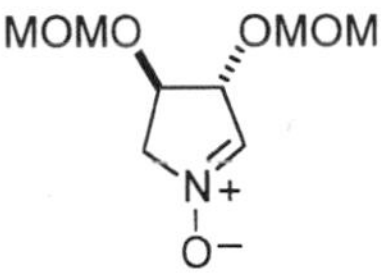 MOMO⟍ ⟋OMOM N+–O⁻	80
9[e]	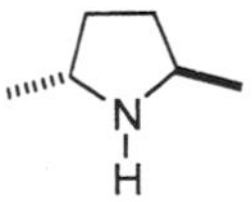 2,5-dimethylpyrrolidine N–H	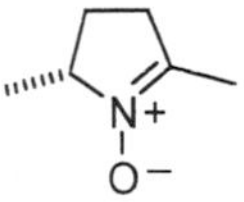 N+–O⁻	51[f]

209

10[g]   MeO...OMe (structure)   MeO...OMe (N-oxide structure)   94[h]

11[i]   (morpholinone structure with Ph, N-H)   (nitrone structure with Ph)   70-80[l]

12[m]   (pyrrolidine structure: Cbz, HN, OTBDPS, OTMS, COOBn, N-H)   (N-oxide structure: Cbz, HN, OTBDPS, OTMS, COOBn)   85

[a]Reference 16.
[b]About 10% of the regioisomeric aldonitrone was observed in the ^{1}H NMR spectrum of the crude mixture.
[c]Reference 18a.
[d]Reference 17a.
[e]Reference 14.
[f]Calculated over two steps starting from the N-benzylamine derivative.
[g]Reference 18b.
[h]Yield of crude product, used without further purification.
[i]Reference 15.
[l]On multigram scale.
[m]Reference 18c.

1. Dipartimento di Chimica Organica "Ugo Schiff", Polo Scientifico, Università degli Studi di Firenze, via della Lastruccia 13, I-50019 Sesto Fiorentino, Firenze, Italy.

2. Cooper, M. S.; Heaney, H.; Newbold, A. J.; Sanderson, W. R. *Synlett* **1990**, 533.

3. (a) Herrmann, W. A.; Fischer, R. W.; Scherer, W.; Rauch, M. U. *Angew. Chem. Int. Ed. Engl.* **1993**, *32*, 1157; (b) Espenson, J. H. *Chem. Commun.* **1999**, 479.

4. Murahashi, S.-I.; Mitsui, H.; Shiota, T.; Tsuda, T.; Watanabe, S. *J. Org. Chem.* **1990**, *55*, 1736.

5. (a) Tufariello, J. J. In *1,3-Dipolar Cycloaddition Chemistry*; Padwa, A.; Ed.; John Wiley & Sons: New York, 1984; (b) Confalone, P. N.; Huie, E. M. *Org. React.* **1988**, *36*, 1; (c) Torssell, K. B. G. *Nitrile Oxides, Nitrones, and Nitronates in Organic Synthesis*; Feuer, H., Ed.; VCH Publishers: New York, 1988; (d) Frederickson, M. *Tetrahedron* **1997**, *53*, 403.

6. For a recent use of nitrones in a different [3+2] cycloaddition, see: Hanessian, S.; Bayrakdarian, M. *Tetrahedron Lett.* **2002**, *43*, 967.

7. (a) Bloch, R. *Chem. Rev.* **1998**, *98*, 1407; (b) Enders, D.; Reinhold, U. *Tetrahedron: Asymmetry* **1997**, *8*, 1895; (c) Lombardo, M.; Trombini, C. *Synthesis* **2000**, 759; (d) Merino, P.; Franco, S.; Merchan, F. L.; Tejero, T. *Synlett* **2000**, 442.

8. (a) Janzen, E. G. In *Free Radicals in Biology*; Pryor, W. A., Ed.; Academic Press: New York, 1980; vol. 4, pp. 115-154; (b) Rosen, G. M.; Finkelstein, E. *Adv. Free Rad. Biol. Med.* **1985**, *1*, 345-375; (c) Janzen, E. G.; Haire, D. L. In *Advances in Free Radical Chemistry*; Tanner, D. D., Ed.; Jai Press: Greenwich, CT, 1990.

9. (a) Cicchi, S.; Corsi, M.; Goti, A. *J. Org. Chem.* **1999**, *64*, 7243; (b) Cicchi, S.; Marradi, M.; Goti, A.; Brandi, A. *Tetrahedron Lett.* **2001**, *42*, 6503.

10. Murray, R. W.; Singh, M. *J. Org. Chem.* **1990**, *55*, 2954.

11. Zajac, W. W., Jr.; Walters, T. R.; Darcy, M. G. *J. Org. Chem.* **1988**, *53*, 5856.

12. Murahashi, S.-I.; Shiota, T.; Imada, Y. *Org. Synth., Coll. Vol. IX* **1998**. 632.

13. Murahashi, S.-I.; Shiota, T. *Tetrahedron Lett.* **1987**, *28*, 2383.

14. Einhorn, J.; Einhorn, C.; Ratajczak, F.; Gautier-Luneau, I.; Pierre, J.-L. *J. Org. Chem.* **1997**, *62*, 9385.

15. Long, A.; Baldwin, S. W. *Tetrahedron Lett.* **2001**, *42*, 5343.

16. Goti, A.; Nannelli, L. *Tetrahedron Lett.* **1996**, *37*, 6025.

17. Related methods employing H_2O_2 instead of UHP have been published: (a) Murray, R. W.; Iyanar, K.; Chen, J.; Wearing, J. T. *J. Org. Chem.* **1996**, *61*, 8099; (b) Yamazaki, S. *Bull. Chem. Soc. Jpn.* **1997**, *70*, 877.

18. (a) Einhorn, J.; Einhorn, C.; Ratajczak, F.; Durif, A.; Averbuch, M.-T.; Pierre, J.-L. *Tetrahedron Lett.* **1998**, *39*, 2565; (b) Shibata, T.; Uemae, K.; Yamamoto, Y. *Tetrahedron: Asymmetry* **2000**, *11*, 2339; (c) Okue, M.; Kobayashi, H.; Shin-ya, K.; Furihata, K.; Hayakawa, Y.; Seto, H.; Watanabe, H. Kitahara, T. *Tetrahedron Lett.* **2002**, *43*, 857.

Appendix

Chemical Abstracts Nomenclature (Registry Number)

N-Benzylidene-benzylamine N-oxide; Benzenemethanamine,
  N-(phenylmethylene)-, N-oxide; (3376-26-9)

Methyltrioxorhenium: Rhenium, methyltrioxo-, (T-4); (70197-13-6)

Urea hydrogen peroxide: Urea, compd. with hydrogen peroxide (H_2O_2) (1:1); (124-43-6)

Dibenzylamine: Benzenemethanamine, N-(phenylmethyl)-; (103-49-1)

A CONVENIENT PREPARATION OF AN ORTHOGONALLY PROTECTED Cα,Cα-DISUBSTITUTED AMINO ACID ANALOG OF LYSINE:
1-tert-BUTYLOXYCARBONYL-4-((9-FLUORENYLMETHYLOXYCARBONYL)AMINO)-PIPERIDINE-4-CARBOXYLIC ACID
[1,4-Piperidinedicarboxylic acid, 4-[[(9H-fluoren-9-ylmethyloxy)carbonyl]amino]-, (1,1-dimethylethyl) ester]

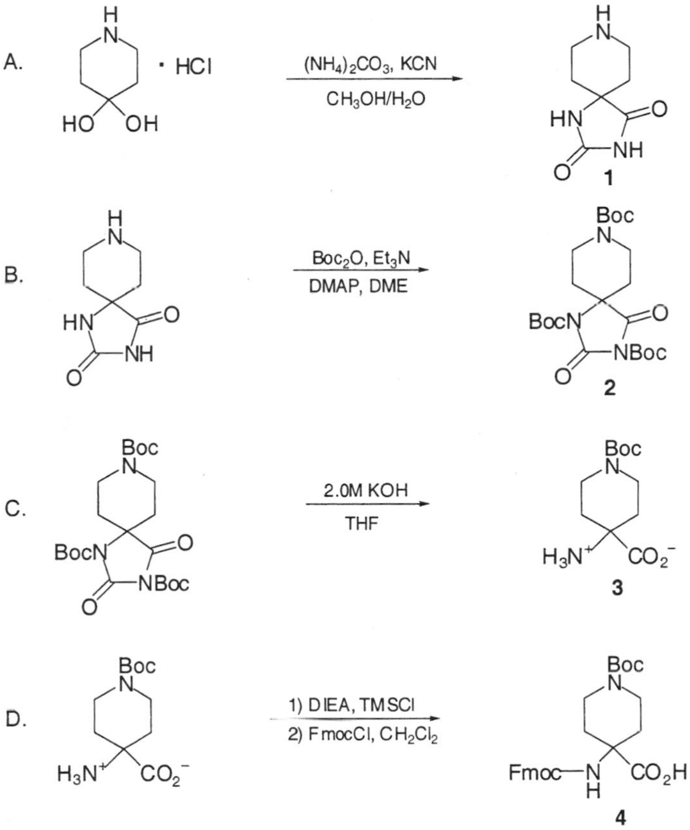

Submitted by Lars G. J. Hammarström, Yanwen Fu, Sidney Vail, Robert P. Hammer, and Mark L. McLaughlin.[1]

Checked by Rick L. Danheiser and Martin E. Hayes.

1. Procedure

Caution: Potassium cyanide is a potent poison, which should always be handled with gloves in a well-ventilated hood. Contact with acid releases toxic hydrogen cyanide gas.

A. *Piperidine-4-spiro-5'-hydantoin* (**1**). A 1000-mL, single-necked, round-bottomed flask equipped with a magnetic stirbar and an addition funnel fitted with an argon inlet is charged with 4-piperidone monohydrate hydrochloride (30.0 g, 195 mmol), ammonium carbonate (41.3 g, 420 mmol), 250 mL of methanol, and 150 mL of deionized water (Note 1). The mixture is allowed to stir at room temperature until all solids dissolve and then a solution of potassium cyanide (26.7 g, 410 mmol) (Note 2) in 100 mL of deionized water is added dropwise to the reaction mixture over 10 min. The reaction flask is sealed with a glass stopper and the reaction mixture is stirred at room temperature for 48 hr. The resulting suspension is concentrated to a volume of 300 mL by rotary evaporation at 40°C, after which the solution is cooled to 10°C. The white solid which precipitates is collected in a Büchner funnel using suction filtration. Concentration of the filtrate to a volume of 200 mL results in precipitation of additional product that is separated by filtration and combined with the first crop. The resulting light yellow solid is washed with four 25-mL portions of deionized water to yield a white solid (Note 3). The product is allowed to dry for 2 hr in the air and then is dried under reduced pressure (85°C; 0.5 mm) for 48 hr (Note 4) to yield 25.7-26.2 g (77-79%) of the desired hydantoin as a white solid (Note 5).

B. *1-tert-Butyloxycarbonylpiperidine-4-spiro-5'-(1',3'-bis(tert-butyloxy-carbonyl))* *hydantoin (2)*. A 2000-mL, three-necked, round-bottomed flask equipped with an argon inlet adapter, glass stopper, and an overhead mechanical stirrer is charged with a suspension of the hydantoin **1** (26.0 g, 154 mmol) in 1000 mL of 1,2-dimethoxyethane (Note 6). Triethylamine (15.7 g, 154 mmol) (Note 7) is added in one portion, and the resulting white suspension is stirred for 30 min. Di-tert-butyl dicarbonate (168.0 g, 770 mmol) is then added by pipette, followed by 4-dimethylaminopyridine (DMAP) (0.2 g, 1.5 mmol) (Note 8). Six additional 0.2 g-portions of DMAP are added at 12 hr intervals during the course of the reaction. The reaction mixture is stirred vigorously for a total of 72 hr, and the resulting light yellow solid is then collected in a Büchner funnel using suction filtration. The filtrate is concentrated to a volume of 60 mL by rotary evaporation, and the resulting solution is cooled to 15°C. The precipitate which appears is collected using suction filtration, added to the first crop, and the combined solids are dissolved in 500 mL of chloroform. This solution is washed with three 200-mL portions of 1.0N HCl, and the combined aqueous phases are extracted with 100 mL of chloroform. The combined organic layers are washed with 100 mL of saturated aq NaHCO$_3$ solution and 100 mL of brine (Note 9), dried over anhydrous MgSO$_4$, filtered, and concentrated by rotary evaporation. The resulting solid is dried at room temperature at 0.01 mm for 24 hr. The resulting finely ground light yellow solid is suspended in 400 mL of diethyl ether in a 1000-mL, round-bottomed flask equipped with a magnetic stirbar, stirred for 2 hr, and filtered on a Büchner funnel washing with four 50-mL portions of diethyl ether (Note 10). The product is dried under vacuum (85°C; 0.5 mm) for 24 hr to give 60.0–65.3 g (83-90%) of **2** as an ivory-colored solid (Note 11).

C. *4-Amino-1-tert-butyloxycarbonylpiperidine-4-carboxylic acid* (**3**). A 2000-mL, round-bottomed flask equipped with a magnetic stirbar is charged with a suspension of the hydantoin **2** (40.0 g, 0.8 mol) in 340 mL of THF (Note 12), and 340 mL of 2.0M potassium hydroxide solution (Note 13) is added in one portion. The flask is stoppered and the reaction mixture is stirred for 4 hr (Note 14) and then poured into a 1000-mL separatory funnel. The layers are allowed to separate over 45 min and the aqueous layer is then drained into a 1000-mL round-bottomed flask. This solution is cooled at 0°C while the pH is adjusted to 8.0 by the slow addition of ca. 100 mL of 6.0N HCl solution. The resulting solution is further acidified to pH 6.5 by slow addition of

215

2.0 N HCl solution (Note 15). The white precipitate which appears is collected by filtration on a Büchner funnel and the filtrate is concentrated to a volume of 60 mL to furnish additional precipitate which is collected by filtration. The combined portions of white solid are dried at room temperature under reduced pressure (65°C; 0.5 mm) for 12 hr and then suspended in 100 mL of chloroform (Note 16) and stirred for 45 min. The white solid is filtered and then dried under reduced pressure (85°C; 0.5 mm) for 24 hr to yield 13.4–14.1 g (64-68%) (Note 17) of the amino acid **3** as a white solid (Note 18).

D.   *1-tert-Butyloxycarbonyl-4-(9-fluorenylmethyloxycarbonylamino)piperidine-4-carboxylic acid* (**4**). A 1000-mL, three-necked, round-bottomed flask is equipped with a magnetic stirbar, argon inlet adapter, rubber septum, and an exit tube submerged in a 6.0M KOH solution. The apparatus is flame-dried under a flow of argon and then charged with finely ground amino acid **3** (17.0 g, 69.6 mmol). Anhydrous dichloromethane (500 mL) is added, followed by diisopropylethylamine (22.5 g, 30.3 mL, 174 mmol) which is added via syringe over 10 min (Note 19). The reaction mixture is stirred for 30 min and then chlorotrimethylsilane (15.1 g, 17.6 mL, 140 mmol) (Note 20) is added dropwise over 6 min (Note 21). After 30 min, the septum is replaced with a condenser and the reaction mixture is heated at reflux for 3 hr during which time the mixture becomes homogeneous. At 30 min intervals, the reaction vessel is flushed with argon for 30 sec to remove HCl formed in the reaction. The solution is then cooled to −10°C and 9-fluorenylmethyl chloroformate (Fmoc-Cl) (16.7 g, 64.6 mmol) (Note 22) is added in one portion. The resulting solution is stirred for 3 hr under a constant slow stream of argon and then concentrated by rotary evaporation. The resulting paste is taken up in 400 mL of diethyl ether and 1000 mL of aqueous 2.5% Na_2CO_3 solution (Note 23), and the aqueous layer is separated (Note 24) and washed with two 100-mL portions of ether. The aqueous layer is transferred to a 1-L beaker and cooled in an ice bath while 2.0N HCl is added to adjust the pH to 2.0. The resulting suspension is transferred to a 2-L separatory funnel (Note 25), and the precipitated acid is extracted with one 300-mL portion and two 150-mL portions of ethyl acetate. The combined organic phases are dried over MgSO$_4$, filtered, and concentrated by rotary evaporation and then at 60°C under reduced pressure (0.5 mm) for 10 hr. The resulting light yellow solid is suspended in 200 mL of hexane, stirred for 45 min, and then collected by filtration on a Büchner funnel (Note 26). The product is dried under reduced pressure (60°C; 0.5 mm) for 24 hr to afford 22.0–24.4 g (73-76%) (Note 27) of the desired product **4** (Note 28) as a white solid.

2. Notes

1.   4-Piperidone monohydrate hydrochloride and ammonium carbonate were purchased from Alfa-Aesar Chemical Company and used without further purification. Methanol was purchased from Mallinckrodt Chemical Company and used as received.

2.   Potassium cyanide was purchased from Alfa Aesar Chemical Company and used without further purification.

3.   Precipitated hydantoin must be washed with sufficient water to remove any residual ammonia, which can interfere with subsequent reactions. If an ammonia odor or yellow color persists after the initial water washes, additional washes should be performed.

4.   The submitters dried their product in a vacuum oven for 8 hr, then ground the resulting solid to a fine powder and returned it to the oven for 40 hr further.

5.   The hydantoin **1** is only slightly soluble in CD$_3$SOCD$_3$ and displays the following spectroscopic properties: ^{1}H NMR (500 MHz, CD$_3$SOCD$_3$) δ: 1.37 (d, J = 12.5 Hz, 2 H), 1.68 (dt, J = 4.6, 12.2 Hz, 2 H), 2.64 (app t, J = 12 Hz, 2 H), 2.83 (app d, J = 12.2 Hz, 2 H), 8.47 (s, 1 H), 10.3-10.6 (s, 2 H). ^{13}C NMR (125 MHz, CD$_3$SOCD$_3$) δ: 33.8, 41.2, 61.1, 156.4, 178.1.

6.   1,2-Dimethoxyethane (Dri-Solv) was purchased from EM Science Chemical Company and used as received.

7.   Triethylamine was purchased from Alfa-Aesar Chemical Company and used without further purification.

8.   Di-tert-butyl dicarbonate and 4-dimethylaminopyridine were purchased from Alfa-Aesar Chemical Company and used without further purification.

9.   The initial HCl wash results in an emulsion and up to 2 hr may be required for phase separation. At that point, any remaining emulsion should be separated and added to 100 mL of chloroform and 100 mL of 1.0N HCl solution. The chloroform layer is then combined with the other organic phases.

10.   Triturating with diethyl ether was found to be necessary to successfully remove residual di-tert-butyl dicarbonate and di-tert-butyl iminodicarboxylate which may be produced as a result of residual traces of ammonia from step A.

11.   Spectroscopic data for **2**: ^{1}H NMR (500 MHz, CDCl$_3$) δ: 1.46 (s, 9 H), 1.53 (s, 9 H), 1.57 (s, 9 H), 1.73 (m, 2 H), 2.67 (dt, J = 5.2, 13.6 Hz, 2 H), 3.32-3.50 (m, 2 H), 4.00-4.23 (m, 2 H). ^{13}C NMR (125 MHz, CDCl$_3$) δ: 27.9, 28.2, 28.6, 29.7, 29.9, 39.2, 40.2, 62.5, 80.1, 85.3, 87.1, 145.2, 147.3, 148.0, 154.5, 169.8.

12.   Tetrahydrofuran was purchased from Mallinckrodt Chemical Company and used as received.

13.   Potassium hydroxide was purchased from Mallinckrodt Chemical Company and used without further purification.

14.   Two phases appear within the first 15 min due to the heavy ionic content of the aqueous layer. Di-tert-butyl iminodicarboxylate, which is produced as a result of the hydrolysis, is selectively soluble in the THF layer, while the amino carboxylate salt of the product remains in solution in the aqueous layer.

218

15. Reaching the equilibrium where the amino acid zwitterion predominates is a slow process. After acidifying to pH 6.5, the solution is allowed to stir at 0°C for 25 min during which time the pH of the solution slowly increases. The pH is readjusted to pH 6.5 by slow addition of 2.0N HCl at 0°C. Repetition of this procedure as many as ten times may be necessary to insure the pH value of the aqueous solution remains at 6.5.

16. Triturating with chloroform is necessary to successfully remove a trace amount of 1-tert-butyloxycarbonylpiperidine-4-spiro-5'-(1'-tert-butyloxycarbonyl)-hydantoin.

17. The submitters report obtaining **3** in 87% yield.

18. Compound **3** is only slightly soluble in MeOH-d_4. Spectroscopic data for **3**: ^{1}H NMR (500 MHz, CD$_3$OD) δ: 1.44 (s, 9 H), 1.58-1.66 (m, 2 H), 2.07-2.14 (m, 2 H), 3.44-3.54 (m, 2 H), 3.64-3.72 (m, 2 H). ^{13}C NMR (125 MHz, CDCl$_3$) δ: 28.8, 33.3, 40.7, 41.7, 59.9, 81.5, 156.5, 175.9.

19. Diisopropylethylamine was purchased from Alfa-Aesar chemical Company and used without further purification. Dichloromethane was purified by pressure filtration through activated alumina.

20. Chlorotrimethylsilane was distilled from calcium hydride immediately prior to use.

21. Upon addition of chlorotrimethylsilane, vigorous HCl production is observed.

22. The checkers purchased Fmoc-Cl from Alfa-Aesar Chemical Company. Impurities are sometimes found in Fmoc-Cl purchased from various sources, and it is

necessary to verify its purity by TLC and ¹H-NMR analysis. If necessary, Fmoc-Cl can be purified by recrystallization from hexane.

23.   Efficient magnetic stirring for up to 2 hr is necessary to dissolve the crude product in the mixture of ether and aqueous 2.5% sodium carbonate solution.

24.   The solution is allowed to stand in a separatory funnel for at least 30 min before the aqueous layer is drained out. It is sometimes observed that a portion of the sodium carboxylate salt of **4** forms a third layer below the aqueous layer, in which case this layer should be combined with the sodium carbonate layer.

25.   The precipitated amino acid forms a gum on the inside walls of the beaker. Magnetic stirring and manual agitation with excess ethyl acetate is necessary to transfer the residual precipitate.

26.   Triturating with hexane is necessary to remove trace amounts of solvents (ethyl acetate and ether) in the product.

27.   The submitters obtained 26.9 g (83%) of the product.

28.   The product displays the following spectroscopic data: ¹H NMR (500 MHz, CDCl₃) δ: 1.46 (s, 9 H), 1.90-2.12 (br s, 4 H), 3.00-3.13 (br s, 2 H), 3.72-3.92 (br s, 2 H), 4.16-4.22 (m, 1 H), 4.38-4.48 (br s, 2 H), 5.06-5.14 (br s, 1 H), 7.30 (t, J = 7.4 Hz, 2 H), 7.39 (t, J = 7.4 Hz, 2 H), 7.56 (d, J = 7.0 Hz, 2 H), 7.75 (d, J = 7.4 Hz, 2 H). ¹³C NMR (125 MHz, (CD₃)₂SO) δ: 28.1, 31.3, 46.7, 56.7, 65.4, 78.8, 120.2, 125.3, 127.1, 127.7, 140.8, 143.8, 153.9, 155.4,175.1.

Waste Disposal Information

All toxic materials were disposed of in accordance with "Prudent Practices in the Laboratory"; National Academy Press; Washington, DC, 1995.

3. Discussion

The effect of C^α, C^α-disubstituted amino acids ($\alpha\alpha$AAs) on peptide secondary structure has been studied in recent years.[2a-d] While longer side-chain C^α, C^α-di-n-alkyl amino acids promote extended peptide conformation,[2a] alicyclic $\alpha\alpha$AAs, in which the C^α carbon forms a cyclic bridge with itself, such a 1-aminocyclopentane-1-carboxylic acid (Ac_5c) and 1-aminocyclohexane-1-carboxylic acid (Ac_6c), have helix-forming characteristics similar to those of 1-aminoisobutyric acid (Aib).[2a,c]

Since most $\alpha\alpha$AAs are hydrophobic in nature, peptides rich in $\alpha\alpha$AAs are generally restricted to study in organic solvents due to their low solubility in aqueous media. There have been very few examples of side-chain functionalized $\alpha\alpha$AAs that would allow for the synthesis of highly water-soluble peptides rich in $\alpha\alpha$AA content.[3] This is primarily due to difficulty of synthesis, since side-chain functionalized derivatives must be orthogonally protected to allow for incorporation into solid-phase peptide synthesis. The harsh conditions, under which standard methods of $\alpha\alpha$AA synthesis are performed, make this a difficult task.

Despite recent advances in synthetic methods to mildly generate polyfunctional $\alpha\alpha$AAs,[4] the Bucherer-Berg approach is still the most convenient method to generate simple $\alpha\alpha$AAs in good yields.[5] However, hydantoins suffer from the limitation of requiring harsh conditions for hydrolysis, thus compromising the ability to generate side-chain protected polyfunctional amino acids. This problem was partially overcome by Rebek and coworkers, who discovered that N,N'-bis-(t-butyloxycarbonyl)hydantoins can be hydrolyzed under much milder conditions.[6] The progress made in ααAA synthesis by mild hydantoin hydrolysis allowed the submitters to develop the synthesis of the orthogonally protected ααAA Fmoc-Api(Boc)-OH. This ααAA analog of lysine allows for the preparation of alicyclic ααAAs, while retaining high water-solubility.[6] As with other alicyclic ααAAs, Api has been found to strongly favor helical conformations of resulting peptides.[7b]

It has been found that the tris(tert-butyloxycarbonyl) protected hydantoin of 4-piperidone **2**, selectively hydrolyses in alkali to yield the N-tert-butyloxycarbonylated piperidine amino acid **3**. The hydrolysis, which is performed in a biphasic mixture of THF and 2.0M KOH at room temperature, cleanly partitions the deprotonated 4-amino-N'-(tert-butyloxycarbonyl)piperidine-4-carboxylic acid into the aqueous phase of the reaction with minimal contamination of the hydrolysis product, di-tert-butyl iminodicarboxylate, which partitions into the THF layer. Upon neutralization of the aqueous phase with aqueous hydrochloric acid, the zwitterion of the amino acid is isolated. The Bolin procedure to introduce the 9-fluorenylmethyloxycarbonyl protecting group efficiently produces **4**.[8] This synthesis is a significant improvement over the previously described method[9] where the final protection step was complicated by contamination of the hydrolysis side-product, di-tert-butyl iminodicarboxylate, which is very difficult to separate from **4**, even by chromatographic means.

1. Department of Chemistry, Louisiana State University, Baton Rouge, LA 70803.

2. (a) Paul, P. K. C.; Sukumar, M.; Bardi, R.; Piazzesi, A. M.; Valle, G.; Toniolo, C.; Balaram, P. *J. Am. Chem. Soc.* **1986**, *108*, 6363-6370; (b) Toniolo, C.; Benedetti, E. *Macromolecules* **1991**, *24*, 4004-4009; (c) Benedetti, E.; DiBlasio, B.; Iacovino, R.; Menchise, V.; Saviao, M.; Pedone, C.; Bonora, G. M.; Ettore, A.; Graci, L.; Formaggio, F.; Crisma, M.; Valle, G.; Toniolo, C. *J.*

Chem. Soc., Perkin Trans. 2 **1997**, 2023-2032; (d) Toniolo, C.; Crisma, M.; Formaggio, F.;Benedetti, E.; Santini, A.; Iacovino, R.; Saviano, M.; DiBlasio, B.; Pendone, C.; Kamphius, J. *Biopolymers* **1996**, *40*, 519-522.

3. Yokum, T. S.; Bursavich, M. G.; Piha-Paul, S. A.; Hall, D. A.; McLaughlin, M. L. *Tetrahedron Lett.* **1997**, *38*, 4013-4016.

4. Fu, Y.; Hammarström, L. G. J.; Miller, T. J.; Fronczek, F. R.; McLaughlin, M. L.; Hammer, R. P. *J. Org. Chem.* **2001**, *66*, 7118-7124.

5. (a) Bucherer, H. T.; Steiner, W. *J. Prakt. Chem.* **1934**, *140*, 291-316; (b) Bergs, H.; German patent 566,094 (May 26, 1929); *Chem Abst.* **1993**, *27*, 1001.

6. Kubik, S.; Meissner, Rebek, Jr., J. *Tetrahedron Lett.* **1994**, *35*, 6635-66

7. (a) Yokum, T. S.; Gauthier, T. J.; Hammer, R. P.; McLaughlin, M. L. *J. Am. Chem. Soc.* **1997**, *119*,

8. 1167-1168; (b) Hammarström, L. G. J.; Gauthier, T. J.; Hammer, R. P.; McLaughlin, M. L. *J. Peptide Res.* **2001**, *58*, 108-116.

8. Bolin, D. R.; Sytwu, I.–I.; Humiec, F.; Meienhofer, J. *Int. J. Pept. Protein Res.* **1989**, 353.

9. Wysong, C. L.; Yokum, T. S.; Morales, G. A.; Gundry, R. L.; McLaughlin, M. L.; Hammer, R. P. *J. Org. Chem.* **1996**, *61*, 7650-7651.

Appendix

Chemical Abstracts Nomenclature (Registry Number)

4-Piperidone monohydrate hydrochloride: 4-Piperidinone, hydrochloride; (41979-39-9).

Ammonium carbonate: Carbonic acid, diammonium salt; (506-87-6).

Potassium cyanide: Potassium cyanide [K(CN)]; (151-50-8)

Piperidine-4-spiro-5'-hydantoin: 1,3,8-Triazaspiro [4.5]decane-2,4-dione; (13625-39-3).

Triethylamine: Ethanamine, N,N-diethyl; (121-44-8).

Di-tert-butyl dicarbonate: Dicarbonic acid, bis(1,1-dimethylethyl) ester; (24424-99-5).

DMAP(4-(Dimethylamino)pyridine: 4-Pyridinamine, N,N-dimethyl-; (1122-58-3).

1-tert-Butyloxycarbonylpiperidine-4-spiro-5'-(1',3'-bis(tert-butyloxycarbonyl) hydantoin:
    1,3,8-Triazaspiro[4.5]decane-1,3,8-tricarboxylic acid, 2,4-dioxo-,
    tris(1-dimethylethyl) ester; (183673-68-9).

4-Amino-1-tert-butyloxycarbonylpiperidine-4-carboxylic acid: 1,4-Piperidinedicarboxylic
    acid, 4-amino-, 1-(1,1-dimethylethyl) ester; (183673-71-4).

1-tert-Butyloxycarbonyl-4-(9-fluorenylmethyloxycarbonylamino)piperidine-4-carboxylic
    acid: 1,4-Piperidinedicarboxylic acid, 4-[[(9H-fluoren-
    9-ylmethyloxy)carbonyl]amino]-,1-(1,1-dimethylethyl) ester; (183673-66-7).

Diisopropylethylamine: 2-Propanamine, N-ethyl-N-(1-methylethyl)-; (7087-68-5).

9-Fluorenylmethyl chloroformate: Carbonochloridic acid, 9H-fluoren-9-ylmethyl ester;
(28920-43-6).

Chlorotrimethylsilane: Silane, chlorotrimethyl-; (75-77-4).

SYNTHESIS AND USE OF GLYCOSYL PHOSPHATES
AS GLYCOSYL DONORS

A.

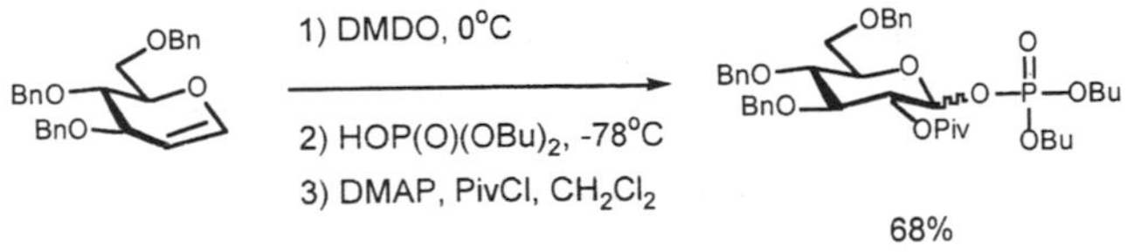

1) DMDO, 0°C

2) HOP(O)(OBu)$_2$, -78°C

3) DMAP, PivCl, CH$_2$Cl$_2$

68%

B.

TMSOTf, CH$_2$Cl$_2$

-60--40°C

90%

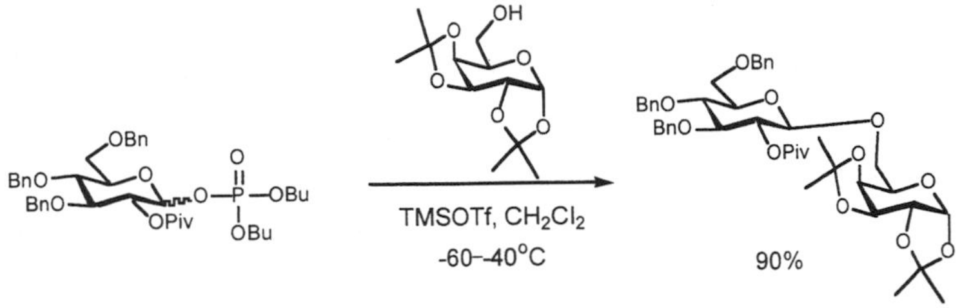

Submitted by Kerry R. Love and Peter H. Seeberger.[1]

Checked by Youseung Shin and Dennis P. Curran.

1. Procedure

A.   *Dibutyl 3,4,6-tri-O-benzyl-2-O-pivaloyl-D-glucopyranosyl phosphate.* A dry
500-mL, round-bottomed flask is charged with 5.00 g (12.0 mmol) of 3,4,6-tri-O-benzyl-
D-glucal (Note 1) and 7 mL of toluene (Note 2). The toluene is removed azeotropically
on a rotary evaporator and this process is repeated two more times. The residue is
dried under vacuum for 30 min. Under a flow of nitrogen, the flask is charged with a
magnetic stirbar and 50 mL of dichloromethane (Note 2) and the resulting solution is
cooled to 0°C. A solution of dimethyldioxirane (200 mL of a 0.08M solution, 16.0 mmol)
(Note 3) is added to the flask via cannula and the reaction mixture is stirred at 0°C for
10 min (Note 4). Volatile materials are then removed by rotary evaporation at 0°C (Note
5) and the resulting white residue is dried at 2 mm for 5 min. The flask is equipped with

a rubber septum fitted with a nitrogen inlet needle and the residue is dissolved in 40 mL of CH_2Cl_2 and the solution is cooled to –78°C. A solution of dibutyl phosphate (2.85 mL, 14.4 mmol) (Note 6) in 10 mL of dichloromethane is added via cannula. After 10 min at –78°C, the reaction mixture is warmed to 0°C and 4-(dimethylamino)pyridine (DMAP) (5.86 g, 48.0 mmol) and pivaloyl chloride (2.96 mL, 24.0 mmol) are added (Note 7). The solution is stirred for 30 min at 0°C, then allowed to warm to room temperature and stirred for 16 hr (Note 8). A solution of 25% ethyl acetate in hexanes is added and the resulting suspension is filtered through a pad of silica gel. The filtrate is concentrated to a clear oil and purified by column chromatography through a short plug of 80 g of silica gel (elution with 35% ethyl acetate-hexanes) (Note 9) to afford 5.86 g (67%) of dibutyl 3,4,6-tri-O-benzyl-2-O-pivaloyl-D-glucopyranosyl phosphate as a colorless oil (Notes 10, 11).

B.   *3,4,6-Tri-O-benzyl-2-O-pivaloyl-β-D-glucopyranosyl-(1→6)-1,2:3,4-di-O-isopropylidene-α-D-galactopyranoside.* A dry 500-mL, round-bottomed flask is charged with dibutyl 3,4,6-tri-O-benzyl-2-O-pivaloyl-D-glucopyranosyl phosphate (5.89 g, 8.11 mmol), 1,2:3,4-di-O-isopropylidene-α-D-galactopyranose (1.92 g, 7.37 mmol) (Note 12), and 7 mL of toluene (Note 2). The toluene is removed azeotropically on a rotary evaporator and this process is repeated two more times. The residue is dried under vacuum for 1 hr. The flask is equipped with a rubber septum fitted with a nitrogen inlet needle and a magnetic stirbar, and 100 mL of CH_2Cl_2 is added. The resulting solution is cooled to –60°C. Trimethylsilyl triflate (TMSOTf) (1.47 mL, 8.11 mmol) (Note 13) is added and the solution is allowed to warm slowly to –40°C (Note 14). After 30 min, the reaction is quenched by addition of 15 mL of triethylamine (Note 15) and allowed to warm to room temperature. The reaction mixture is concentrated by rotary evaporation and the residue is purified by column chromatography on 140 g of silica gel (elution with 30% ethyl acetate-hexanes) to afford 5.09 g (85%) of 3,4,6-tri-O-benzyl-2-O-pivaloyl-D-glucopyranosyl-(1→6)-1,2:3,4-di-O-isopropylidene-α-D-galactopyranoside as a colorless oil (Notes 16,17).

2. Notes

1. Tri-O-benzyl-D-glucal (97%) was purchased from Aldrich Chemical Company, Inc. and was used as supplied without further purification.

2. Toluene and dichloromethane (CH_2Cl_2) were purified by a JT Baker Cycle-Tainer Solvent Delivery System.

3. Dimethyldioxirane was prepared according to the literature procedure[2] as an 0.08M solution and was dried over 4Å beaded molecular sieves for 24 hr prior to use.

4. Analytical thin-layer chromatography was performed on E. Merck silica gel 60 F_{254} plates (0.25 mm) and compounds were visualized by dipping the plates in a cerium sulfate-ammonium molybdate solution followed by heating.

5. Expedient removal of the volatiles after the epoxidation with dimethyldioxirane is crucial to achieve reproducible yields because the epoxide is extremely water sensitive.

6. Dibutyl phosphate (97%) was purchased from Fluka and was used as supplied without further purification.

7. DMAP (99%) and pivaloyl chloride (99%) were purchased from Aldrich Chemical Company, Inc. and were used as supplied without further purification.

8. The submitters recommend that the reaction be worked up immediately upon reaching room temperature, but the checkers found that it was not detrimental to allow the reaction to stir overnight at room temperature prior to workup.

9. Prolonged chromatography when purifying glycosyl phosphates will lead to hydrolysis and decomposition of the product.

10. Early fractions were mixtures of α/β anomers while later fractions contained pure β-anomer. The fractions were combined to give the product as a 9:1 mixture of

227

β/α-anomers. Analytical data for this compound are as follows: (β-phosphate) R_f 0.29 (ethyl acetate:hexane, 1:3); $[\alpha]^{24}_D$ −1.9 (c = 1.50, CH_2Cl_2); IR (thin film) cm^{-1}: 2946, 1740, 1454, 1282, 1016; ^{1}H NMR (500 MHz, $CDCl_3$) δ: 0.96-0.88 (m, 6H), 1.20 (s, 9H), 1.40-1.34 (m, 4H), 1.64-1.59 (m, 4H), 3.64-3.61 (m, 1H), 3.78-3.70 (m, 3H), 3.82 (t, J = 9.5 Hz, 1H), 4.08-4.00 (m, 4H), 4.51 (d, J = 11.0 Hz, 1H), 4.69-4.54 (m, 2H), 4.70 (d, J = 11.0 Hz, 1H), 4.80-4.75 (m, 2H), 5.17 (app t, J = 8.5 Hz, 1H), 5.24 (app t, J = 7.3 Hz, 1H), 7.16-7.14 (m, 2H), 7.33-7.25 (m, 13H); ^{13}C NMR (125 MHz, $CDCl_3$) δ: 14.0, 19.1, 26.9, 32.7, 39.2, 68.1, 68.2, 68.4, 73.3, 73.9, 75.9, 76.2, 83.1, 97.0, 127.6, 128.0, 128.1, 128.2, 128.3, 128.7, 138.1, 138.2, 177.2, (d, J_{C-P} = 5.0 Hz); ^{31}P NMR (200 MHz, $CDCl_3$) δ: -2.2; FAB MS m/z (M)$^+$ calcd 726.3532, obsd 726.3537. (α-Phosphate) R_f 0.36 (ethyl acetate:hexane, 1:3); $[\alpha]^{24}_D$ +50.5 (c = 0.63, CH_2Cl_2); IR (thin film) cm^{-1}: 2960, 2872, 1736, 1454, 1282; ^{1}H NMR (500 MHz, $CDCl_3$) δ: 0.97-0.91 (m, 6H), 1.24 (s, 9H), 1.44-1.36 (m, 4H), 1.86-1.61 (m, 4H), 3.68 (d, J = 11.0 Hz, 1H), 3.86-3.79 (m, 2H), 4.10-4.02 (m, 5H), 4.56-4.50 (m, 3H), 4.63 (d, J = 11.5 Hz, 1H), 4.83-4.80 (m, 3H), 4.99-4.97 (m, 1H), 5.85 (dd, J = 1.8, 6.4 Hz, 1H), 7.18-7.15 (m, 2H), 7.35-7.27 (m, 13H),; ^{13}C NMR (125 MHz, $CDCl_3$) δ: 13.8, 18.8, 27.3, 32.4, 32.5, 39.0, 67.8, 67.9, 68.0, 68.2, 72.6, 72.7, 73.7, 75.4, 75.6, 79.5, 94.7, 127.6, 127.7, 127.9, 128.0, 128.1, 128.2, 128.3, 128.5, 128.6, 138.0, 138.1, 138.3, 177.7, (d, J_{C-P} = 5.5 Hz); ^{31}P NMR (200 MHz, $CDCl_3$) δ: -2.5; FAB MS m/z (M)$^+$ calcd 726.3532, obsd 726.3537.

11. Practical considerations limit the scalability of this reaction due to the highly reactive and water sensitive intermediates formed. Furthermore, the time required for removal of large amounts of solvent *in vacuo* allows for the opening of the intermediate epoxide leading to diol formation.

12. 1,2:3,4-Di-*O*-isopropylidene-α-D-galactopyranoside (97%) was purchased from Aldrich Chemical Company, Inc. and was used as supplied without further purification.

13. TMSOTf (99%) was purchased from Acros Organics and was used as supplied without further purification.

14. The submitters report that the less reactive α-phosphate can be completely consumed by allowing the reaction mixture to warm to –20°C prior to workup.

15. Triethylamine (98%) was purchased from J. T. Baker and was used as supplied without further purification.

16. The product was obtained in 90-91% yield in other runs. Analytical data for this compound are as follows: $[\alpha]^{24}_D$ –45.2 (c = 2.34, CH_2Cl_2); IR (thin film) cm^{-1}: 3029, 2978, 2933, 2904, 1741, 1134, 1028; ^{1}H NMR (500 MHz, $CDCl_3$) δ: 1.21 (s, 9H), 1.31 (s, 3H), 1.32 (s,3H), 1.43 (s, 3H), 1.51 (s, 3H), 3.53-3.50 (m, 1H), 3.63-3.59 (m,1H), 3.76-3.69 (m, 4H), 3.97-3.94 (m, 1H), 4.10-4.07 (m, 1H), 4.29-4.25 (m, 2H), 4.46 (d, J = 8.0 Hz, 1H), 4.58-4.53 (m, 3H), 4.64 (d, J = 8.0 Hz, 1H), 4.79-4.69 (m, 3H), 5.10 (app t, J = 8.5 Hz), 5.49 (d, J = 5.0 Hz, 1H), 1H), 7.19-7.15 (m, 2H), 7.39-7.24 (m, 13H),; ^{13}C NMR (125 MHz, $CDCl_3$) δ: 24.7, 25.4, 26.3, 26.4, 27.5, 39.1, 67.4, 69.0, 70.9, 71.5, 73.3, 73.9, 75.2, 75.7, 78.1, 83.6, 96.6, 101.8, 108.8, 109.5, 127.7, 127.9, 127.8, 128.0, 128.1, 128.2, 128.6, 128.7, 138.4, 138.5, 177.1; FAB MS m/z (M)$^+$ calcd 776.3772, obsd 776.3770.

17. Continued elution gave unreacted α-phosphate (0.125 g, 2%).

Waste Disposal Information

All toxic materials were disposed of in accordance with "Prudent Practices in the Laboratory"; National Academy Press; Washington, D.C., 1995.

3. Discussion

The role of complex carbohydrates in biological processes is now widely appreciated. Oligosaccharides and glycoconjugates are essential for many cellular events, such as recognition, adhesion and signaling between cells.[3] Carbohydrates have also been implicated in a variety of disease states including cancers[4] and in a host of bacterial and viral infections.[5] The widespread biological implications of carbohydrates have rendered them targets of intense study. Microheterogeneity of naturally derived oligosaccharides, however, limits the access to pure materials in appreciable quantity. Chemical synthesis, therefore, remains the best way procure material for biological investigations.

Advances in carbohydrate chemistry, particularly in the development of powerful glycosylating agents, have provided access to molecules of biological interest. In particular, glycosyl trichloroacetimidates,[6] glycosyl fluorides,[7] thioglycosides[8] and n-pentenyl glycosides[9] have each been used in the construction of complex oligosaccharides. Limitations of these glycosyl donors, including lengthy syntheses, difficult activation conditions, and long reaction times, have lead us to develop glycosyl phosphates for the facile construction of a variety of glycosidic linkages.

Glycosyl phosphates can be synthesized from a variety of intermediates; anomeric lactols[10,11] and 1,2-anhydrosugars[12] are most typically used. Additionally, other glycosyl donors, such as glycosyl trichloroacetimidates[13] and n-pentenyl glycosides,[14] may be converted into glycosyl phosphates. The aforementioned procedure generates the 1,2-anhydrosugar from a suitably protected glycal with dimethyldioxirane. Opening of the anhydrosugar at low temperature using a phosphate diester, followed by acylation of the C^2-hydroxyl group leads to fully protected glycosyl phosphates (Figure 1).[12]

This method is attractive for three reasons: First, glycal precursors are desirable starting materials because they require the differentiation of only three hydroxyl groups, each having unique reactivity. Generation of phosphates from anomeric lactols or other glycosyl donors requires many synthetic steps for the differential protection of the five hydroxyl groups present. Second, the procedure can be carried out in one pot providing quick access to the desired glycosyl phosphates in high yields. Finally, choice of

230

solvent for the opening of the anhydrosugar can lead to either alpha- or beta-enriched phosphates.[12]

Initial work by Ikegami and coworkers showed that glycosyl phosphates are highly reactive glycosylating agents.[15] Activation of the glycosyl phosphate by addition of a stoichiometric amount of a Lewis acid in the presence of an acceptor quickly forms the desired disaccharide in good yield (Table 1). Glycosyl phosphates have been successfully employed in the synthesis of both O- and C-glycosides,[16] including biologically relevant compounds such as the Leishmania antigenic tetrasaccharide[17] and the H-type II blood group pentasaccharide.[18] Use of phosphates on solid-support in an automated fashion further demonstrates their utility as glycosyl donors.[19]

Figure 1
Glycosyl Phosphates Prepared from Glycals

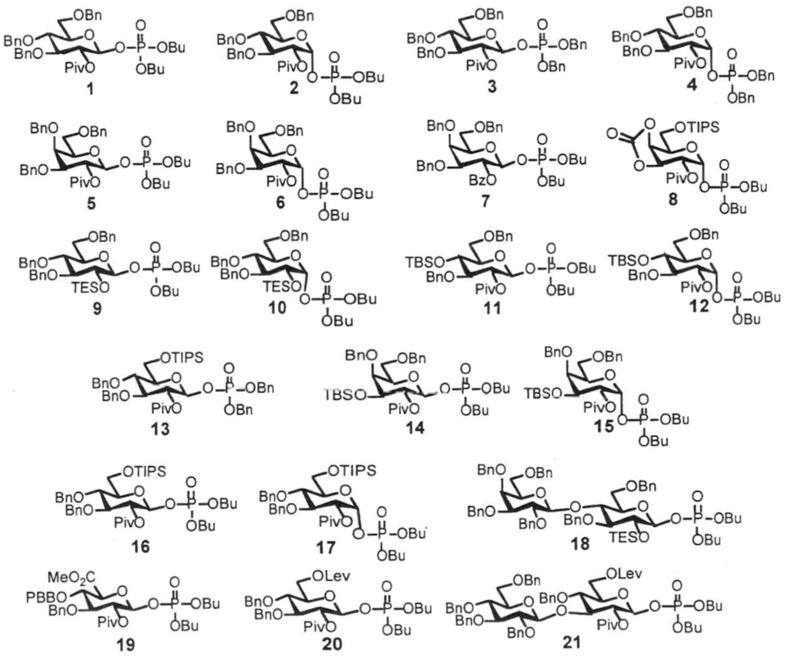

231

Table 1
Glycosylation with Glycal Derived Glycosyl Phosphates

Donor	Acceptor	Product	Yield (%)
1	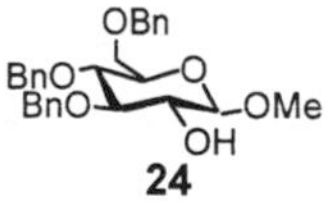 24	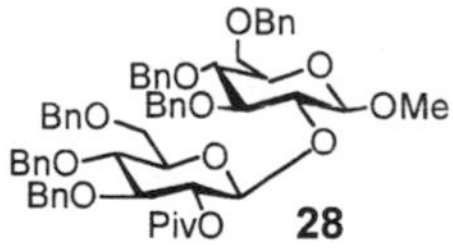 28	83
2	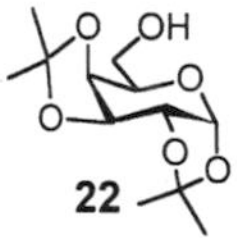 22	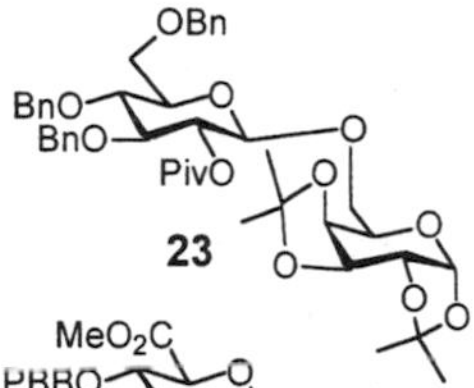 23	87
19	22	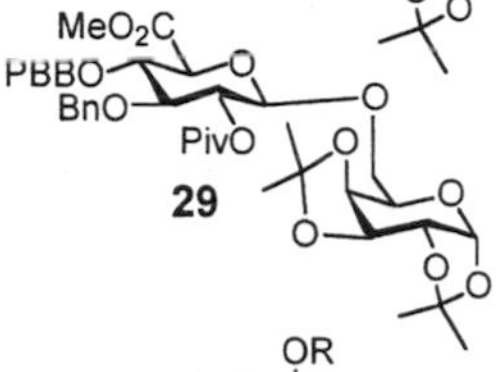 29	72
19	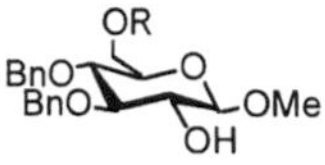 25	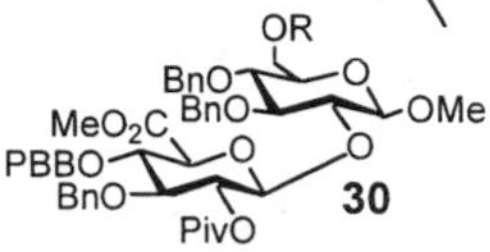 30	84
3	EtSH 26	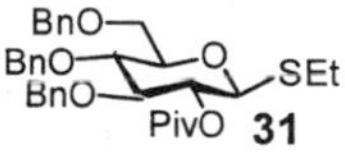 31	90
7	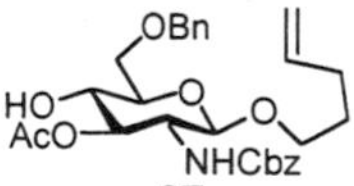 27	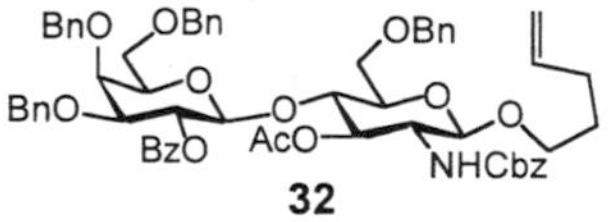 32	96
9	22	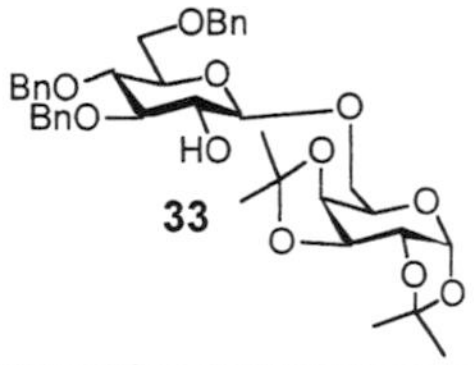 33	71

1. Department of Chemistry, Massachusetts Institute of Technology, Cambridge, MA 02139. Present address: Laboratory of Organic Chemistry, ETH Hönggerberg/ HCI, CH – 8093, Zürich, Switzerland, E- mail:seeberger@org.chem.ehtz.ch

2. Murray, R. W.; Singh, M. *Org. Synth.* **1996**, *73*, 91.

3. Varki, A. *Glycobiology* **1993**, *3*, 97.

4. Hakomori, S. *Adv. Cancer Res.* **1989**, *52*, 257.

5. Rudd, P. M.; Eliott, T.; Cresswell, P.; Wilson, I. A.; Dwek, R. A. *Science* **2001**, *291*, 2370.

6. Schmidt, R. R. *Adv. Carbohydr. Chem. Biochem.* **1994**, *50*, 21.

7. Mukaiyama, T.; Murai, Y.; Shida, S. *Chem. Lett.* **1981**, 431.

8. Garegg, P. J. *Adv. Carbohydr. Chem. Biochem.* **1997**, *52*, 179.

9. Mootoo, D. R.; Date, V.; Fraser-Reid, B. J. *J. Am. Chem. Soc.* **1988**, *110*, 2662.

10. Via the chlorophosphate: (a) Sabesan, S.; Neira, S. *Carbohydr. Res.* **1992**, *223*, 169. (b) Inage, M.; Chaki, H.; Kusumoto, S.; Shiba, T. *Chem Lett.* **1982**, 1281.

11. Via dehydrative glycosylation: Garcia, B. A.; Gin, D. Y. *Org. Lett.* **2000**, *2*, 2135.

12. Plante, O. J.; Palmacci, E. R.; Andrade, R. B.; Seeberger, P. H. *J. Am. Chem. Soc.* **2001**, *123*, 9545.

13. Hoch, M.; Heinz, E.; Schmidt, R. R. *Carbohydr. Res.* **1989**, *191*, 21.

14. Pale, P.; Whitesides, G. M. *J. Org. Chem.* **1991**, *56*, 4547.

15. Hashimoto, S.; Honda, T.; Ikegami, S. *J. Chem. Soc. Comm.* **1989**, 685.

16. Palmacci, E. R.; Seeberger, P. H. *Org. Lett.* **2001**, *3*, 1547.

17. Hewitt, M. C.; Seeberger, P. H. *J. Org. Chem.* **2001**, *66*, 4233.

18. Love, K. R.; Andrade, R. B.; Seeberger, P. H. *J. Org. Chem.* **2001**, *66*, 8165.

19. Plante, O. J.; Palmacci, E. R.; Seeberger, P. H. *Science*, **2001**, *291*, 1523.

233

Appendix

Chemical Abstracts Nomenclature (Registry Number)

Dibutyl 3,4,6-tri-*O*-benzyl-2-*O*-pivaloyl-D-glucopyranosyl phosphate: β-D-Glucopyranose, 3,4,5-tris-*O*-(phenylmethyl)-, 1-(dibutyl phosphate) 2-(2,2-dimethylpropanoate); (223919-63-7)

3,4,6-Tri-*O*-benzyl-D-glucal: D-arabino-Hex-1-enitol, 1,5-anhydro-2-deoxy-3,4,6-tris-*O*-(phenylmethyl)-; (55628-54-1)

Dibutyl phosphate: Phosphoric acid, dibutyl ester; (107-66-4)

4-(Dimethylamino)pyridine: 4-Pyridinamine, N,N-dimethyl-; (1122-58-3)

Pivaloyl chloride: Propanoyl chloride, 2,2-dimethyl-; (3282-30-2)

3,4,6-Tri-*O*-benzyl-2-*O*-pivaloyl-β-D-glucopyranosyl-(1→6)-1,2:3,4-di-*O*-isopropylidene-α-D-galactopyranoside: α-D-Galactopyranose, 6-O-[2-O-(2,2-dimethyl-1-oxopropyl)-3,4,6-tris-*O*-(phenylmethyl)-β-D-glucopyranosyl}-1,2:3,4-bis-*O*-(1-methylethylidene)-; (219122-26-6)

1,2:3,4-Di-*O*-isopropylidene-α-D-galactopyranose: α-D-galactopyranose, 1,2:3,4-bis-*O*-(1-methylethylidene)-; (4064-06-6)

Trimethylsilyl triflate: Methanesulfonic acid, trifluoro-, trimethylsilyl ester; (27607-77-8)

THE USE OF POLYSTYRYLSULFONYL CHLORIDE RESIN AS A SOLID SUPPORTED CONDENSATION REAGENT FOR THE FORMATION OF ESTERS: SYNTHESIS OF N-[(9-FLUORENYLMETHOXY)CARBONYL]-L-ASPARTIC ACID; α-tert-BUTYL ESTER, β-(2-ETHYL[(1E)-(4-NITROPHENYL)AZO]PHENYL]AMINO]ETHYL ESTER

Submitted by Norbert Zander[1] and Ronald Frank.[2]

Checked by Richard V. Coelha, Jr., Klaas Schildknegt, and Sarah E. Kelly.

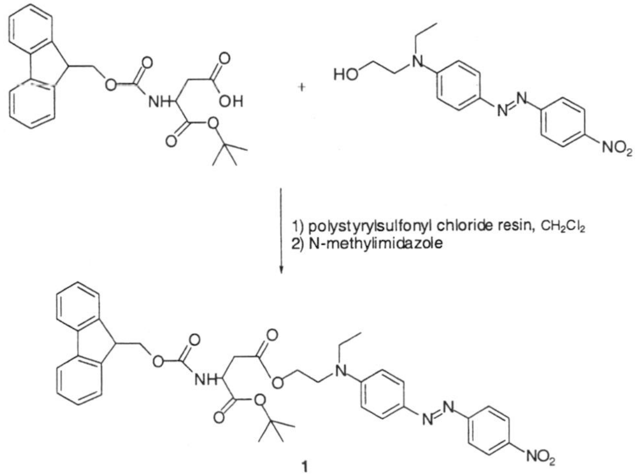

1. Procedure

To a 50-mL polypropylene vial (Note 1) are added 0.839 g (2.67 mmol) of 2-[ethyl[4-[(1E)-(4-nitrophenyl)azo]phenyl]amino]ethanol (Disperse Red 1, Note 2), 0.985 g of (2.39 mmol) N-[(9H-fluoren-9-ylmethoxy)carbonyl]-L-aspartic acid,1-(1,

1-dimethylethyl) ester (Fmoc-L-Asp-OtBu, Note 3), 3.26 g (4.73 mmol) of polystyrylsulfonyl chloride resin (Note 4), and 30 mL anhydrous methylene chloride (Note 5). The vial is capped and the mixture is shaken for five min (Note 6). N-Methylimidazole (0.764 mL, 9.58 mmol) is then added to the deep red mixture (Note 7) and the resulting mixture is shaken for 2 hr (Note 8).

N-Methylimidazole is then removed from the reaction mixture with Amberlyst 15 ion exchange resin (Note 9) using the following procedure. To a 2-cm diameter column equipped with a glass frit (Note 1) is added 7 g of Amberlyst 15, which is rinsed with 25 mL of methylene chloride. The reaction mixture is added and the resin mixture is rinsed with 50 mL of methylene chloride. The filtrate is collected in a 250-mL flask (Note 10) and the solvent is removed on a rotary evaporator to afford 1.63-1.66 g (96-98%) of the desired ester **1** as a deep red foam. HPLC analysis showed a purity of 98% (Notes 11, 12, 13, 14).

2. Notes

1.   Polystyrene resin strongly sticks to glass, and consequently either polypropylene reaction vessels or silanized glassware should be used. Glassware was silanized by the following procedure. The glassware and a small beaker containing neat chlorotrimethylsilane were placed in a desiccator. Vacuum was applied until the chlorotrimethylsilane started to boil and then the desiccator was closed. The next day, the glassware was removed, rinsed with methylene chloride, and dried at 80°C.

2.   Disperse Red 1, dye content ca. 95%, is available from Aldrich, Sigma-Aldrich Chemie GmbH. The checkers obtained Disperse Red 1 from Sigma-Aldrich Corporation.

3.   The submitters obtained Fmoc-L-Asp-OtBu from Calbiochem-Novabiochem AG. The checkers obtained Fmoc-L-Asp-OtBu from EMD Biosciences, Inc.

4.   Polystyrylsulfonyl chloride resin is available from Argonaut Technologies. The batch used by the submitters had a substitution of 1.44 mmol/g (the checker's batch had a substitution of 1.45 mmol/g). The use of polystyrylsulfonyl chloride resin from Novabiochem in this type of reaction resulted in drastically longer reaction times and less pure products.

5.   Methylene chloride was the solvent most compatible with the reagents and resin. Tetrahydrofuran or a 1 : 1 mixture of dimethylformamide (DMF) and methylene chloride could also be employed, but longer reaction times were necessary. 1,4-Dioxane, toluene, N-methylpyrrolidinone, and DMF alone were not suitable as solvents.

6.   Mixtures containing polystyrene resins should either be shaken or stirred with a mechanical stirrer. Stirring with a magnetic stir bar results in destruction of resin beads and the resulting debris can clog frits during filtrations.

7.   In place of N-methylimidazole (MeIm), only dimethylaminopyridine (DMAP) could be substituted. The solid-supported amines piperidinomethyl- or morpholinomethyl polystyrene resins, pyridine, and tertiary amines like triethylamine and N-methylmorpholine were not effective.

8.   HPLC analysis showed complete conversion after 90 min. The use of less base resulted in longer reaction times.

9.   Amberlyst 15 is available from Fluka, Sigma-Aldrich Chemie GmbH. The checkers obtained Amberlyst 15 from Sigma-Aldrich Corporation. The use of only 6 g of Amberlyst 15 resulted in incomplete removal of MeIm.

10.   Further rinsing with methylene chloride yielded trace quantities of additional product, which was less pure by HPLC analysis. The main impurity was **2** (see Discussion) produced by cleavage of the *t*-butyl ester by the Amberlyst 15 resin.

11.   Analytical reversed-phase HPLC was performed using a 50 mm x 2 mm i.d. 3 μm C18(2) column (LUNA, Phenomenex, Germany); solvent A was 0.1% TFA in HPLC-grade water; solvent B was 0.1% TFA in HPLC-grade acetonitrile; UV-detection was at 220 nm; flow rate was 0.8 mL/min; gradient elution: 0 min 95% A, 0-11 min 5% A, 11-12.5 min 5% A, 12.5-13 min 95% A, 13.5-17 min 95% A.

12.   The submitters performed the reaction using a 0.12 mmol excess of Fmoc-L-Asp-O*t*-Bu, under which conditions aminomethylated polystyrene resin was required to remove the excess carboxylic acid (Note 13). The checkers modified the reaction to use 0.28 mmol excess Disperse Red 1. The initial Amberlyst-15 filtration removes this material.

13.   To remove carboxylic acid, the crude product is redissolved in 20 mL of methylene chloride and is shaken for 30 min with 1 g of aminomethylated polystyrene resin with a substitution of 1.02 mmol/g, available from Novabiochem (Note 3). After filtration and washing of the resin with 50 mL of methylene chloride, the filtrates were

collected together in a 250-mL flask and the solvent was removed on a rotary evaporator.

14.   Spectral properties of the product are as follows: IR (KBr): cm^{-1} 2975, 1733, 1600, 1549, 1514, 1433, 1391, 1336, 1253, 1133, 853, 741; ^{1}H NMR (400 MHz, DMSO-d_6): δ 1.11 (t, 3H, J = 6.8 Hz), 1.30 (s, 9H), 2.61 (dd, 1H, J = 16.4, 8.1 Hz), 2.71 (dd, 1H, J = 16.2, 6.6 Hz), 3.50 (q, 2H, J = 6.9 Hz), 3.68 (m, 2H), 4.18 (t, 1H, J = 6.6 Hz), 4.32 – 4.23 (m, 5H), 6.88 (d, 2H, J = 9.5 Hz), 7.28 (t, 2H, J = 7.5 Hz), 7.37 (t, 2H, J = 7.3 Hz), 7.66 (d, 2H, J = 7.5 Hz), 7.74 (d, 1H, J = 8.3 Hz), 7.81 (d, 2H, J = 9.1 Hz), 7.85 (d, 2H, J = 7.5 Hz),), 7.88 (d, 2H, J = 9.1 Hz), 8.32 (d, 2H, J = 9.1 Hz; ^{13}C NMR (100 MHz, DMSO-d_6): δ 12.6, 28.1, 36.1, 36.7, 45.7, 47.2, 48.8, 51.6, 62.6, 66.3, 81.8, 112.4, 120.2, 120.8, 123.1, 123.8, 125.6, 125.8, 126.8, 127.7, 128.3, 136.3, 141.4, 143.4, 144.4, 147.5, 152.3, 156.5, 156.7, 170.5, 170.7; HRMS (ESI): Calcd for $C_{39}H_{41}N_5O_8+Na^+$: 730.2853, found 730.2854.

Waste Disposal Information

All toxic materials were disposed in accordance with "Prudent Practices in the Laboratory"; National Academy Press; Washington, DC 1995.

3. Discussion

Although solid-supported reagents and scavengers have been used in organic synthesis for decades, it was the development of combinatorial and parallel high throughput synthesis techniques that brought this class of reagents to wider attention.

While there are numerous applications of solid-supported reagents and scavengers only a few examples for the formation of esters have been described[3]. Carboxylates, generated with solid-supported organic bases, were alkylated with alkyl halides[4]. Solid-supported organic bases were also used as scavenger resins in the esterification of benzyl alcohol with benzoyl chlorides, giving clean benzyl esters in high yields[5]. This approach requires the acid chloride to be available. A recent report describes the alkylation of carboxylic acids with carbenium ions, generated from polymer-supported triazines[6]. This approach, however, requires the preformation of the polymer-supported triazine for each alkyl group to be transferred and a relatively high excess of the alkylating resin. No examples for the formation of aryl esters are given.

We report here a convenient and general procedure for the formation of esters directly from the easily available carboxylic acids and alcohols or phenols[7]. The reaction uses polystyrylsulfonyl chloride resin as an efficient dehydration reagent and N-methylimidazole as basic catalyst. The method employs only commercially available supported reagents and scavengers and allows compounds to be obtained in excellent yields and high purity by simple filtration. It should therefore be especially suitable for an automated parallel synthesis. The scope of our procedure was examined by treating six carboxylic acids with Fmoc-glycinol and seven alcohols with Fmoc-glycine. The results are summarized in Table 1. The purity of all the different esters derived from Fmoc-glycinol (entry 1-6) was very good (>93%), while the reaction time necessary for a complete conversion varied considerably. All products of the esterification of Fmoc-glycine (entry 7-13) with different alcohols and donor or acceptor substituted phenols were essentially 100% pure, except for *tert*-butyl alcohol, which reacted only after the addition of 0.25 equivalents of DMAP. The conversion was slow, but the resulting ester was of good purity.

Table 1:

Entry	Carboxylic acid	Alcohol	Purity [%][c, d]	Reaction Time
1	isobutyric acid	Fmoc-Gly-ol[a]	100	60 min
2	pivalic acid	"	98 (2)	22 h
3	6-bromohexanoic acid	"	100	30 min
4	benzoic acid	"	94	60 min
5	4-acetamidobenzoic acid	"	93 (7)	22 h
6	3-nitrobenzoic acid	"	100	15 min
7	Fmoc-Gly-OH[b]	ethanol	100	30 min
8	"	isobutyl alcohol	100	30 min
9	"	tert-butyl alcohol	87	22 h[e]
10	"	benzyl alcohol	100	30 min
11	"	phenol	100	30 min
12	"	4-Ethoxycarbonylphenol	98	30 min
13	"	4-methoxyphenol	100	30 min

a) 0.0706 mmol Fmoc-Gly-ol, 1.3 eq. carboxylic acid, 1.3 eq sulfonyl chloride resin, 4 eq. MeIm 1 mL abs. DCM; b) 0.0673 mmol Fmoc-Gly-OH, 1.3 eq. alcohol, 1.3 eq sulfonyl chloride resin, 4 eq. MeIm 1 mL abs. DCM; c) HPLC of the crude product, percent starting material in parentheses; d) all compounds were characterized by HPLC-ESI-MS or ^{1}H-NMR, e) 0.25 eq. DMAP added

Scheme 1:

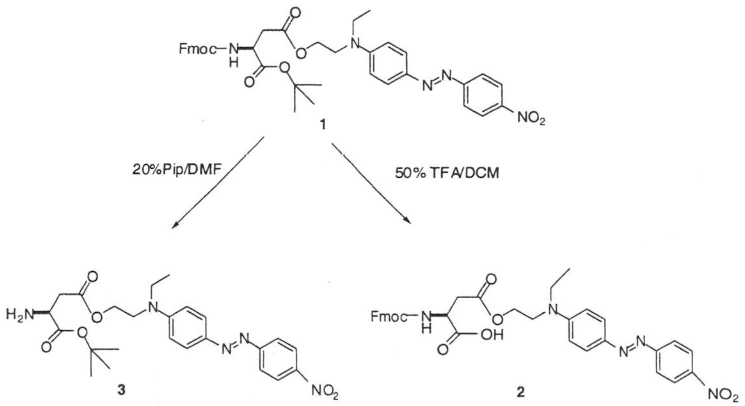

The example described in the experiment illustrates the compatibility of this reaction and the work up procedure with the base-sensitive Fmoc- as well as the acid-sensitive *tert*-Bu-ester protecting group. No by-product **3** due to Fmoc cleavage by N-methylimidazole was found. The small amount of by-product **2** resulting from the cleavage of the tert-butyl ester by the acidic ion exchange resin Amberlyst 15 during the removal of N-methylimidazole is easily separated during this process (Note 13). The product **1** has potential in the synthesis of color-labelled peptides[8] after cleavage of either of the protecting groups as illustrated in Scheme 1. The amine **3** should be useful for peptide synthesis in solution while the carboxylic acid **2** could be applied for solid phase synthesis. The Fmoc group was easily removed in 10 min with 20% piperidine in DMF. LC-MS analysis of **3** showed no by-products resulting from the cleavage of the two ester groups. The *tert*-butyl ester was cleaved in one hour with 50% TFA in methylene chloride without side reactions.

1.   AIMS Scientific Products GmbH, Mascheroder Weg 1b. 38124 Braunschweig, Germany.

2.   German Research Center for Biotechnology, Department of Chemical Biology, Mascheroder Weg 1, 38124 Braunschweig. Germany.

3.   Ley, S. V.; Baxendale, I. R.; Bream, R. N.; Jackson, P. S.; Leach, A. G.; Longbottom, D. A.; Nesi, N.; Scott, J. S.; Storer, R. I.; Taylor, S. J.; *J. Chem. Soc., Perkin Trans. 1*, **2000**, 3815-4195.

4.   Cainelli, G.; Manescalchi, F.; *Synthesis*, **1975**, 723-724.

5.   Gayo, L. M.; Suto, M. J.; *Tetrahedron Lett.*, **1997**, *38*, 513-516.

6.   Rademann, J.; Smerdka, J.; Jung, G.; Grosche, P.; Schmid, D.; *Angew. Chem.* **2001**, *113*, 390-393.

7.   Zander, N.; Frank, R.; *Tetrahedron Lett.* **2001**, *42*, 7783-7785.

8.   For N-α-dye labelled amino acid derivatives see: Sameiro, M.; Gonçalves, T.; Maia, H. L. S.; *Tetrahedron Lett.* **2001**, *42*, 7775-7777.

Appendix

Chemical Abstracts Nomenclature (Registry Number)

Disperse Red 1; Ethanol, 2-[ethyl[4-[4-nitrophenyl)azo]phenyl]amino]-; (2872-52-8).

N-[(9H-fluoren-9-ylmethoxy)carbonyl]-L-aspartic acid, 1-(1,1-dimethylethyl) ester; L-Aspartic acid, N-[(9H-fluoren-9-ylmethoxy)carbonyl]-, 1-(1,1-dimethylethyl ester; (1290460-009-9).

N-Methylimidazole: 1H-Imidazole, 1-methyl-; (616-47-7).

PHENYLSULFENYLATION OF NONACTIVATED δ-CARBON ATOM BY PHOTOLYSIS OF ALKYL BENZENESULFENATES: PREPARATION OF 2-PHENYLTHIO-5-HEPTANOL
(Heptane, 2-phenylthio-5-hydroxy)

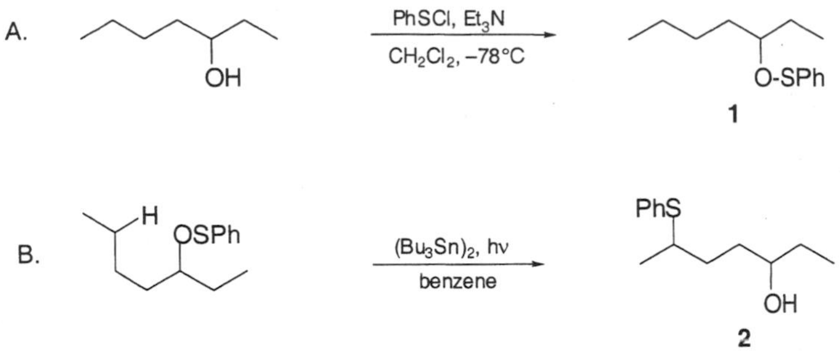

A.

PhSCl, Et$_3$N

CH$_2$Cl$_2$, −78°C

1

B.

(Bu$_3$Sn)$_2$, hv

benzene

2

Submitted by Goran Petrović, Radomir N. Saičić and Živorad Čeković[1]

Checked by Fangzheng Li and Marvin J. Miller.

A. Procedure

A.   *3-Heptyl benzenesulfenate.* A 1-L, three-necked, round-bottomed flask equipped with a 50-mL pressure-equalizing addition funnel fitted with an argon inlet, an internal thermometer, a drying tube, and a magnetic stir bar is charged with 10.0 g (86.1 mmol) of 3-heptanol (Note 1), 420 mL of anhydrous methylene chloride (Note 2), and 30 mL (22 g, 215 mmol) of freshly distilled triethylamine (Note 3). The flask is flushed with argon and the solution is cooled in a acetone/dry ice bath to an internal temperature of -72°C. Benzenesulfenyl chloride (10.93 mL, 13.7 g, 94.73 mmol) (Note 4) is added over 20 min to the efficiently stirred solution. The reaction mixture is stirred at −72°C for 45 min and then protected from light (Note 5) and allowed to warm to room temperature.

The resulting mixture is diluted with 200 mL of methylene chloride and then washed successively with 200 mL of deionized water, 200 mL of 1.5M hydrochloric acid, 200 mL of deionized water, 200 mL of saturated aqueous sodium hydrogen carbonate solution, and 200 mL of deionized water. The organic solution is dried over anhydrous sodium sulfate, filtered, and then concentrated under reduced pressure from a flask wrapped with aluminum foil. The residual oil is purified by vacuum distillation (bp 131-132°C, 3 mm) to give 17.2 g (89% yield) of the title compound **1** as a green-yellow oil (Notes 6, 7, 8).

B. *2-Phenylthio-5-heptanol.* A photochemical reactor consisting of a tubular pyrex flask, a magnetic stirbar, a water-cooled high pressure mercury lamp, and an argon inlet tube (Note 9) is charged with 12.3 g (54.8 mmol) of 3-heptyl benzenesulfenate, 1.6 g (2.8 mmol) of hexabutylditin (Note 10), and 220 mL of benzene (Note 11). The solution is purged with argon for 10 min to remove oxygen and then irradiated for 1 hr with water cooling (Note 12). The solvent is then removed under reduced pressure to afford a pale yellow oil which is purified by dry flash chromatography on silica gel (Note 13). The fractions containing the desired product are combined and concentrated by rotary evaporation under reduced pressure to afford 6.5 g (53%) of 2-phenylthio-5-heptanol as a colorless oil (Note 14).

2. Notes

1. 3-Heptanol was purchased from Fluka Chemika and distilled (bp 155°C) before use. The checkers purchased 3-heptanol from Aldrich Chemical Company and used it without purification.

2.   Methylene chloride was distilled from phosphorus pentoxide and stored over activated molecular sieves (4Å).

3.   Triethylamine was purchased from the Merck & Company, Inc. and distilled from calcium hydride. The checkers used triethylamine purchased from Aldrich Chemical Company.

4.   Benzenesulfenyl chloride was prepared from the reaction of thiophenol with sulfuryl chloride in the presence of triethylamine in petroleum ether (bp 35-50°C) according to the procedure of Barrett, A. G. M.; Dhanak, D.; Graboski, G. G.; Taylor, S. J. *Org. Synth. Coll. Vol. VIII,* **1993**, 550. This reagent was stored under argon in a refrigerator and protected from light, under which conditions it can be stored for one month without significant change (G. Zelčans, *Encyclopedia of Reagents for Organic Synthesis,* Ed. L. A. Paquette, Wiley, New York, 1995, Vol. 1, p. 272).

5.   Alkyl benzenesulfenates are sensitive to light and their preparation should be carried out in a darkened hood.

6.   The crude product is sufficiently pure for use in the next step. The submitters preferred to perform the distillation at a lower pressure (89°C, 0.3 mm) in order to minimize decomposition and obtained the product in 91% yield. The checkers obtained the product in 84-89% yield in several runs.

7.   The preparation of alkyl benzenesulfenates is accompanied by the formation of products of higher oxidation states of sulfur. These products have higher boiling points and are separated by careful distillation of the product under reduced pressure.

8.   The product exhibits the following spectroscopic properties: IR (KBr) cm^{-1}: 3061, 2960, 2934, 2874, 2000-1600, 1583, 1477, 1464, 1439, 1113, 1069, 1024, 943, 916, 893, 811, 737, 690; ^{1}H NMR (300 MHz, CDCl$_3$) δ: 0.88 (t, J = 7.5 Hz, 6H), 1.24-1.31 (m, 4H), 1.56-1.70 (m, 4H), 3.52 (q, J = 5.7 Hz, 1H), 7.16-7.52 (m, 5H); ^{13}C NMR (75 MHz, CDCl$_3$) δ: 9.5, 14.1, 22.9, 26.9, 27.5, 33.5, 89.3, 125.3, 128.9, 141.8.

9.   The reaction apparatus used by the submitters for irradiation on a 12-g scale was 18 cm high and 6 cm in diameter and had a 55/50 joint at the top for an immersion well. The apparatus is fitted with an argon inlet tube and water-cooled condenser which is connected to a mineral oil bubbler. The light source was a 125W high pressure Hanovia mercury lamp. The checkers used a tubular pyrex flask 15 cm high and 7.5 cm in diameter. The light source was a 450W high pressure ACE mercury lamp (7825-34 lamp) with filter to provide 250 nm light.

10.   Hexabutylditin was purchased from Aldrich Chemical Company, Inc. and used without purification.

11.   Anhydrous benzene was purchased from Aldrich Chemical Company and used without purification.

12.   The course of the reaction was followed by TLC (silica gel 60 A, 2.5 x 7.5 cm plates, elution with 95:5 ethyl acetate/hexane, visualization with 50% sulfuric acid followed by heating) by monitoring the disappearance of benzenesulfenate starting material **1**.

13.   A chromatography column of 4.5-cm diameter was charged with 250 g of silica gel (ICN Silica 10/18 60 A). The viscous oily crude product is dissolved in 10 mL

of methylene chloride, 10 g of silica gel is added, and the solvent is evaporated. The resulting dry powder is applied on the top of the column which is then successively eluted with petroleum ether, 95:5 petroleum ether/acetone, and 90:10 petroleum ether/acetone.

14.   The submitters obtained the product in 61% yield; the checkers isolated the purified product in 53-56% yield in several runs. The product exhibits the following spectroscopic properties: IR (KBr) cm[-1]: 3400, 3074, 2961, 2930, 2874, 2000-1600, 1584, 1480, 1458, 1439, 1376, 1303, 1263, 1178, 1092, 1068, 1025, 1000, 971, 923, 746, 692; ^{1}H NMR (300 MHz, CDCl$_3$) δ: 0.92 (t, J = 7.5 Hz, 3H), 1.27 (d, J = 6.9 Hz, 3H), 1.40-1.80 (m, 6H), 3.22 (m, 1H), 3.48 (m, 1H), 7.18-7.29 (m, 3H), 7.37-7.40 (m, 2H); ^{13}C NMR (75 MHz, CDCl$_3$) δ: 9.9, 21.5, 21.5, 30.4, 30.4, 32.9, 33.1, 34.3, 34.5, 43.7, 43.8, 73.2, 73.3, 126.9, 128.9, 132.3, 135.7.

Waste Disposal Information

All toxic materials were disposed of in accordance with "Prudent Practices in the Laboratory"; National Academy Press; Washington, DC, 1995.

3. Discussion

The introduction of various functional groups onto a remote nonactivated carbon atom is of great synthetic importance.[2] The regioselective functionalization of a nonactivated δ-carbon atom involves a free radical 1,5-hydrogen transfer from δ-carbon atom to oxygen or nitrogen centered radicals.[3] Transposition of a radical center from oxygen or nitrogen to the remote carbon atom offers possibilities for introduction of different functional groups onto the nonactivated carbon atom or for formation of a new carbon-carbon bond. Diversity of this type of reaction offers possibilities for introduction of different oxygen,[2] nitrogen,[4] halogen,[5] and sulfur[6] functional groups as well as an olefinic bond[7] onto the δ-carbon atom. The present procedure is based on the original results[5] of the author and co-workers.

Formation of δ-phenylthio alcohols from alkyl benzenesulfenates **3** formally represents a simple interchange of the positions of δ-hydrogen and phenylthio group. The sequence of radical reactions is initiated by tributyltin radical and is based on its thiophilic addition to the sulfur of an alkyl benzenesulfenate.[6] The formed alkoxyl radical **4** upon intramolecular 1,5-hydrogen migration gives the δ-carbon radical **5**. In the absence of any other reactive species the carbon radical undergoes homolytic substitution at the sulfur atom in the alkyl benzenesulfenate **3** to give the δ-phenylthio alcohol **6** and to generate the alkoxyl radical **4** as a transfer radical which continues the chain (see Scheme 1).

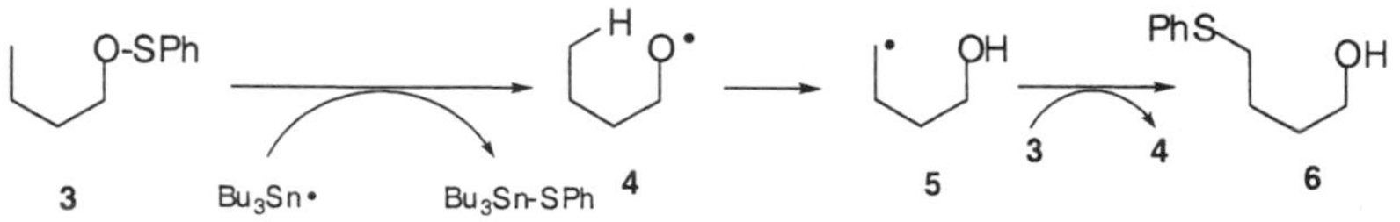

Scheme 1

δ-Phenylsulfenylation is also conceivable in a tin-free variant; however, when alkyl benzenesulfenates were irradiated in the absence of hexabutylditin, the reaction proceeded at a lower rate and gave 10-15% lower yields of δ-phenylthio alcohols.[6]

This method for introduction of the thioether functional group tolerates the presence of a broad range of functional groups, such as alkene, ester, carbonyl, and cyano groups.

Introduction of the phenylthio group onto the δ-carbon atom of alcohols can have valuable synthetic applications. δ-Phenylthio alcohols can be oxidized to the corresponding δ-sulfoxides and sulfones (with their versatile reactivities) or they can be deprotonated by strong base converting the δ-carbon atom to a nucleophilic species. Conversion of δ-phenylthio alcohols to the corresponding δ-carbonyl compounds can be achieved via halogenation followed by subsequent hydrolysis. In this way an inversion of the reactivity of the δ-carbon atom may be accomplished and it can react as an electron acceptor.

The procedure of phenylsulfenylation of δ-carbon atom was applied to a variety of other substrates as summarized in Table 1.

1. Faculty of Chemistry, University of Belgrade, Studentski trg 16, PO Box 158, 11000 Belgrade, Serbia&Montenegro

2. (a) Heusler, K; Kalvoda, J. *Angew. Chem., Int. Ed. Engl.* **1964**, *3*, 525; (b) Mihailovic, M. Lj.; Ceković, Z. *Synthesis,,* **1970**, 209; (c) Barton, D. H. R., *Pure and Appl. Chem.* **1968**, *16*, 1; (d) Feray, N.; Kuznetsov, P.; Renaud, P.,

in *Radicals in Organic Synthesis,* Eds. Renaud, P. and Sibi, M. P., Vol. 2. Ch. 3.6. p. 246, Wiley-VCH, Weinheim, 2001; (e) Čeković, Ž., *Tetrahedron,* **2003**, *59,* 8073

3. Majetich, G.; Wheless, K. *Tetrahedron,* **1995**, *51,* 7095.

4. (a) Barton, D. H. R.; Parakh, S. I. *Half a Century of Free Radical Chemistry,* Cambridge University Press, Cambridge, 1993; (b) Barton, D. H. R.; Beaton, J. M.; Geller, L. E.; Pechet, M. M. *J. Am. Chem. Soc.,* **1960**, *82,* 2640; (c) Barton, D. H. R.; Beaton, J. M.; Geller, L. E.; Pechet, M. M. *J. Am. Chem. Soc.,* **1961**, *83,* 4076.

5. Walling, C.; Padwa, A. *J. Am. Chem. Soc.* **1961**, *83,* 2207;

6. Petrović, G.; Saičić, R. N.; Čeković, Ž. *Tetrahedron Lett.* **1997**, *38,* 7107; (b) Petrović, G.; Saičić, R. N.; Čeković, Ž. *Tetrahedron* **2003**, *59,* 187; (c) Petrović, G.; Saičić, R. N.; Čeković, Ž. *Synlett,* **1999**, 635; (d) Pasto, D. J.; Cottard, F. *Tetrahedron Lett.* **1994**, *35,* 4303.

7. (a) Čeković, Ž.; Green, M. M. *J. Am. Chem. Soc.* **1974**, *96,* 3000; (b) Čeković, Ž.; Dimitrijević, Lj.; Djokić, G.; Srnić, T. *Tetrahedron,* **1979**, *35,* 2021

TABLE 1.
δ-PHENYLTHIO ALCOHOLS PREPARED BY PHOTOLYSIS OF ALKYL BENZENESULFENATES IN THE PRESENCE OF HEXABUTYLDITIN

Alkyl benzenesulfenates	δ-Phenylthio alcohols	Yields (%)
	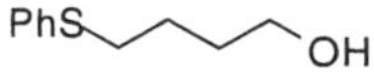	17
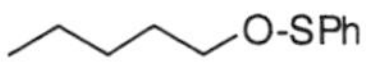	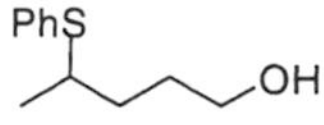	80
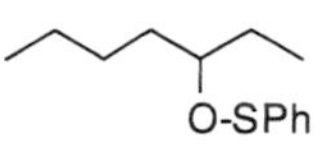	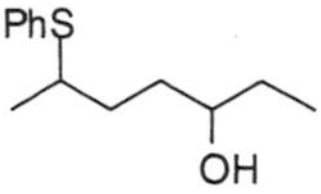	61
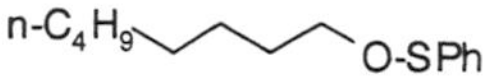	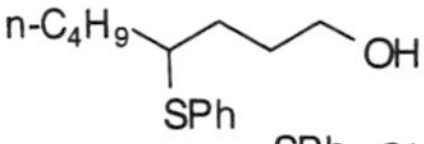	47
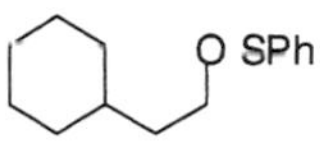	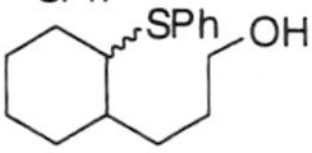	51
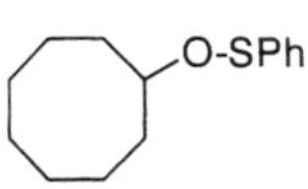	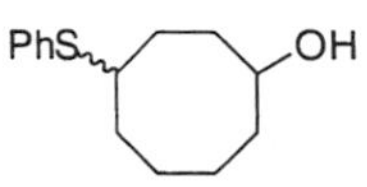	35
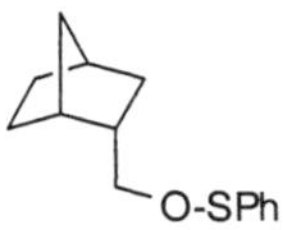	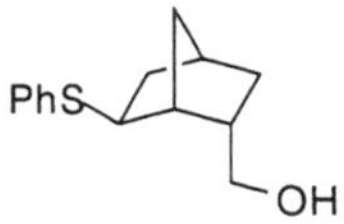	64
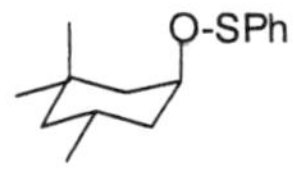	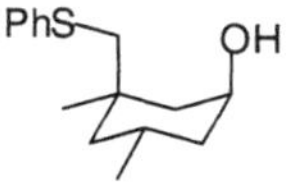	62
	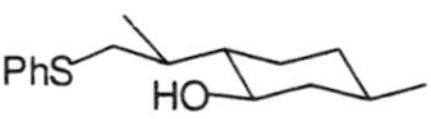	43
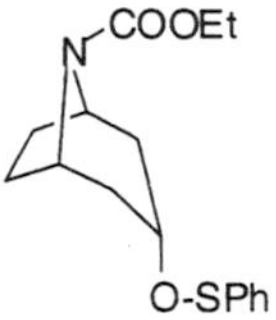	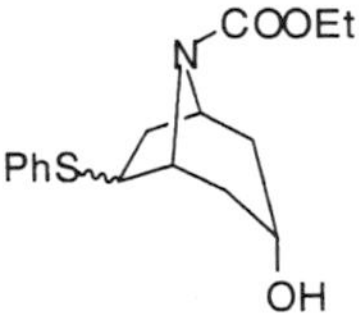	45
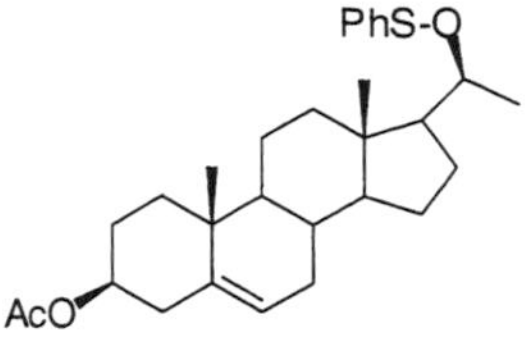	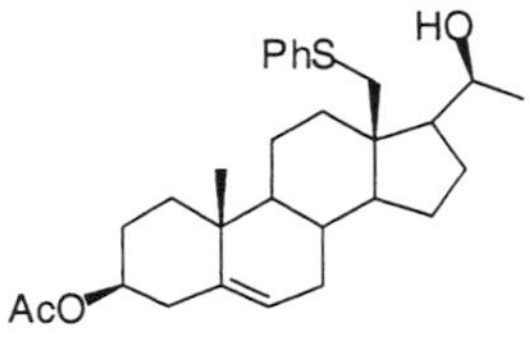	40

Appendix
Chemical Abstracts Nomenclature (Registry Number)

2-Phenylthio-5-heptanol: 3-Heptanol, 6-(phenylthio); (198778-75-5)

3-Heptyl benzenesulfenate: Benzenesulfenic acid, 1-ethylpentyl ester; (198778-69-7)

3-Heptanol; (589-82-2)

Triethylamine: Ethanamine, N,N-diethyl-; (121-44-8)

Benzenesulfenyl chloride; (931-59-9)

Hexabutylditin: Distannane, hexabutyl; (813-19-4)

Thiophenol: Benzenethiol; (108-98-5)

Sulfuryl chloride; (7791-25-5)

Petroleum ether: Ligroine; (8032-32-4)

PREPARATION OF PIVALOYL HYDRAZIDE IN WATER

(Propanoic acid, 2,2-dimethyl-, hydrazide)

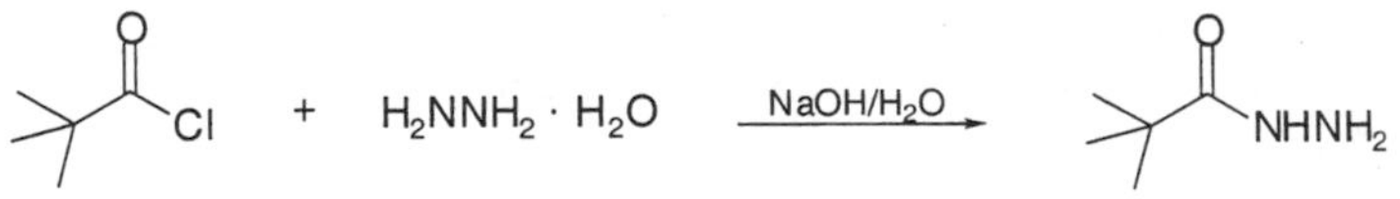

Submitted by Bryan Li,[1] Raymond J. Bemish, David R. Bill, Steven Brenek, Richard A. Buzon, Charles K-F Chiu and Lisa Newell.

Checked by Claire Coleman and Peter A. Wipf.

1. Procedure

Pivaloyl hydrazide. A 1-L, three-necked, round-bottomed flask equipped with a Teflon-coated thermocouple and mechanical stirrer is charged with 400 mL of water and sodium hydroxide (12.87 g, 322 mmol), and the resulting mixture is stirred until all of the solids dissolve (Note 1). Hydrazine (35% aqueous solution, 36.83 g, 400 mmol) is then added in one portion. The mixture is cooled in an ice-water/acetone bath to an internal temperature of −5 to 0°C, and trimethylacetyl chloride (38.6 mL, 320 mmol) is added dropwise (Note 2) over a period of 40-60 min while maintaining the reaction temperature between −5 and 0°C (Note 3). The reaction mixture is then transferred to a 1-L, pear-shaped flask and concentrated to a volume of ca. 100 mL by rotary evaporation (at ca. 100 mm) (Notes 4, 5) and the resulting suspension is filtered (Note 6). The filtrate is further concentrated to a volume of ca. 40 mL (Note 7) and 100 mL of toluene is then added. The resulting solution is transferred to a three-necked, round-bottomed flask equipped with a Dean-Stark water separator, thermometer, and rubber septum. Distillation is continued at atmospheric pressure until a constant boiling point is reached (Note 8) and then the pot volume is further reduced to ca. 40 mL. The resulting heterogenous mixture is filtered (Note 9) and the filtrate is concentrated by rotary evaporation under reduced pressure to provide the desired product as a colorless oil, which on standing solidifies to a white semi-solid. This material is recrystallized from 100 mL of isopropyl ether to afford 18.6-20.4 g (50-55%) of pivaloyl hydrazide (Notes 10, 11).

2.   Notes

1. All reagents were purchased from Aldrich Chemical Company (except for trimethylacetyl chloride, which the checkers obtained from Acros) and were used without further purification. The checkers used a low temperature alcohol thermometer in place of a Teflon-coated themocouple. The third neck of the flask was left open to the atmosphere.

2. A syringe pump was used for the addition of acid chloride in order to achieve a steady flow rate. The tip of the syringe needle (gauge 20) was submerged in the reaction mixture. Dropwise addition of trimethylacetyl chloride at 0 – 5°C resulted in the immediate formation of a precipitate.

3. The reaction was complete at the end of the pivaloyl chloride addition. On 5-L or larger scale, the reaction was conducted at temperatures of 10–15°C without loss of selectivity.

4. A small amount of hydrazine hydrate was present in the reaction mixture at this point, but a safety evaluation indicated the final reaction mixture had a very low thermal potential (_H =15.3 J/g). This poses a minimum thermal hazard for vacuum distillation.

5. The submitters concentrated the reaction mixture by vacuum distillation (100 mm, bath temperature 70°C, vapor temperature 51°C). The weight after concentration was ca. 120 g. The checkers used rotary evaporation with a bath temperature of 65 to 70°C without any problems, and employed an explosion shield as a safety precaution.

6. The bis-acylation byproduct ($Me_3CCONHNHCOCMe_3$) was removed by filtration; 20 mL of water was used for washing the filter cake.

7. The submitters removed solvent by vacuum distillation (100 mm, bath temperature 70°C, vapor temperature 51°C).

8. Azeotropic removal of water was complete when the vapor temperature reached 111°C.

9. Sodium chloride was removed by filtration.

10. The submitters obtained the product in 72% yield without recrystallization and determined the product to be >97% pure by HPLC (by area; conditions: 250 mm Kromasil C4 column using acetonitrile (A)/water (B) and 0.1% TFA in water (C), 0:90:10 A:B:C ramp to 90:0:10 A:B:C over 15 min and hold for 5 min. Flow rate 1 mL/min and

detection wavelength 210 nm. The checkers found the purity prior to recrystallization to be typically 85-90% with the impurity determined to be the bis-acylation product by LC-MS. The checkers employed C8 microsorb-MW 100 (250 mm) and Alltech C18 (100 mm) columns using acetonitrile (A)/water (B) and 0.1% TFA in water (C), 0:90:10 A:B:C ramp to 90:0:10 A:B:C over 15 min and hold for 5 min. Flow rate 1 mL/min and detection wavelength 210 nm.

11. The product exhibits the following physical properties: mp 67.1-69.2°C; IR (NaCl) cm^{-1} 3471, 3327, 2968, 1660; ^{1}H NMR (CDCl$_3$, 400 MHz): δ 1.19 (s, 9H), 4.40-4.85 (br, 3 H); ^{13}C NMR (CDCl$_3$, 75 MHz): δ 27.4, 38.1, 179.3; MS (m/z): 117 ([M + 1]$^+$). Anal. Calcd for C$_5$H$_{12}$N$_2$O: C, 51.70; H, 10.41; N, 24.12. Found C, 51.62; H, 10.78; N, 24.22.

Waste Disposal Information

All toxic materials were disposed of in accordance with "Prudent Practice in the Laboratory"; National Academy Press; Washington, DC, 1995.

3. Discussion

Hydrazides (RCONHNH$_2$) are highly useful starting materials and intermediates in the synthesis of heterocyclic molecules.[2] They can be synthesized by hydrazinolysis of amides, esters and thioesters.[3] The reaction of hydrazine with acyl chlorides or anhydrides is also well known,[4] but it is complicated by the formation of 1,2-

diacylhydrazines, and often requires the use of anhydrous hydrazine which presents a high thermal hazard. Diacylation products predominate when hydrazine reacts with low molecular weight aliphatic acyl chlorides, which makes the reaction impractical for preparatory purposes.[5]

Recently we needed to prepare large amounts of pivaloyl hydrazide (1). A literature survey indicated several approaches: (1) heating pivalic acid with hydrazine hydrate with a Lewis acid catalyst such as activated alumina[6] or titanium oxide;[7] (2) heating hydrazine hydrate at high temperature (140°C) with ethyl pivalate;[8] (3) condensing phthaloyl hydrazine with pivaloyl chloride, followed by deprotection of the phthaloyl group;[9] and (4) reaction of ethyl thiopivalate with hydrazine hydrate. Reaction safety evaluations revealed that hydrazine monohydrate has an onset temperature of *ca.* 125°C in a Differential Scanning Calorimetry (DSC) experiment, and possesses a very high thermal potential ($\Delta H = 2500$ J/g),[10,11] which prompted us to develop a method for the synthesis of 1 that did not require heating. After some experimentation we determined that the reaction of pivaloyl chloride with hydrazine proceeds most efficiently in water to give a 4:1 ratio[12] of 1 to $Me_3CCONHNHCOCMe_3$ (2). The use of organic solvents (MeOH, THF, 2-propanol) with water[13] invariably led to formation of biphasic mixtures and predominant formation of 2.[14] Reaction workup is also simplified using water as solvent. Upon partial concentration the bis-acylhydrazide by-product 2 precipitated out of the reaction mixture and is conveniently removed by filtration. Removal of the remainder of the water by displacement with toluene leads to precipitation of NaCl, which is also easily removed by filtration. The filtrate is then

258

further concentrated to provide **1** in >97% purity, typically in 55–75% yield. This procedure has been employed to prepare 10 Kg batches of **1** with no difficulty.

The protocol is effective in preparation of hydrazides of 5 carbons or less. Cyclopropanecarboxylic acid hydrazide[15] and isobutyric acid hydrazide[16] were prepared from their corresponding acid chlorides in 64% and 71% yields, respectively. However, when this method was applied to cyclohexanecarboxylic acid chloride, the bis-acylhydrazide was the predominant product, and the mono-acylhydrazide[17] was isolated in 25% yield.[18]

1. Chemical Research and Development, Pfizer Global Research and Development, Groton Laboratories, Groton, CT 06340, USA

2. a. Monge, A.; Aldana, I.; Lezamiz, I.; Fernandez-Alvarez, E. *Synthesis*, **1984**, *2*, 160; b. Rao, C. S.; Ramachandra, R.; Kshirsagar, S.; Vashi, D. M.; Murty, V. S. N. *Indian J. Chem. Sect. B*, **1983**, *22*, 230; c. Essassi, E. M.; Fifani, J. *Bull. Soc. Chim. Belg.* **1987**, *96*, 63.

3. Edwards, L. H. US4500539, US Appl. 83-514073.

4. Paul H.; Stoye, D. *Chap. 10 The Chemistry of Hydrazide, In The Chemistry of Amides*; Zabicky J.; Ed.; John Wiley & Sons, 1970; p 515.

5. Smith, P. A. S. *Org. React.* **1946**, *3*, 366.

6. Flowers, W. T.; Robinson, J. F.; Taylor, D. R.; Tipping, A. E. *J. Chem. Soc. Perkin Trans.* 1, **1981**, 349.

7. Verbrugge, P. A.; De Waal, J. *Eur. Pat. Appl.* **1995,** EP 653419.

8. a. Struve, G. *J. Prakt. Chem.* **1894,** *50*, 295; b. Quast, H.; Kees, F. *Chem. Ber.* **1981**, *114*, 802; c. Wieland, H.; Dennstedt, J. *Justus Liebigs Ann. Chem.* **1927**, *16*, 452.

9. Brosse, N.; Pinto, M-F.; Jamart-Gregoire, B. *J. Chem. Soc., Perkin Trans. 1* **1998**, *22*, 3685.

10. Under nitrogen atmosphere, the thermal potential is slightly lower (_H = 2150 J/g), and the onset temperature is 150 °C.

11. For perspective, 1 g of hydrazine monohydrate is equivalent to ~0.5 g of 2,4,6-trinitrotoluene (TNT) in terms of thermal potential; T. Grewer, *Thermal Hazards of Chemical Reactions*; Elsevier:Amsterdam, **1994**, Vol. 4.

12. By integration of the ^{1}H NMR spectrum of the crude reaction product.

13. Water is still introduced from the use of hydrazine hydrate when no additional water is added.

14. Pivaloyl chloride and **1** are both preferentially soluble in the organic phase, which gives rise to **2** as the major product.

15. Szmuszkovicz, J. *Ger. Offen.* (1972), DE 2228044; CAN 78:72230.

16. Galons, H.; Cave, C.; Miocque, M.; Rinjard, P.; Tran, G.; Binet, P. *Eur. J. Med. Chem.* **1990**, *25*, 785.

17. McElvain, S. M.; Starn, R. E., Jr. *J. Am. Chem. Soc.* **1955**, *77*, 4571.

18. Cyclohexanecarboxylic acid hydrazide has low solubility in water. A biphasic mixture developed during the reaction, which led to the predominant formation of the bis-acylated material.

Appendix

Chemical Abstracts Nomenclature (Registry Number)

Hydrazine; (10217-52-4)

Pivaloyl chloride: Propanoyl chloride, 2,2-dimethyl-; (3282-30-2)

Pivaloyl hydrazide: Propanoic acid, 2,2-dimethyl-, hydrazide; (42826-42-6)

BORIC ACID CATALYZED AMIDE FORMATION FROM CARBOXYLIC ACIDS AND AMINES: N-BENZYL-4-PHENYLBUTYRAMIDE

(Benzenebutanamide, N-(phenylmethyl)-)

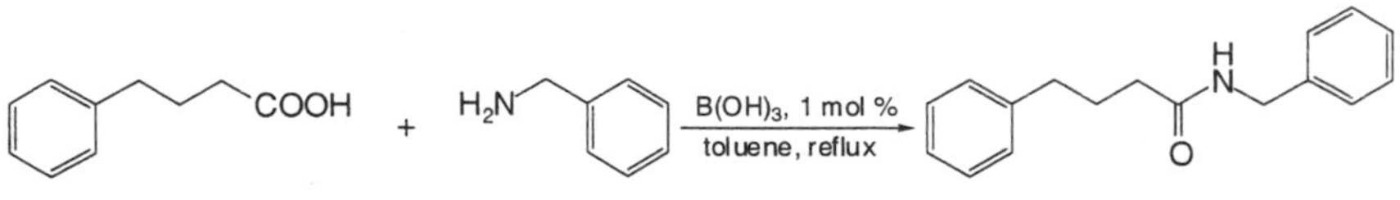

Submitted by Pingwah Tang.[1]

Checked by Helga Krause and Alois Fürstner.

1. Procedure

N-Benzyl-4-phenylbutyramide. A flame-dried, 200-mL, three-necked, round-bottomed flask is equipped with two glass stoppers, a vacuum-jacketed Dean-Stark trap topped with a reflux condenser fitted with a nitrogen inlet, and a Teflon-coated magnetic stirring bar. The reaction vessel is charged with 4-phenylbutyric acid (4.92 g, 30 mmol) (Note 1), boric acid (0.020 g, 0.30 mmol) (Note 2), and 88 mL of toluene (Note 3). To the stirred colorless reaction mixture is added benzylamine (3.32 g, 31 mmol) (Note 4) in one portion. The reaction mixture is heated at reflux for 16 hr and ca. 0.6 mL of water is collected in the Dean-Stark trap (Notes 5, 6, 7, 8). The mixture is allowed to cool to ambient temperature and then is poured with stirring into 500 mL of hexanes leading to the immediate precipitation of a white solid. Stirring is continued for an additional 30 min and then the precipitate is filtered off with suction through a sintered glass filter funnel. The collected solid is successively washed with two 60-mL portions of hexanes and two 60-mL portions of distilled water (Note 9) and then is dried in vacuo at room temperature for 12 hr to afford 6.90 g (91%) of N-benzyl-4-phenylbutyramide as a white solid (Notes 10, 11).

2. Notes

1. 4-Phenylbutyric acid (99%) was purchased from Aldrich Chemical Company and used as received.

2. Boric acid (99.999%) was purchased from Aldrich Chemical Company and used as received.

3. Toluene (99.8% anhydrous, water <0.001%, evaporation residue <0.005%) was purchased from Aldrich Chemical Company and used as received. The capacity of the receiver in the Dean-Stark trap was 28 mL. The initial amount of toluene placed in the flask was 88 mL. Upon heating to reflux, 28 mL of toluene was distilled from the flask and collected in the receiver. The remaining volume of toluene in the reaction flask was ca. 60 mL, corresponding to an approximately 0.50M concentration of the reactants.

4. Benzylamine (redistilled 99.5+ %) was purchased from Aldrich Chemical Company and used without further purification.

5. The theoretical amount of water is 0.54 mL.

6. TLC analysis indicated that the reaction was complete. TLC was performed on pre-coated plates with silica gel 60 F_{254} purchased from EM Sciences (Merck) using hexanes/ethyl acetate (1/1) as the eluent.

7. HPLC analysis indicated that the reaction was complete. HPLC was conducted utilizing Perkin-Elmer's Series 200 autosampler and pump with either a Perkin-Elmer Series 200 or Applied Biosystems 785A UV/Vis detector (both monitoring at $\lambda = 220$ nm). Phenomenex's Kromasil 5 micron C18 column (50 mm X 4.6 mm) fitted with Phenomenex's C18 (ODS, Octadecyl) security guard cartridges (4 mm X 3 mm) were utilized with a 3 mL/min flow-rate. The solvent system consisted of mixtures of acetonitrile and water, containing 0.1% (v/v) trifluoroacetic acid, with the gradient starting at 5% acetonitrile and increasing to 95% over 10.5 min.

8. GC analysis indicated that the reaction was complete. Gas chromatographic analyses were performed on a Agilent 6890N GC system equipped with a 30-m 5% polyphenyl methyl siloxane capillary column (0.32 mm i.d.). Helium was used as the carrier gas, and the flow rate was kept constant at 7.7 mL/min. The retention time was measured under the following conditions: injector temperature: 250°C; initial temperature of the column: 150°C; increment rate 8°C/min to 200°C. After being kept at 200°C for 8.0 min, the temperature was raised to 290°C at an increment rate of 8°C/min.

9. The water wash removes residual boric acid. Elemental analysis of the final product indicated that boron content was less than 0.005%.

10. The filtrate (wash liquor) contained some white solid which was collected, triturated with ca. 5 mL of heptane, and dried to provide additional 0.35-0.50 g of pure product as a white solid (0.4-0.7%). Analytical data (TLC, GC, HPLC and elemental analysis) indicated that this crop has a purity of ≥ 95%.

11. The amide was obtained in 88-91% yield in different runs. The product has the following physico-chemical properties: GC: R_t: 10.31 min (Note 8); HPLC: R_t: 5.63 min (Note 7); TLC R_f: 0.40 (Note 6); mp 78-80°C; IR (Kbr.) cm^{-1} 3293, 3085, 3029, 2944, 1632, 1550, 1453, 1236, 1155, 1080, 1027, 724, 695; ^1H NMR (300 MHz, DMSO-d_6): δ 1.76-1.87 (m, 2 H), 2.15 (t, 2 H, J = 7.2), 2.55 (t, 2 H, J = 7.5), 4.26 (d, 2 H, J = 6.0), 7.16-7.33 (m, 10 H), 8.30 (br, 1 H); ^{13}C NMR (75.6 MHz, DMSO-d_6): δ 27.1, 34.6, 34.7, 41.9, 125.7, 126.6, 127.1, 128.20, 128.23, 128.27, 139.7, 141.7, 171.7. Anal. calcd for $C_{17}H_{19}NO$: C, 80.60; H, 7.56; N, 5.53. Found: C, 80.29; H, 7.53; N, 5.47; B < 0.005.

Waste Disposal Information

All toxic materials were disposed in accordance with "Prudent Practices in the Laboratories"; National Academy Press; Washington, DC, 1965

3. Discussion

Many procedures for the formation of carboxylic acid amides are known in the literature. The most widely practiced method employs carboxylic acid chlorides as the electrophiles which react with the amine in the presence of an acid scavenger. Despite its wide scope, this protocol suffers from several drawbacks. Most notable are the limited stability of many acid chlorides and the need for hazardous reagents for their preparation (thionyl chloride, oxalyl chloride, phosgene etc.) which release corrosive and volatile by-products. Moreover, almost any other functional group in either reaction partner needs to be protected to ensure chemoselective amide formation.[2] The procedure outlined above presents a convenient and catalytic alternative to this standard protocol.

Although there are several reports in the literature on boron-mediated amide formations, the boron reagents had to be used in stoichiometric amounts.[3-9] Recently, Yamamoto et al. presented the first truly catalytic method allowing for a direct amide formation from free carboxylic acids and amines as the reaction partners.[10-12] Best results were obtained by using phenylboronic acids bearing electron withdrawing substituents in the *meta*- and/or *para*-positions such as 3,4,5-trifluorophenylboronic acid or 3,5-bis(trifluoromethyl)boronic acid as the catalysts.

During the course of our discovery program directed to small molecules for drug delivery, it was discovered that cheap, readily available, non-toxic, and environmentally benign boric acid, $B(OH)_3$, also constitutes a highly effective catalyst for direct amide formation. Benzylamines and cyclic aliphatic amines such as piperidines react smoothly. In most cases, the use of 5 mol% of $B(OH)_3$ is sufficient for obtaining excellent yields. Likewise, aniline derivatives afford the corresponding amides without incident even if they are hardly nucleophilic due to the presence of electron withdrawing substituents on the arene ring; in such cases, however, the amount of $B(OH)_3$ has to be increased to ca. 25% to ensure complete conversion. The scope of the method is illustrated by the examples compiled in Tables 1 and 2. Particularly noteworthy are the operational simplicity of this new method which might therefore qualify for large-scale preparations, as well as the excellent chemoselectivity profile that can make protection/deprotection sequences obsolete. This is illustrated by the examples shown below.

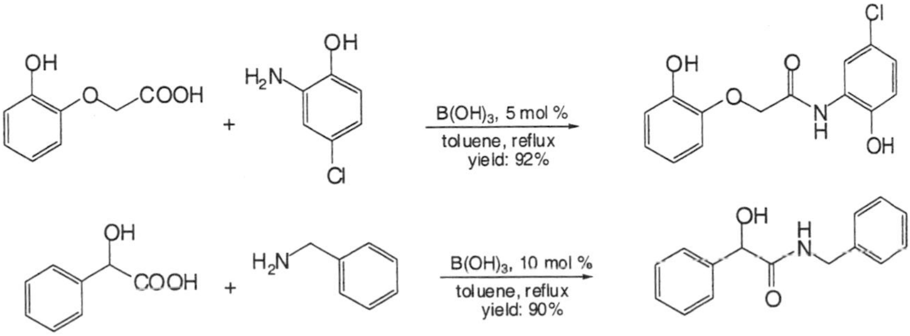

It is proposed that the boric acid reacts with the carboxylic acid to form a mixed anhydride as the actual acylating agent.[9,13] Upon reaction with an amine, this intermediate forms the desired carboxamide and regenerates the catalytically active boric acid.

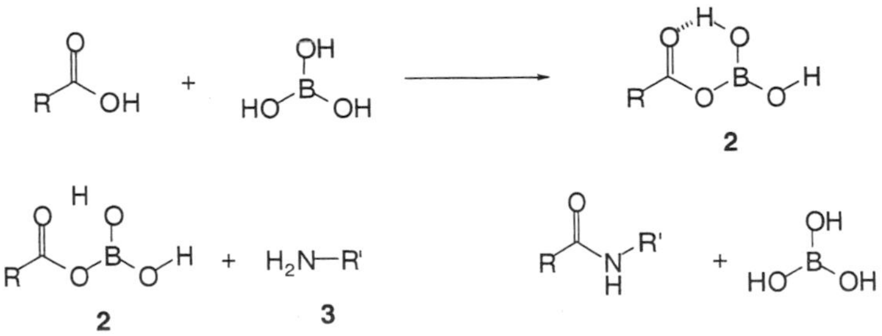

Table 1

$$\text{Carboxylic acid} \quad + \quad \text{Amine} \quad \xrightarrow[\text{reflux}]{\text{B(OH)}_3} \quad \text{Carboxamide}$$

Entry	Carboxylic acid	Amine	Carboxamide	Solvent	Time (hr)	Yield % (a)
1				Heptane	15 (b)	99
2				Toluene	15	92
3				Toluene	15	95
4				Toluene	15	95
5				Toluene	15	83
6				Toluene	16	92
7				Toluene	4	90
8				Xylene	15 (c)	90
9				o-Xylene	20	85
10			No carboxamide produced	Toluene	16 (d)	0
11				Toluene	16	80
12				Toluene	16 (e)	60

268

Table II

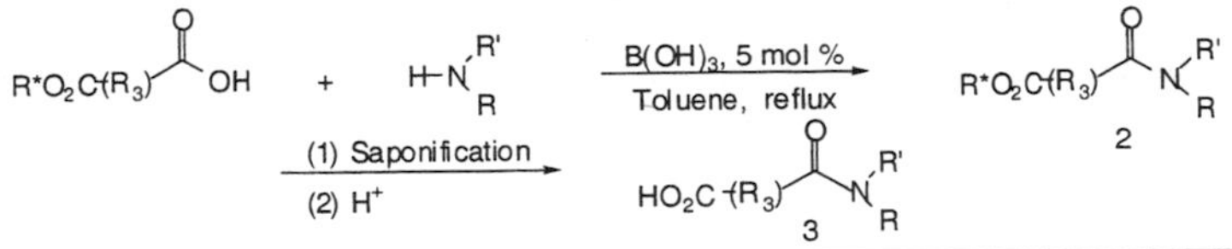

$$R^*O_2C\text{-}(R_3)\text{-}COOH \ + \ H\text{-}N\overset{R'}{\underset{R}{}} \ \xrightarrow{\text{B(OH)}_3, \ 5 \ mol \%, \ \text{Toluene, reflux}} \ R^*O_2C\text{-}(R_3)\text{-}C(O)N\overset{R'}{\underset{R}{}} \quad 2$$

$$\xrightarrow[\text{(2) H}^+]{\text{(1) Saponification}} \ HO_2C\text{-}(R_3)\text{-}C(O)N\overset{R'}{\underset{R}{}} \quad 3$$

Entry	Carboxylic acid	Amine	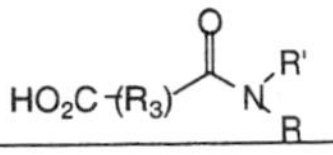 HO₂C-(R₃)-C(O)N(R')(R)	Reaction Time (hr)	Yield % of 3
1	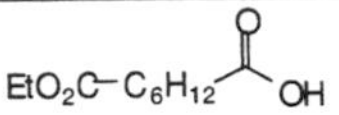 EtO₂C–C₆H₁₂–COOH	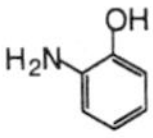 H₂N–C₆H₄(OH)	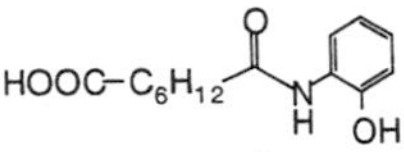 HOOC–C₆H₁₂–C(O)NH–C₆H₄(OH)	16	75
2	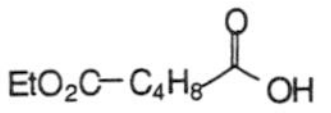 EtO₂C–C₄H₈–COOH	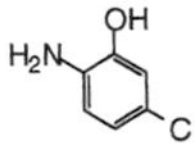 H₂N–C₆H₃(OH)(CH₃)	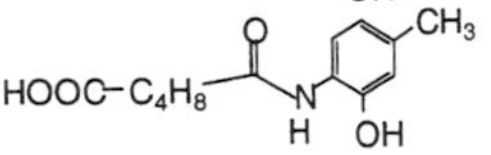 HOOC–C₄H₈–C(O)NH–C₆H₃(CH₃)(OH)	4	77
3	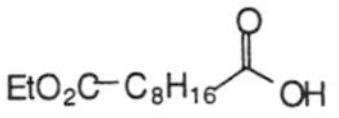 EtO₂C–C₈H₁₆–COOH	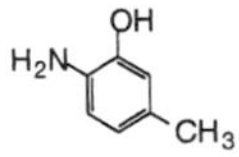 H₂N–C₆H₃(OH)(CH₃)	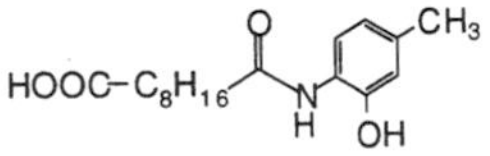 HOOC–C₈H₁₆–C(O)NH–C₆H₃(CH₃)(OH)	4	78
4	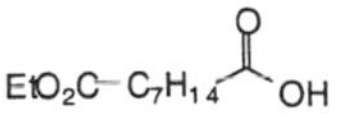 EtO₂C–C₇H₁₄–COOH	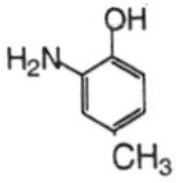 H₂N–C₆H₃(OH)(CH₃)	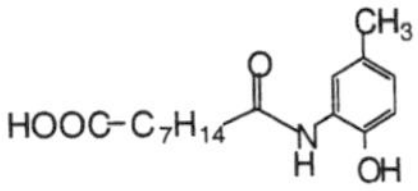 HOOC–C₇H₁₄–C(O)NH–C₆H₃(CH₃)(OH)	4	77
5	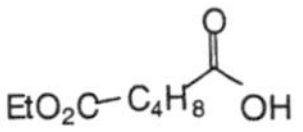 EtO₂C–C₄H₈–COOH	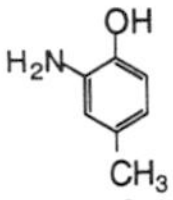 H₂N–C₆H₃(OH)(CH₃)	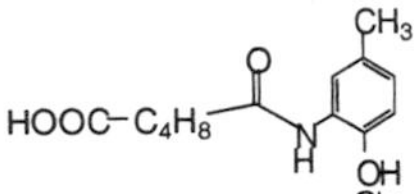 HOOC–C₄H₈–C(O)NH–C₆H₃(CH₃)(OH)	4	78
6	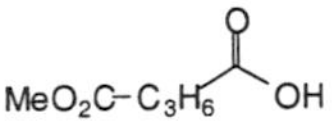 MeO₂C–C₃H₆–COOH	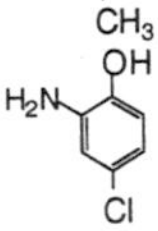 H₂N–C₆H₃(OH)(Cl)	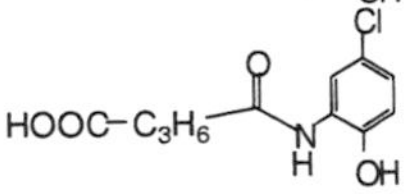 HOOC–C₃H₆–C(O)NH–C₆H₃(Cl)(OH)	16	72
7	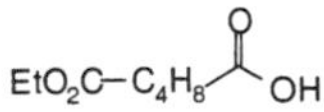 EtO₂C–C₄H₈–COOH	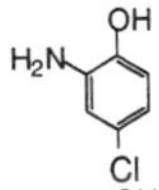 H₂N–C₆H₃(OH)(Cl)	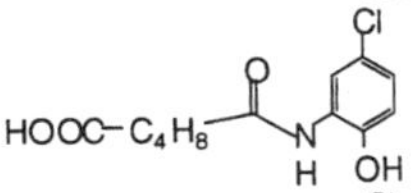 HOOC–C₄H₈–C(O)NH–C₆H₃(Cl)(OH)	16	70
8	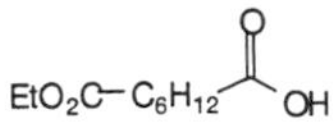 EtO₂C–C₆H₁₂–COOH	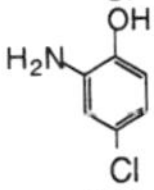 H₂N–C₆H₃(OH)(Cl)	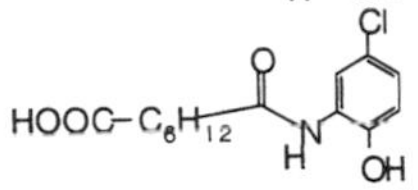 HOOC–C₆H₁₂–C(O)NH–C₆H₃(Cl)(OH)	16	57
9	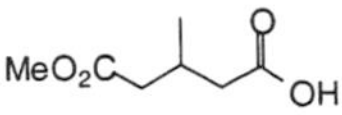 MeO₂C–CH₂CH(CH₃)CH₂–COOH	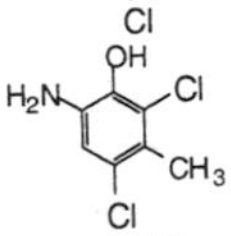 H₂N–C₆H(OH)(Cl)(CH₃)(Cl)	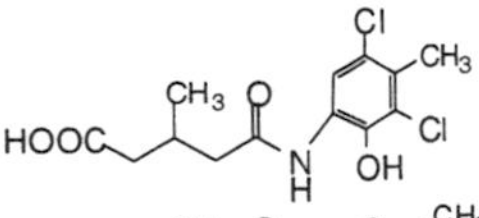 HOOC–CH₂CH(CH₃)CH₂–C(O)NH–C₆(Cl)(CH₃)(Cl)(OH)	16	68
10	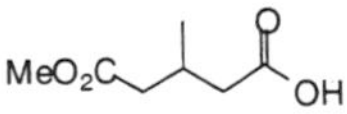 MeO₂C–CH₂CH(CH₃)CH₂–COOH	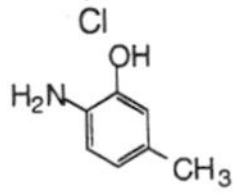 H₂N–C₆H₃(OH)(CH₃)	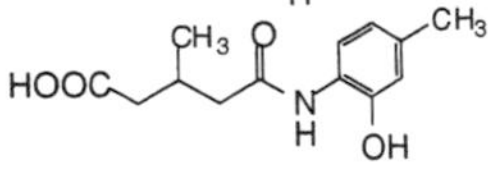 HOOC–CH₂CH(CH₃)CH₂–C(O)NH–C₆H₃(CH₃)(OH)	16	70

269

Several important generalizations emerge from the study of the boric acid catalyzed amidation of carboxylic acids and amines.

(1)   In most cases, 5 mol% of the catalyst is sufficient to catalyze the amidation. In all cases, the reactions proceed cleanly in high yields to the expected carboxamides.

(2)   No or little side reactions either with the unprotected hydroxy group present in the phenylamines, or with the unprotected hydroxyl group present in the carboxylic acids are observed. The catalytic amidation of unprotected α-hydroxycarboxylic acids also proceeds well under similar conditions.

(3)   The present method is successfully applicable to aromatic amines, which are less nucleophilic and less reactive. It works well with phenylamines, even bearing deactivating groups such as carboxylic acid ester moieties; in these cases we use 25 mol% of boric acid to drive the reaction to completion.

(4)   The catalyst employed in the reactions, boric acid, is inexpensive and commercially available. Boric acid is a "green" catalyst. By virtue of the simplicity of the process, the operation is easy to conduct. Therefore, it is amenable for large-scale preparations.

(5)   The catalytic amidation is an atom-economical process because it maximizes the incorporation of all materials used in the process into the

270

final product. Therefore, it allows organic molecule architects for the quick building of molecular complexity.

1. Department of Chemistry, Emisphere Technologies, Inc., Tarrytown, NY 10591-6715

2. Kelly, S. E.; LaCour, T. G. *Synth. Commun.* **1992,** *22*, 859.

3. Tani, J.; Oine, T.; Inoue, I. *Synthesis* **1975**, 714.

4. Trapani, G.; Reho, A.; Latrofa, A. *Synthesis* **1983**, 1013.

5. Pelter, A.; Levitt, T. E.; Nelson, P. *Tetrahedron* **1970**, *26*, 1539.

6. Collum, D. B.; Chen, S.-C.; Ganem, B. *J. Org. Chem.* **1978**, *43*, 4393.

7. Pelter, A.; Levitt, T. E. *Tetrahedron* **1970**, *26*, 1545.

8. Carlson, R.; Lundstedt, T.; Nordahl, A.; Prochazka, M. *Acta Chem. Scand., Ser. B* **1986**, *40*, 522.

9. For mixed anhydride: Srinivas, P.; Gentry, E. J.; Mitscher, L. A. Poster session 237, 223rd ACS National Meeting, Division of Organic Chemistry, Orlando, FL, U.S.A., April 7-11, 2002.

10. Ishihara, K.; Ohara, S.; Yamamoto, H. *J. Org. Chem.* **1996,** *61*, 4196.

11. Ishihara, K.; Kondo, S.; Yamamoto, H. *Synlett* **2001**, *9*, 1371.

12. Ishihara, K.; Ohara, S.; Yamamoto, H. *Org. Synth., Coll. Vol. X* **2004**, 80.

13. Ishihara, K. In *Lewis Acids in Organic Synthesis*; Yamamoto, H., Ed.; Wiley-VCH: Weinheim, Germany, **2000**, Vol 1, p. 89.

Appendix

Chemical Abstracts Nomenclature (Registry Number)

N-Benzyl-4-phenylbutyramide: Benzenebutanamide, N-(phenylmethyl)-; (179923-27-4)

4-Phenylbutyric acid: Benzenebutanoic acid; (1821-12-1)

Boric acid: Boric acid (H_3BO_3); (10043-35-3)

Benzylamine: Benzenemethanamine; (100-46-9)

Accepted for checking during the period October 1, 2002 through

May 1, 2004. An asterisk(*) indicates that the

procedure has been subsequently checked.

Previously, *Organic Syntheses* has supplied these procedures upon request. However, because of the potential liability associated with procedures which have not been tested, we shall continue to list such procedures but requests for them should be directed to the submitters listed.

3001   Chiral Tridentate Schiff Base Cr(III) Catalyst for Highly Diastereo- and
            Enantioselective Catalytc Hetero-Diels-Alder Reactions
            David E. Chavez and Eric N. Jacobsen, Harvard University,
            Department of Chemistry, 12 Oxford Street, Cambridge, MA 02138

3002   Preparation of (S)-Methyl Glycidate via Hydrolytic Kinetic Resolution
            Christian P. Stevenson and Eric N. Jacobsen, Harvard University,
            Department of Chemistry, 12 Oxford Street, Cambridge, MA 02138

3006*   (S,S)-1,2-Bis(tert-butylmethylphosphino)-ethane
            K. V. L. Crépy and T. Imamoto, Department of Chemistry
            Faculty of Science. Chiba University, Yayoi-cho, Inage-ku,Chiba
            263-8522 JAPAN

3011   Optically Active (*R*,*R*)-Hydrobenzoin from Benzoin or Benzil
            Takao Ikariya, Shoshei Hashiguchi, Kunihiko Muata, and Ryoji Noyori,
            Department of Chemistry, Faculty of Science, Nagoya University,
            Chikusa-ku, Nagoya 464-01, JAPAN

3016   Tetrabutylammonium Triphenyldifluorosilicate for C-Si Bond Cleavage
            P. DeShong, Department of Chemistry, University of Maryland, College
            Park, MD 20742

3018   Flexible Synthesis of (*S*)-Isoserine in Protected Form
            J. Robertson and P. M. Stafford, Dyson Perrins Laboratory, University
            of Oxford, South Parks Road, Oxford OX1 3QY UK

3019   Synthesis of (2S)-(–)-exo-Morpholinoisoborneol [(–)-MIB]
Y. Chen, S-j. Jeon, P. Walsh and W. Nugent, University of
Pennsylvania, Department of Chemistry, 231 South 34th Street,
Philadelphia, PA 19104-6323

3022R   D-Ribonolactone and 2,3-Isopropylidene (D-Ribonolactone).
L. B. Townsend and J. D. Williams, Department of Medicinal
Chemistry, 4567 College of Pharmacy, 428 Church St., Ann Arbor, MI
48109-1065

3026   Synthesis of 2α-Benzyloxy-8-oxabicyclo[3.2.1]oct-6-en-3-one by [4+3]
Cycloaddition
Maria Vidal-Pascual, Carolina Martínez-Lamenca, and H. M. R.
Hoffmann, Universität Hannover, Institut für Organische Chemie,
Schneiderberg 1 B, D-30167 Hannover GERMANY

3030   2,2-Diethoxy-1-isocyanoethane
Francesco Amato and Stefano Marcaccini, Departimento di Chimica
Organica, "Ugo Schiff", Università di Firenze, Via della Lastruccia, 13,
I-50019 Sesto Fiorentino (FI) ITALY

3032   Catalytic Asymmetric Acylatioin of Alcohols using a 1,2-Diamine
Derived from (S)-Proliine: (1S-, 2S)-trans-1-Benzoyloxy-2-
bromocyclohexane
Dai Terakado and Takeshi Oriyama, Faculty of Science, Ibaraki
University, 2-1-1 Bunkyo, Mito 310-8512 JAPAN

3034*   Preparation of Hexakis (4-Bromophenyl)benzene(HBB)
Rajenda Rathore and Carrie L. Burns, Department of Chemistry,
Marquette University, P.O. Box 1881, Milwaukee, WI 53201

3035   ortho-Formylation of Phenols: Preparation of 3-Bromosalicylaldehyde
Trond Vidar Hansen and Lars Skattebøl, University of Oslo,
Department of Chemistry, P.O. Box 1033 Blindern, Oslo, Norway

3036   Preparation of 4-Acetylamino- 2,2,6,6-tetramethylpiperidine-1-
oxoammonium Tetrafluoroborate and its use as an Oxidant for
Alcohols
James M. Bobbitt and Nabyl Merbouh, University of Connecticut,
Department of Chemistry, 55 North Eagleville Road, Unit 3060, Storrs,
Connecticut 06269-3060

3038   Iridium-Catalyzed C-H Borylation of Arenes and Heteroarenes:
        1-Chloro-3-Iodo-5-(4,4,5,5-tetramethyl-1,3,2-dioxaborolan-2-
        yl)benzene and 2-(4,4,5,5-tetramethyl-1,3,2-dioxaborolan-2-yl)indole

Tasuo Ishiyama, Jun Takagi, Yusuke Nobuta and Norio Miyaura,
Division of Molecular Chemistry, Graduate School of Engineering,
Hokkaido University, Sapporo, 060-8628 JAPAN

3039   Preparation of 1- Methoxy-2-(4-methoxyphenoxy)benzene.

Zhiguo J. Song and Elizabeth Buyck, Department of Process
Research,, Merck Research Labs., P.O. Box 2000, Rahway, NJ
07065

This index comprises the names of contributors to Volume **80** and **81**. For authors of previous volumes, see either indices in \Collective Volumes I through X, or the single volume entitled *Organic Syntheses, Collective Volumes I-VIII, Cumulative Indices,* edited by J. P. Freeman.

Adams, N. D., **81**,157
Amos, D. T., **80**, 133
Armstrong, J., **81**, 178
Asahara, M., **81**, 26

Bailey, W. F., **81**, 121
Banik, B. K., **81**, 188
Banik, I., **81**, 188
Becker, F. F., **81**, 188
Begtrup , M., **81**, 134
Bégué, J.-P., **80**, 184
Bemish, R. J., **81**, 254
Bhupathy, M., **80**, 219
Bill, D. R., **81**, 254
Bonnet-Delpon, D., **80**, 184
Brenek, S., **81**, 254
Brenner, M., **80**, 57
Brickner, S. J., **81**, 112
Brown, W. D., **81**, 98
Buson, R. A., **81**, 254

Cai, D , **81**, 89
Cardona, F., **81**, 204
Cekovic, Z., **81**, 244
Cha, J. K., **80**, 111
Chen, D., **80**, 18
Chen, Q.-Y., **80**, 172
Chiu, C. K.-F., **81**, 105, 254
Chobanian, H., **81**, 157
Corey, E. J. **80**, 38
Crocker, L. S., **80**, 219
Crousse, B., **80**, 184
Curran, D. P., **80**, 46

Danheiser, R. L., **80**, 133, 160
Davies, I. W., **80**, 200

de Meijere, A., **81**, 26
DeMong, D. E., **80**, 18, 31
Denmark, S. E., **81**, 42, 54
Dixon, D., **80**, 129
Dolbier, Jr., W. R., **80**, 172
Duan, J.-X., **80**, 172

Ellis, K. C., **81**, 157

Ferguson, M. L., **80**, 85
Frank, R., **81**, 235
Frohn, M., **80**, 1, 9
Fu, G. C., **81**, 63
Fu, Y., **81**, 215
Fukuyama, T., **80**, 207
Fürstner, A., **81**, 33

Goti , A., **81**, 204
Gouliaev, A. H., **81**, 98
Grebe, T., **81**, 134
Grubbs, R. H., **80**, 85

Hagiwara, H., **80**, 195
Hammer, R. P., **81**, 213
Hammarström, L. G. J., **81**, 213
Hank, R. F., **81**, 106
Hirao, T., **81**, 26
Hoerrner, R. S., **81**, 89
Hoshi, T., **80**, 195
Hsung, R. P., **81**, 147

Isaka, M., **80**, 144
Iserloh, U., **80**, 46

Jackson, R. F. W., **81**, 77
Jacobs, K., **81**, 140
Jensen, M. S., **81**, 89

Kakiuchi, F., **80**, 104
Kanemasa, S., **80**, 46
Katz, T. J., **80**, 227, 233
Kesavan, V., **80**, 184
Keller, J., **81**, 178
Kim, S.-H., **80**, 111

King, S. A., **81**, 178
Kopecky, D. J., **80**, 177
Kristensen, J. L., **81**, 134
Kuboyama, T., **80**, 207

Larsen, R. D., **81**, 89
LaVecchia, L., **80**, 57
Lawlor, M. D., **80**, 160
Lee, T. W., **80**, 160
Leitner, A., **81**, 33
Leutert, T., **80**, 57
Ley, S. V., **80**, 129
Li, B., **81**, 105, 254
Li, J., **81**, 195
Li, W., **81**, 89
Littke, A. F., **81**, 63
Livinghouse, T., **80**, 93
Longbottom, D. A., **80**, 129
Love, K. R., **81**, 225
Luderer, M. R., **81**, 121
Lysén, M., **81**, 134

Maligres, P., **80**, 190
Manninen, P. R., **81**, 112
Mano, E., **81**, 195
Marcoux, J.-F., **80**, 200
Marshall, J. A., **81**, 157
McLaughlin, M. L., **81**, 213
McNamara, J., **80**, 219
Mealy, M.J., **81**, 121
Moradei, O. M., **80**, 66
Mori, M., **81**,1
Muguruma, Y., **81**, 26
Mulder, J. A.., **81**, 147
Murai, S., **80**, 104
Murry, J., **81**, 105

Nakamura, E., **80**, 144
Nakamura, M., **80**, 144
Nelson, D. P., **81**, 89
Nelson, T. D., **80**, 219
Newell, L.., **81**, 254
Noe, M. C., **80**, 38
Nugiel, D. A., **81**, 140

Oderaotoshi, Y., **80**, 46
Ogawa, A., **81**, 26
Okamoto, I., **80**, 160
O'Leary, D. J., **80**, 85
Ono, H., **80**, 195

Pagendorf, B. L., **80**, 93
Paquette, L. A., **80**, 66
Paruch, K., **80**, 227, 233
Patel, M. C., **80**, 93
Perez-Gonzalez, M., **81**, 77
Petrovic, G., **81**, 244
Porinchu, M., **81**, 173
Punzalan, E. R., **81**, 123

Ravikumar, K. S., **80**, 184
Renslo, A. R., **80**, 133
Rose, J. D., **80**, 219
Roth J., **81**, 106
Rush, C., **80**, 219
Rychnovsky, S, D., **80**, 177

Sahu, P. K., **81**, 171
Saicic, R. N., **81**, 244
Sakurai, H., **81**, 26
Sato, F., **80**, 120
Seebach, D., **80**, 57
Seeberger, P. H., **81**, 225
Seidel, G., **81**, 33
Shi, Y., **80**, 1, 9
Shriver, J. A., **80**, 75
Shu, L., **80**, 9
Sinclair, P. J., **80**, 18, 31
Singh, V., **81**, 171
Snelgrove, K., **80**, 190
Söderberg, B. C., **80**, 75
Soldaini, G., **81**, 204
Song, Z. J., **81**, 195
Sowa, M. J., **80**, 219
Stecker, B., **81**, 14
Sung, M. J., **80**, 111
Suzuki, D., **80**, 120

Tabaka, A. C., **81**, 140
Tang, P., **81**, 262
Taylor, J., **80**, 200
Teleha, C. A., **81**, 140
Tian, F., **80**, 172
Tobiassen, H., **81**, 105
Tokuyama, H., **80**, 207
Tonogaki, K., **81**, 1
Tracey, M. R., **81**, 147
Tschaen, D. M., **81**, 195
Tu, Y., **80**, 1, 9

Urabe, H., **80**, 120

Vail, S., **81**, 213
Vedantham, P., **81**, 171
Vedsø, P., **81**, 134
Vyklicky, L., **80**, 227, 233

Wallace, J. A., **80**, 75
Wang, X. Q., **80**, 144
Wang, Z., **81**, 42, 54
Wang, Z.-X., **80**, 1, 9
Waters, M. S., **80**, 190
Williams, R. M., **80**, 18, 31
Winsel, H., **81**, 14
Wright, G. T., **80**, 133

Xiong, H., **81**, 147

Yamago, S., **80**, 144
Yanik, M. M., **81**, 157

Zander, N., **81**, 235
Zhai, D., **80**, 18
Zhao, M. M., **81**, 195

This index comprises subject matter for Volumes **80** and **81.** For subjects in previous volumes, see either the indices in Collective Volumes I through X or the single volume entitled *Organic Syntheses, Collective Volumes I-VIII, Cumulative Indices,* edited by J. P. Freeman.

The index lists the names of compounds in two forms. The first is the name used commonly in procedures. The second is the systematic name according to **Chemical Abstracts** nomenclature, accompanied by its registry number in parentheses. Also included are general terms for classes of compounds, types of reactions, special apparatus, and unfamiliar methods.

Most chemicals used in the procedure will appear in the index as written in the text. There generally will be entries for all starting materials, reagents, intermediates, important by-products, and final products. Entries in capital letters indicate compounds, reactions, or methods appearing in the title of the preparation.

ACETIC ACID 2-METHYLENE-3-PHENETHYLBUT-3ENYL ESTER::
    Benzenepentanol, β,γ-bis(methylene)-, acetate; (445234-76-4), **81**, 1
Acetic acid 5-phenylpent-2-ynyl ester: 2-Pentyn-1-ol, 5-phenyl-, acetate;
    (445234-71-9), **81**, 3
Acetic anhydride: Acetic acid, anhydride; (108-24-7), **80**, 177; **81**, 3
α-ACETOXY ETHER SYNTHESIS, **80**, 177
Acetylation, of alcohols, **81**, 3
        reductive, **80**, 177
Acetyl chloride; (75-36-5), **80**, 227
Acyl hydrazide synthesis, **81**, 254
Alkyl iodide synthesis, **81**, 121
Alkyne-cobalt complex, for Pauson-Khand reaction, **80**, 95
Alkyne-titanium alkoxide complex, **80**, 120
Alkyne synthesis, **81**, 2, 162
Alkynyl iodide synthesis, **81**, 42
Allene synthesis, **81**, 147
Allenylamide synthesis, **81**, 147
Allylamine: 2-Propen-1-amine; (107-11-9), **80**, 93
O-ALLYL-N-(9-ANTHRACENYLMETHYL)CINCHONIDINIUM BROMIDE:
    Cinchonanium, 1-(9-anthracenylmethyl)-9-(2-propenyloxy)-, bromide, (8α,9R)-;
    (200132-54-3), **80**, 38

Allylbenzene: Benzene, 2-propenyl-; (300-57-2), **80**, 111

Allyl bromide:1-Propene, 3-bromo-; (106-95-6), **80**, 38

Allyl iodide: 1-Propene, 3-iodo-; (556-56-9), **80**, 31

Aluminum chloride (AlCl$_3$); (7446-70-0), **80**, 227

Amide synthesis, **81**, 262

Amino acid synthesis, **81**, 213

4-Amino-1-tert-butyloxycarbonylpiperidine-4-carboxylic acid: 1,4-Piperidine dicarboxylic acid, 4-amino-, 1-(1,1-dimethylethyl)ester (9); (183673-71-4), **81**, 216

Ammonia; (7664-41-7), **80**, 31

Ammonium carbonate: Carbonic acid, diammonium salt (8,9); (506-87-6), **81**, 215

Ammonium chloride (NH$_4$Cl) (9); (12125-02-9), **81**, 188

p-Anisidine: Benzenamine, 4-methoxy-; (104-94-9), **80**, 160

Arylboronate synthesis, **81**, 89, 134

Asymmetric hydrogenation, **81**, 178

Benzaldehyde; (100-52-7), **80**, 160

Benzamidine hydrochloride: Benzenecarboximidamide, monohydrochloride (9); (1670-14-0), **81**, 105

Benzenesulfenyl chloride; (931-59-9), **81**, 244, 246

Benzenethiol; Thiophenol; (108-98-5), **80**, 184

(4S)-2-(BENZHYDRYLIDENAMINO)PENTANEDIOIC ACID, 1-tert-BUTYL ESTER-5-METHYL ESTER: L-Glutamic acid, N-(diphenylmethylene)-, 1-(1,1-dimethylethyl) 5-methyl ester; (212121-62-5), **80**, 38

Benzonitrile; (100-47-0), **81**, 123

1,4-Benzoquinone: 2,5-Cyclohexadiene-1,4-dione; (106-51-4), **80**, 233

Benzylamine: Benzenemethanamine; (100-46-9), **81**, 262

N-Benzylidene-p-anisidine: Benzenamine, 4-methoxy-N-(phenylmethylene)-; (783-08-4), **80**, 160

N-Benzylidene-benzylamine N-oxide; Benzenemethanamine, N-(phenylmethylene)-, N-oxide; (3376-26-9), **81**, 204

trans-2-BENZYL-1-METHYLCYCLOPROPAN-1-OL; 1-Cyclopropanol, 1-methyl-2-phenylmethyl-; **80**, 111

N-BENZYL-4-PHENYLBUTYRAMIDE: Benzenebutanamide, N-(phenylmethyl)-; (179923-27-4), **81**, 262

BICYCLO[3.1.0]HEXAN-1-OL; (7422-09-5), **80**, 111

(R)-BINAP: Phosphine, (1R)-[1,1: binaphthalene]-2,2'-diayhdris (diphenyl)-; (76189-55-4). **81**, 178

2,2-Bis(chloromethyl)-5,5-dimethyl-1,3-dioxane: 1,3-Dioxane, 2,2-bis(chloromethyl)-5,5-dimethyl-; (133961-12-3), **80**, 144

Bis (tri-tert-butylphosphine)palladium: Palladium, bis[tris(1,1,-dimethylethyl)phosphine]- (9); (53199-31-8), **81**, 63

Bis(tricyclohexylphosphine)benzylidine ruthenium(IV) dichloride: Ruthenium, dichloro(phenylmethylene)bis(tricyclohexylphosphine)-; (172222-30-9), **80**, 85

6,13-BIS(TRIISOPROPYLSILOXY)-9,10-DIMETHOXY[7]HELICENEBISQUINONE: Dinaphtho[2,1-c:1',2'-g]phenanthrene-1,4,15,18-tetrone, 9,10-dimethoxy-

6,13-bis[[tris(1-methylethyl)silyl]oxy]-; (310899-14-0), **80**, 233

3,6-Bis[1-(triisopropylsiloxy)ethenyl]-9,10-dimethoxyphenanthrene: Silane,
[(9,10-dimethoxy-3,6-phenenthrenediyl)bis(ethenylideneoxy)]tris(1-methylethyl)-;
80, 233

Boc-diallylamine: Carbamic acid, di-2-propenyl-, 1,1-dimethylethyl ester; (151259-38-0),
80, 85

N-Boc hydroxyproline methyl ester; 1,2-Pyrrolidinedicarboxylic acid, 4-hydroxy, 1-(1,1-dimethylethyl)
2-methyl ester, (2S,4R); (74844-91-0), **81**, 178

N-Boc-3-PYRROLINE: 1H-Pyrrole-1-carboxylic acid, 2,5-dihydro-, 1,1-dimethylethyl
ester; (73286-70-1), **80**, 85

endo-1-BORNYLOXYETHYL ACETATE: Ethanol, 1-[[(1R,2S,4R)-1,7,7-trimethylbicyclo[2.2.1]-
hept-2-yl]oxy]-, acetate; (284036-61-9), **80**, 177

Borane-dimethylsulfide complex: Boron, trihydro[thiobis[methane]]-(T-4)-; (13292-87-0), **81**, 43

Boric acid: Boric acid (H$_3$BO$_3$); (10043-35-3), **81**, 262

BORONIC ACID SYNTHESIS, **81**, 89

Bromination, **80**, 75; **81**, 98, 99

Bromine; (7726-95-6), **80**, 75

Bromoacetonitrile; Acetonitrile, bromo-; (590-17-0), **80**, 207

Bromobenzene: Benzene, bromo-; (100-86-1), **80**, 57

1-Bromo-2-propyne: 1-propyne, 3-bromo-; (106-96-7), **80**, 93

4-Bromo-1-butene: 1-Butene, 4-bromo-; (5162-44-7), **81**, 19

5-BROMOISOQUINOLINE: Isoquinoline, 5-bromo- (8,11); (34784-04-8), **81**, 98

5-BROMO-8-NITROISOQUINOLINE: Isoquinoline, 5-bromo-8-nitro-; (63927-23-1), **81**, 98

1-Bromononane: Nonane, 1-bromo-; (693-58-3), **81**, 33

3-Bromopyridine: Pyridine, 3-bromo- (8, 9); (626-55-1), **81**, 89

3-Bromoquinoline: Quinoline, 3-bromo- (8, 9); (5332-24-1), **81**, 90

N-Bromosuccinimide: Succinimide, N-bromo- (8); 2,5-Pyrrolidinedione, 1-bromo-; (128-08-5)
81, 98, 99

3-Butenylmagnesium bromide: Magnesium, bromo-3-butenyl-(8,9); (7013-09-5), **81**, 17

(2S-trans)-2-tert-Butoxycarbonylacetyl-4-hydroxypyrrolidine-1-carboxylic acid, tert-butyl ester:
2-Pyrrolidinepropanoic acid, 1-[(1,1-dimethylethoxy)carbonyl]-4-hydroxy-β-oxo-,1,1-dimethylethyl
ester, [2S-trans]- (9); (167963-29-3), **81**, 178

(R)-(N-tert-BUTOXYCARBONYL)ALLYLGLYCINE: 4-Pentenoic acid,
2-[[(1,1-dimethylethoxy)carbonyl]amino]-, (2R)-; (170899-08-8), **80**, 31

(3R,5R,6S)-4-tert-Butoxycarbonyl-5,6-diphenyl-3-(1'-prop-2'-enyl)morpholin-2-one:
4-Morpholinecarboxylic acid, 2-oxo-5,6-diphenyl-3-(2-propenyl)-, 1,1-dimethylethyl
ester, [3R-(3α,5β,6β)]-; (143140-32-3), **80**, 31

2(S)-(β-tert-BUTOXYCARBONYL-α-(R)-HYDROXY)ETHYL-4(R)-
HYDROXYPYRROLIDINE-1-CARBOXYLIC ACID, tert-BUTYL ESTER: (2-Pyrrolidinepropanoic
acid, 1-[(1,1-dimethylethoxy)carbonyl]-β,4-dihydroxy-, 1,1-dimethylethyl ester, [2S-[2α(S*),4β]]-;
(167963-30-6), **81**, 178

2(S)-(β-tert-BUTOXYCARBONYL-α-(S)-HYDROXY)ETHYL-4(R)-HYDROXYPYRROLIDINE-
1-CARBOXYLIC ACID, tert-BUTYL ESTER, **81**, 178

N-(tert-BUTOXYCARBONYL)-L-IODOALANINE METHYL ESTER: L-Alanine,
N-[(1,1-dimethylethoxy)carbonyl]-3-iodo-, methyl ester; (93267-04-0), **81**, 77

N-(tert-BUTOXYCARBONYL)-L-4-(METHOXYCARBONYL)PHENYL]ALANINE METHYL ESTER:
L-Phenylalanine,N-[(1,1-dimethylethoxy)carbonyl]-4-(methoxycarbonyl)-, methyl ester (9);

(160168-19-4), **81**, 77

N-(tert-Butoxycarbonyl)-L-serine methyl ester: L-Serine, N-[(1,1-dimethylethoxy)carbonyl]-, methyl
    ester (9); (2766-43-0), **81**, 77

(S)-N-(tert-Butoxycarbonyl)valine methyl ester: Valine, N-[(1,1-dimethylethoxy)-
    carbonyl]-, methyl ester, (S)-; (58561-04-9), **80**, 57

(S)-N-(tert-Butoxycarbonyl)valine: Valine, N-[(1,1-dimethylethoxy)carbonyl]-;
    (13734-41-3), **80**, 57

tert-Butyl acetate: Acetic acid, 1,1-dimethylethyl ester; (540-88-5), **81**, 179

n-Butyl acrylate: 2-Propenoic acid, butyl ester; (141-32-2), **80**, 172

n-BUTYL 2,2-DIFLUOROCYCLOPROPANECARBOXYLATE: Cyclopropanecarboxylic
    acid, 2,2-difluoro-, butyl ester; (260352-79-2), **80**, 172

tert-Butyldimethylsilyl chloride; Silane, chloro(1,1-dimethylethyl)dimethyl- (9); (18162-48-6), **81**, 157

(S)-4-(tert-BUTYLDIMETHYLSILYLOXY)-2-PENTYN-1-OL, **81**, 157
    ee determination, **81**, 164

(S)-2-(tert-Butyldimethylsilyloxy)propanal: Propanal, 2-[[(1,1-dimethylethyl)dimethyl silyl]oxy]-, (2S)-;
    (87727-28-4), **81**, 159

n-Butyllithium: Lithium, butyl-; (109-72-8) **80**, 160; **81**, 2, 43, 44, 89, 112, 134, 159

sec-Butyllithium: Lithium, (1-methylpropyl)-; (598-30-1), **80**, 46

tert-Butyllithium; Lithium, (1,1-dimethylethyl)-; (594-19-4), **81**, 122

n-Butylmagnesium chloride: Magnesium, butylchloro-; (693-04-9), **80**, 111

Butyl methacrylate: 1-Propenoic acid, 2-methyl-, butyl ester; (97-88-1), **81**, 63

(5S,6R)-4-tert-BUTYLOXYCARBONYL-5,6-DIPHENYLMORPHOLIN-2-ONE:
    4-Morpholinecarboxylic acid, 6-oxo-2,3-diphenyl-, 1,1-dimethylethyl ester, (2R-cis)-;
    (173397-90-5), **80**, 18, 31

(5R,6S)-4-tert-BUTYLOXYCARBONYL-5,6-DIPHENYLMORPHOLIN-2-ONE:
    4-Morpholinecarboxylic acid, 6-oxo-2,3-diphenyl-, 1,1-dimethylethyl ester, (2S-cis)-;
    (112741-50-1), **80**, 18

1-tert-BUTYLOXYCARBONYL-4-(9-FLUORENYLMETHYLOXYCARBONYLAMINO)PIPERIDINE-
    4-CARBOXYLIC ACID: 1,4,Piperidinedicarboxylic acid, 4-[[(9H-fluoren-9-ylmethyloxy)-
    carbonyl]amino]-, 1-(1,1-dimethylethyl)ester; (183673-66-7), **81**, 213

1-tert-Butyloxycarbonylpiperidine-4-spiro-5'-(1',3'-bis(tert-butyloxycarbonyl) hydantoin:
    1,3,8-Triazaspiro[4.5]decane-1,3,8-tricarboxylic acid, 2,4-diiso-, tris(1-dimethylethyl)ester;
    (183673-68-9), **81**, 214

D-(+)-Camphorsulfonic acid: Bicyclo[2.2.1]heptane-1-methanesulfonic acid,
    7,7-dimethyl-2-oxo-, (1S,4R)-; (3144-16-9), **80**, 66

(2-Carbomethoxy-6-nitrobenzyl)triphenylphosphonium bromide: Phosphonium,
    [[2-(methoxycarbonyl)-6-nitrophenyl]methyl]triphenyl-, bromide; (195992-09-7),
    80, 75

Carbon dioxide; (124-38-9), **80**, 46

Carbon monoxide; (630-08-0), **80**, 75, 93

Carbon tetrabromide: Methane, tetrabromo-; (558-13-4), **81**, 1

Carbonyldihydridotris(triphenylphosphine)ruthenium(II): Ruthenium,
    carbonyldihydridotris(triphenylphosphine); (25360-32-1), **80**, 104

Cesium hydroxide: Cesium hydroxide (CsOH); (21351-79-1), **80**, 38

CHIRAL LEWIS ACID TRIDENTATE LIGAND, **80**, 46

Chloroacetyl chloride: Acetyl chloride, chloro-; (79-04-9), **80**, 200

Chlorobenzene: Benzene, chloro-(8,9); (108-90-7), **81**, 64

4-Chlorobenzoic acid methyl ester: Benzoic acid, 4-chloro-, methyl ester; (1126-46-1), **81**, 34

4-Chlorobenzonitrile: Benzonitrile, 4-chloro- (9); (623-03-0), **81**, 64

2-CHLORO-1,3-BIS(DIMETHYLAMINO)TRIMETHINIUM HEXAFLUOROPHOSPHATE: Methanaminium, N-[2-chloro-3-(dimethylamino)- 2-propenylidene]-N-methyl-, hexafluorophosphate(1-); (249561-98-6), **80**, 207

m-Chloroperbenzoic acid; Benzenecarboperoxoic acid, 3-chloro-; (937-14-4), **80**, 207

9-Chloromethylanthracene: Anthracene, 9-(chloromethyl)-; (24463-19-2), **80**, 38

N-Chlorosuccinimide: 2,5-Pyrrolidinedione, 1-chloro-; (128-09-6), **80**, 133

Chlorotitanium triisopropoxide: Titanium, chlorotris(2-propanolato)-, (T-4)-; (20717-86-6), **80**, 111

Chlorotrimethylsilane: Silane, chlorotrimethyl- (9); (75-77-4) **80**, 172; **81**, 26, 216

Cinchonidine: Cinchonan-9-ol, (8α, 9R)-; (485-71-2), **80**, 38

R-(+)-Citronellal: 6-Octenal, 3,7-dimethyl-, (3R)-; (2385-77-5), **80**, 195

Claisen condensation, **81**, 178

Copper(I) Iodide: Copper iodide (CuI); (7681-65-4), **80**, 129

Cross coupling, **81**, 33

Cross enyne metathesis, **81**, 1

Cycloaddition, of difluorocarbene, **80**, 172

[2+2} Cycloaddition, **80**, 160

[3+2} Cycloaddition, **80**, 144

Cyclohexanecarboxaldehyde; (2043-61-0), **81**, 26

Cyclohexene; (110-83-8), **81**, 43

Cyclohexylamine: Cyclohexanamine; (108-91-8), **80**, 93

Cyclooctadienyl)ruthenium dichloride polymer, **81**, 178

(Cyclooctadienyl)ruthenium dichloride polymer-(R)-BINAP catalyst, [Et₂NH₂]⁺[Ru₂Cl₅(BINAP)₂]⁻, **81**, 178

Cyclopentadiene: 1,3-Cyclopentadiene (8,9); (542-92-7), **81**, 171

2-Cyclopenten-1-one; (930-30-3), **80**, 144

2-CYCLOPENTYLACETOPHENONE: Ethanone, 2-cyclopentyl-1-phenyl-; (23033-65-0), **81**, 121

Cyclopentylmagnesium chloride: Magnesium, chlorocyclopentyl-; (32916-51-1), **80**, 111

Cyclopropane synthesis, **80**, 111; **81**, 15

Cyclopropene synthesis, **80**, 145

Cyclopropanol synthesis, **80**, 111

Cyclopropylamine synthesis, **81**, 14

3,6-DIACETYL-9,10-DIMETHOXYPHENANTHRENE: Ethanone, 1,1'-(9,10-dimethoxy-3,6-phenanthrenediyl)bis-; (310899-08-2), **80**, 227, 233

1,2:5,6-DIANHYDRO-3,4-O-ISOPROPYLIDENE-L-MANNITOL: L-MANNITOL,1,2:5,6-DIANHYDRO-3,4-O-(1-METHYLETHYLIDENE)-; (153059-37-1), **81**, 140

Deuterium oxide: Water-d2; (7789-20-0), **80**, 120

Diazo transfer reaction, **80**, 161

DIBAL-H: Diisobutylaluminum hydride; Aluminum, hydrobis(2-methylpropyl)-; (1191-15-7), **81**, 158

Dibenzofuran; (132-64-9), **80**, 46

Dibenzofuran-4,6-dicarbonyl chloride: 4,6-Dibenzofurandicarbonyl dichloride;
(151412-73-8), **80**, 46

Dibenzofuran-4,6-dicarboxylic acid: 4,6-Dibenzofurandicarboxylic acid;
(88818-47-7), **80**, 46

(R,R)-Dibenzofuran-4,6-dicarboxylic acid bis(2-hydroxy-1-phenylethyl)amide:
4,6-Dibenzofurandicarboxamide, N,N'-bis[(1R)-2-hydroxy-1-phenylethyl]-;
(247097-79-6), **80**, 46

(R,R)-4,6-DIBENZOFURANDIYL-2,2'-BIS(4-PHENYLOXAZOLINE) (DBFOX/PH):
Oxazole, 2,2'-(4,6-dibenzofurandiyl)bis(4,5-dihydro-4-phenyl-, (4R,4'R)-;
(195433-00-2), **80**, 46

Dibenzoyl peroxide: Peroxide, dibenzoyl; (94-36-0), **80**, 75

Dibenzylamine: Benzenemethanamine, N-(phenylmethyl)-; (103-49-1), **81**, 18, 204

N,N-DIBENZYL-*N*-(2-ETHENYLCYCLOPROPYL)AMINE: BENZENEMETHANAMINE,
N-(2-ETHENYLCYCLOPROPYL)-*N*-(PHENYLMETHYL)-; (220247-75-5), **81**, 14

N,N-Dibenzylformamide: Formamide, *N,N*-bis(phenylmethyl)-; (5464-77-7), **81**, 14

Dibenzylideneacetone: 1,4-Pentadien-3-one, 1,5-diphenyl-; (538-58-9), **81**, 56

Dibenzyl phosphate: Phosphoric acid, bis(phenylmethyl) ester; (1623-08-1), **80**, 219

(4,4-Dibromobut-3-enyl)benene: Benzene, (4,4-dibromo-3-butenyl)-; (119405-97-9), **81**, 1

Di-tert-butyl dicarbonate: Dicarbonic acid, bis(1,1-dimethylethyl) ester; (24424-99-5),
80, 18; **81**, 215

Dibutyl phosphate: Phosphoric acid, dibutyl ester (8,9); (107-66-4), **81**, 226

DIBUTYL 3,4,6-TRI-*O*-BENZYL-2-*O*-PIVALOYL-D-GLUCOPYRANOSYL PHOSPHATE:
β-D-GLUCOPYRANOSE, 3,4,5-TRIS-*O*-(PHENYLMETHYL)-, 1-(DIBUTYL PHOSPHATE)
2-(2,2-DIMETHYLPROPANOATE); (223919-63-7), **81**, 225

1,3-Dichloroacetone: 2-Propanone, 1,3-dichloro-; (534-07-6), **80**, 144

1,1-Dichloro-(3S)- (tert-butyldimethylsilyloxy)-2-butanol p-toluenesulfonate: 2-Butanol, 1,1-dichloro-
3-[[(1,1-dimethylethyl)dimethylsilyl]oxy]-, 4-methylbenzene sulfonate, (3S)-; (329914-17-2),
81, 159

Dichlorodicyclopentadienylvanadium; Vanadium, dichlorobis (η^5-2,4-cyclopentadien-1-yl)-;
(12083-48-6), **81**, 26

Dicobalt octacarbonyl: Cobalt, di-μ-carbonylhexacarbonyl di-, (Co-Co); (10210-68-1),
80, 93

1,3-Dicyclohexylcarbodiimide: Cyclohexanamine, N,N'-methanetetraylbis-;
(538-75-0), **80**, 219

dl-1,2-DICYCLOHEXYLETHANEDIOL: 1,2-ETHANEDIOL, 1,2-DICYCLOHEXYL- (9); (92319-61-4)
81, 26

Dicyclopentadiene: 4,7-Methano-1H-indene, 3a,4,7,7a-tetrahydro-; (77-73-6), **81**, 173

(Z)-1,2-DIDEUTERIO-1-(TRIMETHYLSILYL)-1-HEXENE: Silane, (1,2-dideuterio-
1-hexenyl)trimethyl-, (Z)-; **80**, 120

Diels-Alder reaction, **80**, 133, 234; **81**, 171

1, 3-Diene synthesis, **81**, 1

Diethylaminosulfur trifluoride (DAST): Sulfur, (N-ethylethanaminato)trifluoro-, (T-4);
(38078-09-0), **80**, 46

Diethylaminotrimethylsilane: Silanamine, N,N-diethyl-1,1,1-trimethyl-; (996-50-9),
80, 195

Diethyl azodicarboxylate: Azodicarboxylic acid diethyl ester; (1972-58-3), **81**, 142

DIFLUOROCARBENE REAGENT, **80**, 172

DIBAL-H (Diisobutylaluminum hydride): Aluminum, hydrobis (2-methylpropyl)- (9); (1191-15-7), **80**, 177; **81**, 157

Diisopropylamine: 2-Propanamine, N-(1-methylethyl)-; (109-72-8), **81**, 159

Diisopropylethylamine: 2-Propanamine, N-ethyl-N-(1-methylethyl)-; (7087-68-5) **80**, 207; **81**, 216

1,2:4,5-DI-O-ISOPROPYLIDENE-D-erythro-2,3-HEXODIULO-2,6-PYRANOSE: β-D-erythro-2,3-Hexodiulo-2,6-pyranose, 1,2:4,5-bis-O-(1-methylethylidene)-; (18422-53-2), **80**, 1, 9

1,2:4,5-Di-O-isopropylidene-β-D-fructopyranose: β-D-Fructopyranose, 1,2;4,5-bis-O-(1-methylethylidene)-; (25018-67-1), **80**, 1

1,2:3,4-Di-O-isopropylidene-α-D-galactopyranose: α-D-Galactopyranose, 1,2:3,4-bis-O-(1-methylethylidene)-; [4064-06-6], **81**, 226

1,2-Dimethoxyethane: Ethane, 1,2-dimethoxy-; (110-71-4), **80**, 93

Dimethoxymethane: Methane, dimethoxy-; (109-87-5), **80**, 9

9,10-DIMETHOXYPHENANTHRENE: Phenanthrene, 9,10-dimethoxy-; (13935-65-4), **80**, 227

2,2-Dimethoxypropane: Propane, 2,2-dimethoxy-(9); (77-76-9), **80**, 1; **81**, 141

4-(Dimethylamino)pyridine (DMAP): 4-Pyridinamine, N,N-dimethyl-; (1122-58-3), **80**, 177; **81**, 78, 217

Dimethyl biphenyl-4,4'-dicarboxylate: [1,1'-Biphenyl]-4,4'- dicarboxylic acid, dimethyl ester; (792-74-5), **81**, 81

2-(5,5-DIMETHYL-1,3,2-DIOXABORINAN-2-YL)BENZOIC ACID ETHYL ESTER: BENZOIC ACID, 2-(5,5-DIMETHYL-1,3,2-DIOXABORINAN-2-YL)-, ETHYL ESTER; (346656-34-6), **81**, 134

cis-5-(5,5-DIMETHYL-1,3-DIOXAN-2-YLIDENE)HEXAHYDRO-1(2H)-PENTALENONE, **80**, 144

N,N-Dimethylformamide: Formamide, N,N-dimethyl-; (68-12-2), **80**, 46, 200

(Z)-1,1-Dimethyl-1-heptenylsilanol: Silanol, (1Z)-1-heptenyldimethyl-; (261717-40-2), **81**, 42

6,6-DIMETHYL-1-METHYLENE-4,8-DIOXASPIRO[2.5]OCTANE: 4,8-Dioxaspiro [2.5]octane, 6,6-dimethyl-1-methylene; (122968-05-2), **80**, 144

3R,7-DIMETHYL-2-(2-OXOBUTYL)-6-OCTENAL: 6-Octenal, 3,7-dimethyl-2-(3-oxo-butyl)-, (3R)-; (131308-24-2), **80**, 195

Dimethyl sulfate: Sulfuric acid, dimethyl ester; (77-78-1), **80**, 227

Diol synthesis, **81**, 26

DIPHENYL DISULFIDE: Disulfide, diphenyl; (882-33-7), **80**, 184

(1S,2R)-1,2-Diphenyl-2-hydroxyethylamine: Benzeneethanol, β-amino-α-phenyl-, (αR,βS)-rel-; (23412-95-5), **80**, 18

(1R,2S)-1,2-Diphenyl-2-hydroxyethylamine: Benzeneethanol, β-amino-α-phenyl-, (αS,βR)-; (23364-44-5), **80**, 18

N-(Diphenylmethylene)glycine tert-butyl ester: Glycine, (diphenylmethylene)-, 1,1-dimethylethyl ester; (81477-94-3), **80**, 38

Directed metallation, **81**, 134

Disodium ethylenediaminetetraacetate: Glycine, N,N'-1,2-ethanediylbis [N-(carboxymethyl)-, disodium salt; (139-33-3), **80**, 9

Disperse Red 1; Ethanol, 2-[ethyl[4-[4-nitrophenyl)azo]phenyl]amino]-; (2872-52-8), **81**, 235

4-Dodecylbenzenesulfonyl azide: Benzenesulfonyl azide, 4-dodecyl-; (79791-38-1), **80**, 160

Electrophilic aromatic substutution, **80**, 228; **81**, 98
Enol silane synthesis, **80**, 233
Epoxide synthesis, **80**, 9; **81**, 140
Ester hydrolysis, **81**, 34
Esterification, **81**, 3, 235
ETHYL 4-AMINOBENZOATE: BENZOIC ACID,4-AMINO-2, ETHYL ESTER; (94-09-7), **81**, 188
Ethyl benzoate: Benzoic acid ethyl ester; (93-89-0), **81**, 134
Ethyl bromoacetate: Acetic acid, bromo-, ethyl ester; (105-36-2), **80**, 18
(S)-Ethyl 2-(tert-butyldimethylsilyloxy)lactate: Propanoic acid, 2-[[(1,1-dimethylethyl) dimethyl]oxy], ethyl ester; (106513-42-2), **81**, 157
Ethyl (1'R,2'S)-N-tert-butyloxycarbonyl-N-(1',2'-diphenyl-2'-hydroxyethyl) glycinate: Glycine, N-[(1,1-dimethylethoxy)carbonyl]-N-(2-hydroxy-1,2-diphenylethyl)-, ethyl ester, [S-(R*,S*)]-; (112741-73-8), **80**, 18
Ethyl (1'S,2'R)-N-tert-butyloxycarbonyl-N-(1',2'-diphenyl-2'-hydroxyethyl)glycinate: Glycine, N-[(1,1-dimethylethoxy)carbonyl]-N-(2-hydroxy-1,2-diphenylethyl)-, ethyl ester, [R-(R*,S*)]-; (112741-70-5), **80**, 18
Ethyl (1'S,2'R)-N-(1',2'-diphenyl-2'-hydroxyethyl)glycinate: Glycine, N-(2-hydroxy-1,2-diphenylethyl)-, ethyl ester, [R-(R*,S*)]-; (100678-82-8), **80**, 18
Ethyl (1'R,2'S)-N-(1',2'-diphenyl-2'-hydroxyethyl)glycinate: Glycine, N-(2-hydroxy-1,2-diphenylethyl)-, ethyl ester, [S-(R*,S*)]-; (112835-62-8), **80**, 18
Ethylene: Ethene; ((74-85-1), **81**, 4
    purification of, **81**, 8
    reaction apparatus for addition of, **81**, 9
Ethylenediaminetetraacetic acid (EDTA): Glycine, N,N'-1,2-ethanediylbis[N-(carboxymethyl)-; (60-00-4), **80**, 9
(S)-Ethyl lactate: Propanoic acid, 2-hydroxy-, ethyl ester, (2S)-; (687-47-8), **81**, 157
Ethyl 4-nitrobenzoate: Benzoic acid, 4-nitro-, ethyl ester (9); 99-77-4), **81**, 188
Ethyl vinyl ether: Ethene, ethoxy-; (109-92-2), **81**, 4

Ferric acetylacetonate: Tris (2,4-pentanedionato)iron (III); (14024-18-1), **81**, 34
Ferric nitrate nonahydrate: Nitric acid, iron(3+) salt, nonahydrate; (7782-61-8), **80**, 144
Finkelstein reaction, **81**, 121
9-Fluorenylmethyl chloroformate: Carbonochloridic acid, 9H-fluoren-9-yl-, methyl ester; (28920-43-6), **81**, 216
2-Fluorosulfonyl-2,2-difluoroacetic acid: Acetic acid, difluoro(fluorosulfonyl)-; (1717-59-5), **80**, 172
Formic acid; (64-18-6), **81**, 18
D-Fructose; (57-48-7), **80**, 1
Furfural: 2-Furancarboxaldehyde; (98-01-1), **80**, 66

(R)-(–)-Glycidyl butyrate: Butanoic acid, (2R)-oxiranylmethyl ester; (60456-26-0), **81**, 112
Glycosylation, **81**, 225
Grignard reagents, **81**, 33

Grubbs catalyst, **80**, 85; **81**, 4

Heck reaction, **81**, 63
HELICENEBISQUINONES, **80**, 233
3-Heptanol; (589-82-2), **81**, 244
(Z)- 1-HEPTENYLDIMETHYLSILANOL: Silanol, (1Z)-1-heptenyldimethyl-; (261717-40-2), **81**, 42
(Z)-1-HEPTENYL-4-METHOXYBENZENE: BENZENE, 1-(1Z)-1-HEPTENYL-4-METHOXY-;
    (80638-85-3), **81**, 42, 54
3-Heptyl benzenesulfenate: Benzenesulfenic acid, 1-ethylpentyl ester; (198778-69-7), **81**, 244
1-Heptyne: 1-Heptyne; (628-71-7), **81**, 42, 54
Hexabutylditin: Distannane, hexabutyl; (813-19-4), **81**, 245
Hexacarbonyl[μ[(3,4-η:3,4-η)-2-methyl-3-butyn-2-ol]]dicobalt: Cobalt, hexacarbonyl
    [μ-[(3,4-η:3,4-η)-2-methyl-3-butyn-2-ol]]di-, (Co-Co); (40754-33-4), **80**, 93
Hexafluorophosphoric acid: Phosphate (1-), hexafluoro-, hydrogen; (16940-81-1),
    80, 200
Hexamethylcyclotrisiloxane: Cyclotrisiloxane, hexamethyl-; (54-05-9), **81**, 44
1,1,1,3,3,3-Hexamethyldisilazane: Silanamine,1,1,1-trimethyl-N-(trimethylsilyl)-;
    (999-97-3), **80**, 160
1-Hexene, 6-iodo-; (18922-04-8), **81**, 121
5-Hexen-1-ol; (821-41-0). **81**, 121
5-Hexen-1-ol, methanesulfonate; (64818-36-6), **81**, 121
Hydantoin synthesis, **81**, 213
Hydrazine; (10217-52-4), **81**, 254
Hydriodic acid; (10034-85-2), **80**, 129
Hydrogen peroxide (H_2O_2); (7722-84-1), **80**, 9, 184
Hydrosilylation, **81**, 54
(S)-[1-(Hydroxydiphenylmethyl)-2-methylpropyl]carbamic acid, tert-butyl ester:
    Carbamic acid, [1-(hydroxydiphenylmethyl)-2-methylpropyl]-, 1,1-dimethylethyl ester,
    (S)-; (157035-82-0), **80**, 57
2-Hydroxy-2(5H)-furanone: 2(5H)-Furanone, 5-hydroxy-; (14032-66-7), **80**, 66
    Hydroxylamine hydrochloride: Hydroxylamine, hydrochloride; (5470-11-1), **80**, 207
Hydroxylaminolysis, **80**, 209
N-HYDROXY-(S)-1-PHENYLETHYLAMINE OXALATE; Benzenemethanamine,
    N-hydroxy-α-methyl-, (αS)-, ethanedioate (1:1) salt; (78798-33-1), **80**, 207
1-HYDROXY-3-PHENYL-2-PROPANONE: 2-Propanone, 1-hydroxy-3-phenyl-;
    (4982-08-5), **80**, 190

Imidazole: 1H-Imidazole; (288-32-4), **81**, 157
Imidazole synthesis, **81**, 105
Imine formation, **80**, 160
Indium; (7440-74-6), **81**, 188
Iodination of alkynes, **81**, 42
Iodine; (7553-56-2), **81**, 43, 77
4-Iodoanisole: Benzene, 1-iodo-4-methoxy-; (696-62-8), **81**, 45, 55

(Z)-1-Iodo-1-Heptene: 1-Heptene, 1-iodo-, (1Z)- (9); (63318-29-6), **81**, 43

1-Iodo-1-heptyne: 1-Heptyne, 1-iodo-; (54573-13-6), **81**, 42

(E)-3-IODOPROP-2-ENOIC ACID: 2-Propenoic acid, 3-iodo-, (2E)-; (6372-02-7), **80**, 129

(R)-Isobutyl 2-(tert-butyldimethylsilyloxy)propanoate, **81**, 161

Isomerization, of alkynes to allenes, **81**, 147

       of a methylcyclopropene to a methylenecyclopropane, **80**, 147

Isoprene: 1,3-Butadiene, 2-methyl-; (78-79-5), **80**, 133

Isopropyl acetate: Acetic acid, 1-methylethyl ester; (108-21-4), **80**, 219

3,4-O-Isopropylidene-L-mannitol: L-Mannitol, 3,4-O-(1-methylethylidene)-; (153059-36-0), **81**, 141

Isopropylmagnesium chloride: Magnesium, chloro(1-methylethyl)-; (1068-55-9), **80**, 120

Isoquinoline; (119-65-3), **81**, 98

Ketal formation, **80**, 1, 66, 144

    hydrolysis, **81**, 140

Kulinkovich reaction, **80**, 111; **81**, 17

Lactone formation, **80**, 20

Lithium; (7439-93-2), **80**, 31

Lithium borohydride: Borate(-1), tetrahydro-, lithium; (16949-15-8), **81**, 142

Lithium diisopropylamide: 2-Propanamine, N-(1-methylethyl)-, lithium salt; (4111-54-0), **81**, 160

Lithium bis(trimethylsilyl)amide; Lithium hexamethyldisilazide: Silanamine, 1,1,1-trimethyl-N-(trimethylsilyl)-, lithium salt; (4039-32-1), **80**, 31, 160; **81**, 179

Magnesium; (7439-95-4) **80**, 57 (3009)

L-Mannonic acid δ-lactone: L-Mannonic acid, δ-lactone; (22430-23-5), **81**, 140

Meldrum's acid: 1,3-Dioxane-4,6-dione, 2,2-dimethyl-; (2033-24-1), **80**, 133

d-Menthol: Cyclohexanol, 5-methyl-2-(1-methylethyl)-, (1S,2R,5S)-; (15356-60-2), **80**, 66

(5S)-(d-MENTHYLOXY)-2(5H)-FURANONE: 2(5H)-Furanone, 5-[[(1S,2R,5S)-5-methyl-2-(1-methylethyl)cyclohexyl]oxy]-, (5S)-; (122079-41-8), **80**, 66

Metathesis, **80**, 85; **81**, 4

Methanesulfonyl chloride; (124-63-0), **81**, 121

4-Methoxyphenacyl bromide; Ethanone, 2-bromo-1-(4-methoxyphenyl)-; (2632-13-5), **81**, 105

4-Methoxyphenethyl alcohol: Benzeneethanol, 4-methoxy-; (702-23-8), **81**, 195

4-METHOXYPHENYLACETIC ACID: BENZENEACETIC ACID, 4-METHOXY-; (104-01-8) **81**, 195

4-(4-METHOXYPHENYL)-2-PHENYL-1H-IMIDAZOLE: 1H-IMIDAZOLE, 4-(METHOXYPHENYL)-2-PHENYL-; (53458-08-5), **81**, 105

trans-1-(4-METHOXYPHENYL)-4-PHENYL-3-PHENYLTHIO)AZETIDIN-2-ONE: 2-Azetidinone, 1-(4-methoxyphenyl)-4-phenyl-3-(phenylthio)-trans-; (94612-48-3), **80**, 160

Methyl acrylate: 2-Propenoic acid, methyl ester; (96-33-3), **80**, 38

Methyl 2-bromomethyl-3-nitrobenzoate: Benzoic acid, 2-bromomethyl-3-nitro-, methyl ester; (98475-07-1), **80**, 75

Methyl tert-butyl ether; Propane, 2-methoxy-2-methyl-; (1634-04-4), **81**, 195

2-Methyl-3-butyn-2-ol: 3-Butyn-2-ol, 2-methyl-; (115-19-5), **80**, 93

N-Methyldicyclohexylamine: Cyclohexanamine, N-cyclohexyl-N-methyl-; (7560-83-0), **81**, 63, 64

Methyl 2-ethenyl-3-nitrobenzoate: Benzoic acid, 2-ethenyl-3-nitro-, methyl ester; (195992-04-2), **80**, 75

(4S)-(1-METHYLETHYL)-5,5-DIPHENYLOXAZOLIDIN-2-ONE: 2-Oxazolidinone, 4-(1-methylethyl)-5,5-diphenyl-, (4S)-; (184346-45-0), **80**, 57

Methyl 5-hexenoate: 5-Hexenoic acid, methyl ester; (2396-80-7), **80**, 111

N-Methylimidazole: 1H-Imidazole, 1-methyl-; (616-47-7), **81**, 236

METHYL INDOLE-4-CARBOXYLATE: 1H-Indole-4-carboxylic acid, methyl ester; (39830-66-5), **80**, 75

Methyl iodide: Methane, iodo-; (74-88-4), **80**, 57, 144

Methyl 4-iodobenzoate: Benzoic acid, 4-iodo-, methyl ester; (619-44-3), **81**, 78

Methyllithium: Lithium, methyl-; (917-54-4), **81**, 16

Methyl 2-methyl-3-nitrobenzoate: Benzoic acid, 2-methyl-3-nitro-, methyl ester; (59382-59-1), **80**, 75

METHYL 5-METHYLPYRIDINE-2-CARBOXYLATE: 2-Pyridinecarboxylic acid, 5-methyl-, methyl ester; (260998-85-4), **80**, 133

(E)-2-METHYL-3-PHENYLACRYLIC ACID BUTYL ESTER: 2-PROPENOIC ACID, 2-METHYL-3-PHENYL-, BUTYL ESTER,(2E)-; (21511-00-5), **81**, 63

METHYL PHENYL SULFOXIDE: Benzene, (methylsulfinyl)-; (1193-82-4), **80**, 184

N-Methylpyrrolidinone: 2-Pyrrolidinone, 1-methyl-; (872-50-4), **81**, 34

trans-β-Methylstyrene: Benzene, (1E)-1-propenyl-; (873-66-5), **80**, 9

(R,R)-trans-β-METHYLSTYRENE OXIDE: Oxirane, 2-methyl-3-phenyl-, (2R,3R)-; (14212-54-5), **80**, 9

Methyltrioxorhenium: Rhenium, methyltrioxo-, (T-4)-; (70197-13-6), **81**, 204

Methyl tris(isopropoxy)titanium: Titanium, methyltris2-propanolato)-, (T-4)-; (18006-13-8), **81**, 16

Methyl vinyl ketone: 3-Buten-2-one; (78-94-4), **80**, 195

Mitsunobu reaction, **81**, 141

Michael addition, **80**, 40
    of aldehydes to vinyl ketones, **80**, 195

Morpholinone glycine synthon, **80**, 35
    reductive removal of auxiliarry, **80**, 32

Negishi coupling, **81**, 77

Neopentyl glycol: 1,3-Propanediol, 2,2-dimethyl-; (126-30-7), **80**, 144; **81**, 134

Nitration, **81**, 98

Nitrone synthesis, **81**, 204

4-NONYLBENZOIC ACID: Benzoic acid, 4-nonyl-; (38289-46-2), **81**, 33

(Nonylmagnesium bromide: Magnesium, bromononyl-; (39691-62-8), **81**, 33

Organolithium cyclization, **81**, 121

Oxalic acid: Ethanedioic acid; (144-62-7), **80**, 207

Oxazolidinone synthesis, **81**, 112

Oxazoline formation, **80**, 46

Oxidation, of alcohols to carboxylic acids, **81**, 195
    of alcohols to ketones, **80**, 2
    of amines, **80**, 207; **81**, 204
    of phenols, **81**, 171
    of sulfur compounds, **80**, 184
    photolytic, **80**, 66
Oximation, **80**, 130
Oxone: Peroxymonosulfuric acid, monopotassium salt, mixture with dipotassium sulfate
    and potassium hydrogen sulfate; (37222-66-5), **80**, 9
Oxygen; (7782-44-7), **80**, 66

Pd(dba)$_2$: Palladium, bis[(1,2,4,5-η)-1,5-diphenyl-1,4-pentadien-3-one]- (9); (32005-36-0), **81**, 45, 55
Palladium (II) acetate: Acetic acid, palladium(2+) salt; (3375-31-3)), **80**, 75; **81**, 90
Paladium chloride: Palladium chloride (PdCl$_2$); (7647-10-1), **81**, 56

Paraformaldehyde; (30525-89-4), **80**, 75; **81**, 3, 160
PAUSON-KHAND REACTION, **80**, 93
Perchloric acid; (7601-90-3), **80**, 1
Phase transfer catalyst for asymmetric alkylation, **80**, 38
9,10-Phenanthrenequinone; (84-11-7), **80**, 227
4-Phenylbutyric acid: Benzenebutanoic acid; (1821-12-1), **81**, 262
N-Phenylcarbamic acid methyl ester: Carbamic acid, phenyl-methyl ester; (2603-10-3), **81**, 112
1-Phenylcyclohexene: Benzene, 1-cyclohexen-1-yl-; (771-98-2), **80**, 9
(R,R)-1-PHENYLCYCLOHEXENE OXIDE: 7-Oxabicyclo[4.1.0]heptane, 1-phenyl-,
    (1R,6R)-; (17540-04-4), **80**, 9
S-Phenyl diazothioacetate: Ethanethioic acid, diazo-, S-phenyl ester; (72228-26-3),
    80, 160
(E)-4-(2-PHENYLETHENYL)BENZONITRILE: BENZONITRILE, 4-[(1E)-2ˉPHENYLETHENYL]- (9);
    (13041-79-7), **81**, 63
 (S)-1-Phenylethylamine: Benzenemethanamine, α-methyl-, (αS)-; (2627-86-3),
    80, 207
(S)-[(1-Phenylethyl)amino]acetonitrile; Acetonitrile, [[(1S)-1-phenylethyl]amino]-;
    (35341-76-5), **80**, 207
[(1S)-1-Phenylethyl]imino]acetonitrile N-oxide; Acetonitrile, [oxido[(1S)-
    1-phenylethyl]imino]-; (300843-73-6), **80**, 207
 (R)-(–)-2-Phenylglycinol: Benzeneethanol, β-amino-, (βR)-; (56613-80-0), **80**, 46
N-PHENYL-5R-HYDROXYMETHYL-2-OXAZOLIDINONE: 2-Oxazolidinone, 5-(hydroxymethyl)-
    3-phenyl-, (5R); (875080-42-7), **81**, 112
Phenylmagnesium bromide: Magnesium, bromophenyl-; (100-58-3), **80**, 57
 (R)-4-Phenyl-2-oxazolidinone: 2-Oxazolidinone, 4-phenyl-, (4R)-; (90319-52-1), **81**, 147
5-PHENYLPENT-2-YN-1-OL: 2-Pentyn-1-ol, 5-phenyl-; (16900-77-9), **81**, 2
R-4-PHENYL-3-(1,2-PROPADIENY)-2-OXAZOLIDINONE:: 2-OXAZOLIDINONE, 4-PHENYL-
    3-(1,2-PROPADIENYL)-, (4R)-; (256382-50-0), **81**, 147
3-Phenylpropionaldehyde: Benzenepropanal; (104-53-0), **81**, 2

3-Phenyl-2-propylthio-2-propen-1-ol, **80**, 190
3-Phenyl-2-propyn-1-ol: 2-Propyn-1-ol, 3-phenyl-; (1504-58-1), **80**, 190
R-4-Phenyl-3-(2-propynyl)-2-oxazolidinone: 2-Oxazolidinone, 4-phenyl-3-(2-propynyl); (4R)-; (256382-74-8), **81**, 147
S-Phenyl thioacetate: Ethanethioic acid, S-phenyl ester; (934-87-2), **80**, 160
2-PHENYLTHIO-5-HEPTANOL: 3-Heptanol, 6-(phenylthio); (198778-75-5), **81**, 244
Phosphorus oxychloride: Phophoric trichloride; (10025-87-3), **80**, 200
Photolysis, **81**, 245
Pinacol: 2,3-Butanediol, 2,3-dimethyl-; (76-09-5), **81**, 89
Pinacolic coupling, **81**, 26
Piperidine-4-spiro-5'-hydantoin: 1,3,8-Triazaspiro [4.5]decane-2,4-dione; (13625-39-3), **81**, 214
4-Piperidone monohydrate hydrochloride: 4-Piperidinone, hydrochloride; (41979-39-9), **81**, 214
Pivaloyl chloride: Propanoyl chloride, 2,2-dimethyl-; (3282-30-2), **81**, 226, 254
Platinum(0)-1,3-divinyl-1,1,3,3-tetramethyldisiloxane complex: Platinum, 1,3-diethenyl-1,1,3,3-tetramethyldisiloxane complex; (68478-92-2), **81**, 55
Potassium tert-butoxide: 2-Propanol, 2-methyl-, potassium salt; (865-47-4) **80**, 57, 144; **81**, 141
Potassium cyanide: Potassium cyanide [K(CN)]; (151-50-8), **81**, 214
Potassium nitrate:; (7757-79-1), **81**, 99
Propanethiol; (79869-58-2), **80**, 190
Propanoic acid, 2-hydroxy-, ethyl ester, (2S)-; (687-47-8), **81**, 157
Propargyl bromide: 1-Propyne, 3-bromo-; (106-96-7), **81**, 148
Propargylic alcohol synthesis, **81**, 2, 157
N-(2-Propenyl-4-methylbenzenesulfonamide: Benzenesulfonamide, 4-methyl-N-2-propenyl-; (50487-71-3), **80**, 93
N-(2-Propenyl)-N-(2-propynyl)-4-methylbenzenesulfonamide: Benzenesulfonamide, 4-methyl-N-2-propenyl-N-2-propynyl-; (133886-40-5), **80**, 93
Propiolic acid: 2-Propynoic acid; (471-25-0), **80**, 129
Pyridine; (110-86-1), **80**, 93, 177; **81**, 3
Pyridinium chlorochromate: Chromate(1-), chlorotrioxo-, (T-4)-, hydrogen compound with pyridine (1:1); (26299-14-9), **80**, 1
3-PYRIDIN-3-YLQUINOLINE: Quinoline, 3-(3-pyridinyl)-; (96546-80-4), **81**, 89
3-PYRIDYLBORONIC ACID: Boronic acid, 3-pyridinyl-; (1692-25-7), **81**, 89
Pyrophosphate synthesis, **80**, 219

Reduction, of alkynes, **81**, 43
        of aromatic nitro compounds, **81**, 188
        of esters to aldehydes, **81**, 157
        of quinones, **80**, 227
Reductive deuteration, **80**, 120
        N-heteroannulation, **80**, 75
Regioselective alkylation of aromatic ketones, **80**, 104
Remote functionalization, **81**, 246
Rhodium(II) acetate dimer: Acetic acid, rhodium(2+) salt; (5503-41-3), **80**, 160
Rochelle salt: Butanedioic acid, 2,3-dihydroxy-(2R,3R)-, monopotassium monosodium salt; (304-59-6), **81**, 159

Rose Bengal; (11121-48-5), **80**, 66

Salicyl alcohol: Benzenemethanol, 2-hydroxy-; (90-01-7), **81**, 171
Silylation of alcohols, **81**, 157
Sodium; (7440-23-5), **80**, 144
Sodium amide: Sodium amide (NaNH$_2$); (7782-92-5), **80**, 144
Sodium chlorite: Chlorous acid, sodium salt; (7758-19-2), **81**, 195
Sodium dithionite: Dithionous acid, disodium salt; (7775-14-6), **80**, 227
Sodium fluoride: Sodium fluoride (NaF); (7681-49-4), **80**, 172
Sodium hydride: Sodium hydride (NaH); (7646-69-7), **81**, 147
Sodium hypochlorite: Hypochlorous acid, sodium salt (8,9); (7681-52-9), **81**, 195
Sodium iodide: Sodium iodide (NaI); 7681-82-5, **81**, 77, 121
Sodium metaperiodate: Periodic acid (HIO$_4$), sodium salt; (7790-28-5), **81**, 171
Sodium methoxide: Methanol, sodium salt; (124-41-4), **80**, 133
Sodium nitrite: Nitrous acid, sodium salt; (7632-00-0), **80**, 133
Sodium thiosulfate: Thiosulfuric acid (H$_2$S$_2$O$_3$), disodium salt; (7772-98-7), **81**, 78
Solid supports, **81**, 235
9-SPIROEPOXY-endo-TRICYCLO[5.2.2.0^{2,6}]UNDECA-4,10-DIEN-8-ONE: Spiro[4,7-ethano-
    1H-indene-8,2'-oxiran]-9-one, 3a,4,7,7a-tetrahydro ; (146924-02-9), **81**, 171
Styrene: Benzene, ethenyl-; (100-42-5), **81**, 65
Sulfenate synthesis, **81**, 244
Sulfide synthesis, **81**, 244
Sulfuryl chloride; (7791-25-5), **81**, 247
Suzuki synthesis, **81**, 89

TETRABENZYL PYROPHOSPHATE: Diphosphoric acid, tetrakis(phenylmethyl) ester;
    (990-91-0), **80**, 219
Tetrabutylammonium bromide: 1-Butanaminium, N,N,N-tributyl-, bromide;
    (1643-19-2), **80**, 227
Tetrabutylammonium fluoride trihydrate: 1-Butanaminium, N, N, N-tributyl, fluoride, trihydrate;
    (87749-50-6), **81**, 45, 55
Tetrabutylammonium hydrogen sulfate: 1-Butanaminium, N,N,N-tributyl-, sulfate (1:1);
    (32503-27-8), **80**, 9
2,3,3α,4-TETRAHYDRO-2-[(4-METHYLBENZENE)SULFONYL]CYCLOPENTA-
    [C]PYRROL-5(1H)-ONE: Cyclopenta[b]pyrrol-5(1h)-one, 2,3,3a,4-tetrahydro-
    1-[(4- methylphenyl)sulfonyl]-; (205885-50-3), **80**, 93
Tetrakis(hydroxymethyl)phosphonium sulfate ("Pyroset–TKOW"): Phosphonium,
    tetrakis(hydroxymethyl)-, sulfate (2:1); (55566-30-8), **80**, 85
1-Tetralone: 1(2H)-Naphthalenone, 3,4-dihydro-; (529-34-0), **80**, 104
3-(4,4,5,5-Tetramethyl-[1,3,2]dioxaborolan-2-yl)-pyridine: Pyridine, 3-(4,4,5,5-tetramethyl-
    1,3,2-dioxaborolan-2-yl)-; (329214-79-1), **81**, 89
1,1,3,3,-Tetramethyldisiloxane: Disiloxane, 1,1,3,3,-tetramethyl-; (3277-26-7), **81**, 54
N,N,N',N'-Tetramethylethylenediamine: 1,2-Ethanediamine, N,N,N',N'-tetramethyl-;

(110-18-9), **80**, 46

2,2,6,6-Tetramethylpiperidine; Piperidine, 2,2,6,6-tetramethyl-; (768-66-1), **81**, 134

2,2,6,6-Tetramethyl-1-piperidinyloxy (TEMPO): 1-Piperidinyoxy, 2,2,6,6-tetramethyl-; (2564-83-2), **81**, 195

"Thia-Wolff" rearrangement, **80**, 166

Thioanisole: Benzene, (methylthio)-; (100-68-5), **80**, 184

Thiophenol: Benzenethiol; (108-98-5), **81**, 246

Thionyl chloride; (7719-09-7), **80**, 46

Titanium tetrachloride: Titanium chloride ($TiCl_4$)(T-4); (7550-45-0), **81**, 16

Titanium tetraisoproproxide: 2-Propanol, titanium (4+)salt; (546-68-9) **80**, 120; **81**, 16

p-Toluenesulfonyl chloride: Benzenesulfonyl chloride, 4-methyl-; (98-59-9) **80**, 93, 133; **81**, 77, 159

5-(Tosyloxyimino)-2,2-dimethyl-1,3-dioxane-4,6-dione: 1,3-Dioxane-4,5,6-trione, 2,2-dimethyl-, 5-O-[(4-methylphenyl)sulfonyl]oxime; (215436-24-1), **80**, 133

3,4,6-Tri-*O*-benzyl-D-glucal: D-*arabino*-Hex-1-enitol, 1,5-anhydro-2-deoxy-3,4,6-tris-*O*-(phenylmethyl)-; (55628-54-1), **81**, 225

3,4,6-TRI-*O*-BENZYL-2-*O*-PIVALOYL-β-D-GLUCOPYRANOSYL-(1→6)-1,2:3,4-DI-*O*-ISOPROPYLIDENE- α-D-GALACTOPYRANOSIDE: α-D-Galactopyranose, 6-O-[2-O-(2,2-dimethyl-1-oxopropyl)-3,4,6-tris-*O*-(phenylmethyl)-β-D-glucopyranosyl}-1,2:3,4-bis-*O*-(1-methylethylidene)-; [219122-26-6], **81**, 226

Tri-tert-butylphosphiine: Phosphine, tris(1,1-dimethylethyl)-; (13716-12-6), **81**, 55, 65

8-[2-(TRIETHOXYSILYL)ETHYL-1-TETRALONE: 1(2H)-Naphthalenone, 3,4-dihydro-8-[2-(triethoxysilyl)ethyl]-; (154735-94-1), **80**, 104

Triethoxyvinylsilane: Silane, ethenyltriethoxy-; (78-08-0), **80**, 104

Triethylamine; Ethanamine, *N*,*N*-diethyl-; (121-44-8) **80**, 18, 46, 75, 85, 161, 233; **81**, 77, 121, 214, 244

Triethylsilane: Silane, triethyl-; (617-86-7), **80**, 93

Trifluoroethanol: Ethanol, 2,2,2-trifluoro-; (75-89-8), **80**, 184

2,2,2-Trifluoroethyl trifluoroacetate: Acetic acid, trifluoro-, 2,2,2-trifluoroethyl ester; (407-38-5), **80**, 160

Triisopropyl borate: Boric acid (H_3BO_3), tris(1-methylethyl) ester; (5419-55-6), **81**, 89, 134

1,2:3,4:5,6-Tri-O-isopropylidene-L-mannitol: L-Mannitol,1,2:3,4:5,6-tris-O-(1-methylethylidene)-; 153059-35-9), **81**, 140

Triisopropylsilyl triflate: Methanesulfonic acid, trifluoro-, tris(1-methylethyl)silyl ester; (80522-42-5), **80**, 233

Trimethylamine hydrochloride: Methanamine, N,N-dimethyl-, hydrochloride; (593-81-7), **81**, 77

1,6,6-Trimethyl-4,8-dioxaspiro[2.5]oct-1-ene: 4,8-Dioxaspiro[2.5]oct-1-ene, 1,6,6-trimethyl-; (122762-81-6), **80**, 144

TRIMETHYLSILYL 2-FLUOROSULFONYL-2,2-DIFLUOROACETATE: Acetic acid, difluoro(fluorosulfonyl)-, trimethylsilyl ester; (120801-75-4), **80**, 172

(Z)-1-(Trimethylsilyl)-1-hexene: Silane, 1-hexenyltrimethyl-, (Z)-; (52835-06-0), **80**, 120

1-(Trimethylsilyl)-1-hexyne: Silane, 1-hexynyltrimethyl-; (3844-94-8), **80**, 120

Trimethylsilyl triflate: Methanesulfonic acid, trifluoro-, trimethylsilyl ester; [27607-77-8], **81**, 226

Triphenylphosphine: Phosphine, triphenyl-; (603-35-0), **80**, 75; **81**, 2, 91,142

Tris(dibenzylideneacetone)dipalladium: Palladium,tris[μ-[(1,2-η:4,5-η)-(1E,4E)-1,5,-diphenyl-1,4-pentadien-3-one]]di-; (51364-51-3), **81**, 64, 78

Tris(hydroxymethyl)phosphine: Methanol, phosphinidynetris-; (2767-80-8), **80**, 85
Tris(3-pyridyl)boroxin: Pyridine, 3,3',3''-(2,4,6-boroxintriyl)tris-; (160688-99-3), **81**, 89
Tri-*o*-tolylphosphine: Phosphine, tris(2-methylphenyl)-; (6163-58-2), **81**, 78

Urea hydrogen peroxide: Urea, compd. with hydrogen peroxide (H_2O_2) (1:1); (124-43-6), **81**, 204

Vinamidinium salts, **80**, 203
Vinyl iodide synthesis, **81**, 43
Vinylsilane synthesis, **81**, 44
Vinyl sulfide, synthesis, **80**, 190
          hydrolysis, **80**, 191

Wittig reaction, **80**, 77; **81**, 1

Zinc; (7440-66-6), **81**, 26, 78